Springer Tracts in Modern Physics
Volume 194

Springer
Berlin
Heidelberg
New York
Hong Kong
London
Milan
Paris
Tokyo

Springer Tracts in Modern Physics

Springer Tracts in Modern Physics provides comprehensive and critical reviews of topics of current interest in physics. The following fields are emphasized: elementary particle physics, solid-state physics, complex systems, and fundamental astrophysics.
Suitable reviews of other fields can also be accepted. The editors encourage prospective authors to correspond with them in advance of submitting an article. For reviews of topics belonging to the above mentioned fields, they should address the responsible editor, otherwise the managing editor.
See also springeronline.com

S. Ossicini L. Pavesi F. Priolo

Light Emitting Silicon for Microphotonics

With 206 Figures and 10 Tables

Springer

Stefano Ossicini

Università degli Studi di Modena e Reggio Emilia
NanoStructures and BioSystems at Surfaces
Istituto Nazionale per
la Fisica della Materia (S3 INFM)
and Dipartimento di Scienze
e Metodi dell'Ingegneria
Via Allegri 13
42100 Reggio Emilia, Italy
E-mail: ossicini@unimore.it

Francesco Priolo

Università degli Studi di Catania
Istituto Nazionale per la Fisica della Materia (INFM)
and Dipartimento di Fisica e Astronomia
via S. Sofia 64
95123 Catania, Italy
E-mail: priolo@ct.infn.it

Lorenzo Pavesi

Università di Trento
Istituto Nazionale per la Fisica della Materia (INFM)
and Dipartimento di Fisica
Via Sommarive 14
38050 Povo (Trento), Italy
E-mail: pavesi@science.unitn.it

Library of Cataloging-in-Publication Data

Ossicini, Stefano.
Light emitting silicon for microphotonics / S. Ossicini, L. Pavesi, F. Priolo.
p. cm. -- (Springer tracts in modern physics, ISSN 0081-3869 ; v. 194)
Includes bibliographical references and index.
ISBN 3-540-40233-0 (alk. paper)
1. Optoelectronics--Materials. 2. Porous silicon. 3. Nanostructure materials. I. Pavesi, Lorenzo. II. Priolo, F. (Francesco) III. Title. IV. Springer tracts in modern physics ; 194.

QC1.S797
[TA1750]
621.381'045--dc22 2003060990

Physics and Astronomy Classification Scheme (PACS):
61.46.+w, 78.67.-n, 81.07.-b, 85.60.-q, 42.70.-a

ISSN print edition: 0081-3869
ISSN electronic edition: 1615-0430
ISBN 3-540-40233-0 Springer-Verlag Berlin Heidelberg New York

Springer-Verlag Berlin Heidelberg New York
is a part of Springer Science+Business Media

springeronline.com

Printed in Germany

Typesetting: Authors and LE-TEX GbR, Leipzig using a Springer LATEX macro package
Production: LE-TEX Jelonek, Schmidt & Vöckler GbR, Leipzig
Cover concept: eStudio Calamar Steinen
Cover production: *design &production* GmbH, Heidelberg

Printed on acid-free paper SPIN: 10835554 56/3141/YL 5 4 3 2 1 0

To our parents:
Vittoria and Alessandro,
Annalisa and Vittorio,
Nellina and Angelo.

Preface

Mehr Licht – More Light

Johann Wolfgang von Goethe, 1832

Silicon microelectronics devices have revolutionized our life in the second half of the last century. Integration and economy of scale are the two key ingredients for silicon technological success. Silicon has a band gap of 1.12 eV which is ideal for room temperature operation and an oxide that allows the processing flexibility to place more than 10^8 transistors on a single chip. The continuous improvements in silicon technology have made it possible to grow routinely 300 mm wide single silicon crystals at low cost and even large crystals are now in transition to production and under development. The extreme integration levels reached by the silicon microelectronics industry have permitted high speed performance and unprecedented interconnection levels. The present interconnection degree is sufficient to cause interconnect propagations delays, overheating and information latency between single devices. To overcome this bottleneck, photonic materials, in which light can be generated, guided, modulated, amplified and detected, need to be integrated with standard electronics circuits to combine the information processing capabilities of electronics with data transfer at the speed of light. In particular, chip to chip or even intrachip optical communications all require the development of efficient optical functions and their integration with state-of-the-art electronic functions.

Silicon is the desired material, because silicon optoelectronics will open the door to faster data transfer and higher integration densities at very low cost. Silicon microphotonics has boomed these last years. Almost all the various photonic devices have been demonstrated: e.g., silicon based optical waveguides with extremely low losses and small curvature radii, tuneable optical filters, fast switches (ns) and fast optical modulators (GHz), fast CMOS photodetectors, integrated Ge photodetectors for 1.55 μm radiation. Micromechanical systems or photonic crystals have been demonstrated and switching systems are already commercial. On the other hand, the main limitation of silicon photonics is the lack of any practical Si-based light sources: either efficient light emitting diodes or a Si laser. Several attempts have been employed to engineer luminescing transitions in an otherwise indirect band gap material.

The aim of this book is to give a comprehensive account of the efforts made in the last ten years towards light emitting silicon devices. The subjects covered in detail include fundamental considerations about the inability of silicon to emit light efficiently plus the ways to circumvent this problem (Chap. 1) and an overview of the theoretical methods and related results on the electronic and optical properties of low-dimensional silicon structures (Chap. 2). The significant advances, that have been made in order to produce silicon light emitting devices are reviewed. Efficient room temperature visible emission can now be achieved in various forms, such as porous silicon (Chap. 3), silicon superlattices, nanowires or nanocrystals (Chap. 4), erbium doped silicon and erbium doped silicon nanocrystals (Chap. 5). Silicon based light emitting diodes are now reaching the required power efficiency of the severe optoelectronics market and can compete with III-V semiconductors based devices. The main future challenge for silicon photonics is the demonstration of laser action in silicon based materials; two elements are keys to a laser: the amplifying medium and the optical cavity. Silicon microcavities have been fabricated and together with photonics crystals based on silicon are discussed in Chap. 6. Optical gain in silicon nanostructures has been recently demonstrated (Chap. 4). Ten years of research have pushed silicon very near to laser applications, and the future outlook is sketched in Chap. 7. We hope a silicon based laser will emerge in the next decade and silicon will achieve in microphotonics the same supremacy he has in microelectronics.

The list of friends and colleagues who helped us is too long to quote here. We are grateful to all of them.

Modena-Reggio Emilia
Povo-Trento
Catania
August 2003

Stefano Ossicini
Lorenzo Pavesi
Francesco Priolo

Contents

1 Introduction: Fundamental Aspects

CMOS (Complementary Metal Oxide Semiconductor) circuitry dominates the current semiconductor market due to the astonishing power of silicon electronic integration technology. In contrast to the dominance of silicon in electronics, photonics utilizes a diversity of materials for emitting, guiding, modulating and detecting light. In the last ten years much research effort was aimed at rendering Si an optically active material so that it can be turned from an electronic material to a photonic material. For some the future of Si-based photonics lies in "hybrid" solutions, for others in the utilization of more photonic functions by silicon itself. Many breakthroughs in the field have been made recently. Nevertheless, the major deficiency in Si based optoelectronic devices remains the lack of suitable light emitters. In this chapter starting from the analysis of the electronic and optical properties of silicon, compared to other semiconductors, we present the strategies that have been employed to overcome the physical inability of bulk silicon to emit light efficiently.

1.1 Electronic and Optical Properties of Bulk Silicon

Silicon, the most important elemental semiconductor, crystallizes in the diamond structure. The diamond lattice consists of two interpenetrating face-centered cubic Bravais lattices displaced along the body diagonal of the cubic cell by one quarter of the length of the diagonal. The symmetry group is $O^7{}_h$-Fd3m. The lattice constant is 0.5341 nm.

The energy band structure of solids depends not only on the crystal structure but also on the chemical species, the bonding between the atoms and the bond lengths. This results in several differences in the electronic band structure of the various semiconductors. As an example the band structures of Si, Ge and GaAs are compared in Fig. 1.1, which shows the results of electron energy band calculations within the empirical pseudopotential method [1, 2] (for a survey on band structure calculations see Sect. 2.1). Note that whereas Ge has the same diamond lattice structure of Si, the compound semiconductor GaAs crystallizes in the zinc-blend structure. In this lattice the atoms are arranged as in the diamond case but the two species alternate: the Ga atom

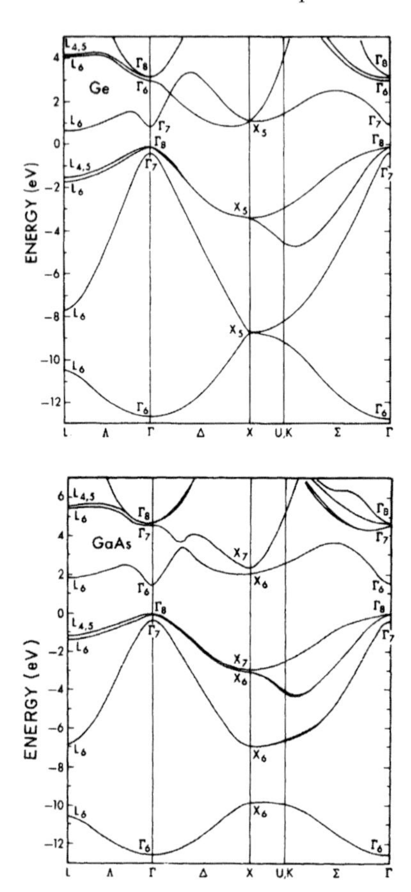

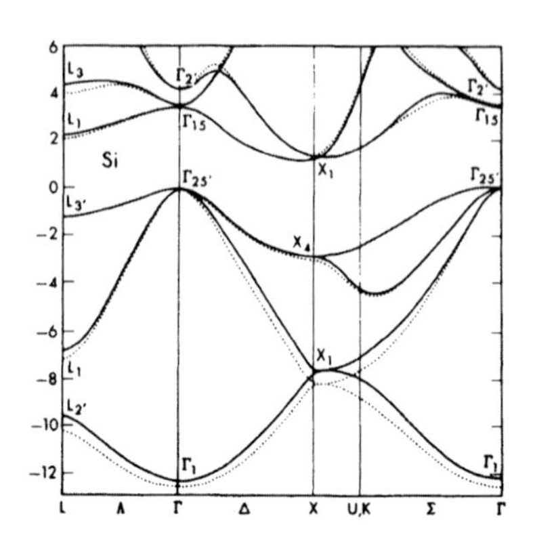

Fig. 1.1. *Top left*: Band structure of Ge calculated within the empirical pseudopotential method. Spin–orbit coupling is included in the calculations. *Top right*: Band structure of Si calculated within the empirical pseudopotential method. Two results are shown: nonlocal pseudopotential (*solid line*) and local pseudopotential (*dashed line*). Spin–orbit coupling is included in the calculations. *Bottom*: Band structure of GaAs calculated within the empirical pseudopotential method. Spin–orbit coupling is included in the calculations. After [1]

occupies the original sites of the face centered cubic lattice, while the As atom is located at the tetrahedral site.

Qualitatively the bands of all three materials are similar in shape, as expected from their common crystal structure and close positions in the periodic table. The main difference in the valence band structure is the degeneracy at the top of the valence band at the Γ point, the center of the Brillouin zone (BZ). The $\Gamma_{25'}$ point of the Si case is split by the spin–orbit interaction at the Γ_8 and Γ_7 points in the case of Ge and GaAs. The spin–orbit splitting Δ_0 at the zone center is a relativistic effect and increases with the atomic number of the element. In Si it is negligible (0.044 eV), while in Ge it has the value of about 0.29 eV and in GaAs of 0.34 eV. The two Γ_8 and Γ_7 bands have different effective masses, termed light and heavy holes, for small wavevector. Moreover, at the X point, some of those states that are doubly degenerate in Si and Ge are split in GaAs, both in the valence and in the conduction band.

The main striking difference is related to the peculiar properties of the conduction band extrema in the three cases. The valence bands exhibit a maximum at the Γ point of the Brillouin zone for the three semiconductors; while the conduction band for GaAs has an absolute minimum at the Γ point, for Si it lies away from the high symmetry points near the X point along the $\langle 001 \rangle$ directions and for Ge it occurs at the zone boundaries in the $\langle 111 \rangle$ directions.

The dispersion of the conduction band in GaAs is parabolic near the band edges. The effective mass for the electron is about 0.067 m_0, where m_0 is the electron mass in vacuum. The low value of the electron effective mass is one of the reasons for the high mobility in GaAs and of its use in high-frequency transistors. In Si the conduction band has six symmetry-related minima in the $\langle 100 \rangle$ directions. Each of these minima gives rise to a conduction band valley. The constant-energy surfaces near the conduction band minima are ellipsoids elongated in the $\langle 100 \rangle$ directions (see Fig. 1.2). For each valley a parabolic approximation can be used. Due to symmetry considerations this leads to different effective masses for the electron, one for the longitudinal x-direction, the other for the transverse y- and z-directions. The high anisotropy of the valleys for Si results in very different masses for the longitudinal and transverse mass, 0.916 and 0.19 m_0.

For Ge the minima, which are located at the L points along the $\langle 111 \rangle$ directions, originate in four equivalent valleys or eight half-valleys. Also here the electron effective masses are highly anisotropic, 1.64 and 0.092 m_0 (see Fig. 1.2).

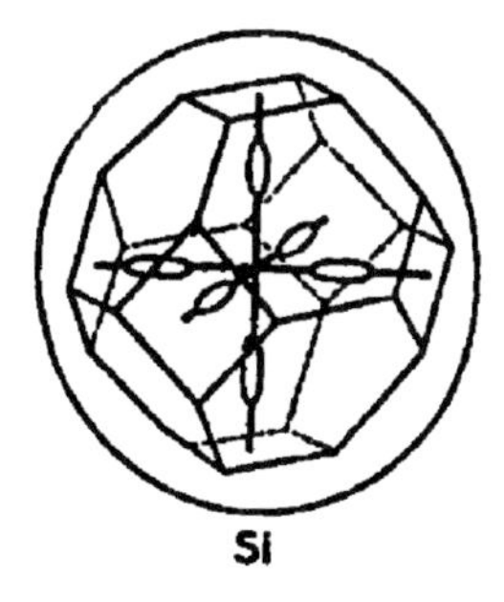

Fig. 1.2. *Left*: Brillouin zone of Ge with constant-energy surfaces for the four equivalent L valleys. *Right*: Brillouin zone of Si with constant-energy surface for the six equivalent X valleys

The energy gap (the energy difference between the conduction band minimum and the valence band maximum) is termed *direct* in the case of GaAs, because an electronic transition between the two extrema can occur without change in wavevector. The energy gap is termed *indirect* when a change in wavevector is needed, as for Si and Ge. The value of the energy gaps ranges from the infrared (Ge) to the near infrared (Si) to the visible (GaAs). Some general properties of the three semiconductors are given in Table 1.1 [4, 5].

Table 1.1. Properties of Si, Ge, and GaAs

Property	Si	Ge	GaAs	Units	Name
a	0.5431	0.5658	0.5653	nm	Lattice constant
ρ	2.329	5.323	5.318	g cm^{-3}	Density
$\hbar\omega_{\mathrm{LO}}$	64	37	36	meV	Optical phonon energy
E_{g}	1.12	0.66	1.42	eV	Energy gap at 300 K
E_{g}	1.17	0.74	1.52	eV	Energy gap at 0 K
$E_{\mathrm{c}}^{(\min)}$	X	L	Γ		Conduction band minimum
E_{c}^{Γ}	3.5	0.80	1.42	eV	Energy gap at Γ
Δ_0	0.044	0.296	0.341	eV	Spin–orbit splitting
χ	4.05	4.01	4.07	eV	Electron affinity
m_{hh}	0.49	0.33	0.51		Heavy hole mass
m_{lh}	0.16	0.043	0.082		Light hole mass
m_{Γ}			0.063		Electron mass
m_{L}	0.916	1.64			Longitudinal electron mass
m_{T}	0.19	0.092			Transverse electron mass
μ_{n}	0.15	0.39	0.85	$\mathrm{m}^2\mathrm{V}^{-1}s^{-1}$	Electron mobility at 300 K
μ_{p}	0.045	0.19	0.04	$\mathrm{m}^2\mathrm{V}^{-1}s^{-1}$	Hole mobility at 300 K
ϵ_{b}	11.7	16.2	12.9		Dielectric constant
n	3.42	4.0	3.3		Static refractive index
R	1.1	6.4	7×10^4	$\times10^{-14}$ $\mathrm{cm}^3\mathrm{s}^{-1}$	Radiative recombination coeff.
C_{11}	16.6	12.6	11.9	$\times10^{-11}$ dyn cm^{-2}	Elastic constant at 300 K
C_{12}	6.4	4.4	5.34	$\times10^{-11}$ dyn cm^{-2}	Elastic constant at 300 K
C_{44}	7.96	6.77	5.96	$\times10^{-11}$ dyn cm^{-2}	Elastic constant at 300K
T_{m}	1685	1210	1513	K	Melting temperature

The optical properties reflect the electronic ones. Figure 1.3 shows the real and imaginary parts of the dielectric function of the three considered semiconductors. They have been determined from reflectivity measurements and the use of Kramers–Kronig relations [3]. They are directly connected to the optical parameters (see Sect. 2.1); the imaginary part is correlated with the absorption coefficient, the real part with the dielectric constant. In the case of a direct band gap semiconductor like GaAs the energy gap is directly given by the optical threshold. This is not true in the case of Si and Ge where the onset of absorption is related to the first direct transition: 0.80 eV in Ge

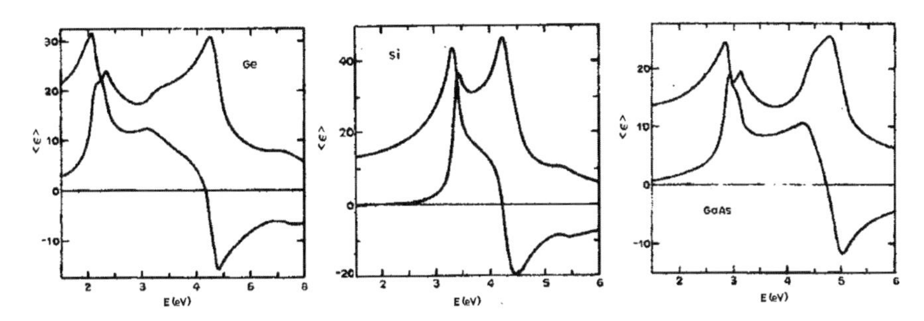

Fig. 1.3. Real ϵ_1 (the curve that also shows negative values) and imaginary parts ϵ_2 of the dielectric function ϵ between 1.5 and 6 eV. *Left*: Crystalline Ge. *Middle*: Crystalline Si. *Right*: Crystalline GaAs. After [3]

and 3.5 eV in Si (see Table 1.1). Moreover for a direct semiconductor the absorption coefficient α near the band edge has a square root dependence on the energy E, whereas a linear dependence of $(\alpha\mathrm{E})^{1/2}$ versus E is indicative of an indirect band gap semiconductor.

In GaAs an electron which has been excited in the conduction band at Γ can easily decay radiatively at the same Γ point in the valence band producing a photon of the same energy as the band gap. Under these conditions the radiative recombination rate is large, of the order of 2×10^7 s^{-1}, and the radiative lifetime is short, of the order of nanoseconds. In the case of Ge and Si, where the minima and maxima occur at different points of the BZ, a phonon must participate in the process in order to conserve the crystal momentum, hence the term indirect transition.

The three-particle process (electron, hole and phonon) has for Si a very low rate, about 10^2 s^{-1}, and the relative radiative lifetimes are in the millisecond range. The luminescence efficiency, defined as the ratio of the number of photons emitted to the number of electron–hole pairs excited, is typically of the order 10^{-4}–10^{-5} in bulk Si; for a direct semiconductor it can be of the order of 10^{-1}. Thus in order to overcome the inability of Si to emit light efficiently one must defeat the tyranny of the $\boldsymbol{k}$ vector, e.g. by trying to reverse the indirect nature of the Si band gap into a direct one.

The presence of an indirect gap is not, however, the only reason for the poor luminescence efficiency [6, 7]. The luminescence efficiency, due to the low radiative recombination rate, is determined by the competition among alternate electron–hole recombination processes. One can define the radiative efficiency of a semiconductor η as

$$\eta = \frac{\tau_{\mathrm{nr}}}{\tau_{\mathrm{r}} + \tau_{\mathrm{nr}}} , \tag{1.1}$$

where τ_{r} is the lifetime of the radiative transition and τ_{nr} is the lifetime relative to the recombination mechanisms that do not involve the creation of a photon. From this equation it is clear that in order to have a large η, τ_{r} must be much shorter than τ_{nr}. As stated above, τ_{r} for direct gap semiconductors is very short (in the range of ns); for Si, τ_{r} is much longer (of the order of ms). Extremely pure Si is needed to have a long nonradiative lifetime. There are several possible nonradiative recombination processes [8]. Just to cite a few:

- Intrinsic defect or impurity recombination that involves the trapping of an electron (or hole) by a deep trap level and the subsequent capture of a hole (or electron), thus originating a recombination process. This mechanism is usually termed Shockley–Read–Hall recombination. For this, the lifetime depends on the concentrations of deep states and is of the order of nanoseconds [9].
- Auger recombination where an electron recombines with a hole transferring its energy and momentum as kinetic energy to another electron and/or hole. Depending on the doping level the radiative lifetime associated with the Auger process ranges between 1 ns and 0.5 ms [10].
- Finally dislocations and surface mediated recombinations are possible. In these cases it is difficult to establish characteristic lifetimes because they vary depending on the quality of the surface. Good surface passivation via a naturally grown oxide removes almost completely these nonradiative recombinations.

The total recombination rate for all the different nonradiative pathways, τ_{nr}, can be obtained by the sum of the individual rates $\tau_{\mathrm{nr}i}$:

$$\frac{1}{\tau_{\mathrm{nr}}} = \sum_i \frac{1}{\tau_{\mathrm{nr}i}} \,. \tag{1.2}$$

For Si, τ_{nr} is at least of the order of μs, three orders of magnitude smaller than the radiative lifetime τ_{r}. Bulk silicon has a very low quantum efficiency and thus it is very unattractive as a light emitting device.

1.1.1 Light Emitting Diode Efficiency

In this and the next sections we will discuss light emitting diodes (LEDs) based on bulk Si. Recently a review on Si based LEDs has been published [14]. The performances of light emitting diodes could be given with different parameters: maximum brightness, quantum efficiency, wall plug-in efficiency, peak power, on-set voltage, etc. A classification of the various efficiencies is reported in Table 1.2.

The practical performance of a light source is primarily evaluated through its *power efficiency* η_{P}, which is the ratio between the radiant flux P emitted by the LED and the input electrical power W_{e} flowing into the LED.[1] The external power efficiency is sometimes reported in lumens per watt. η_{P} is also

[1] The *radiant flux* P is the power carried by the light beams. Given a flux of photons Φ, at a given photon frequency ν, $P = \Phi \times h\nu = \Phi \times hc/\lambda$, where c is the speed of light in vacuum and λ is the wavelength of light in vacuum. The *luminous*

Table 1.2. Definition of the various efficiencies. τ_{rad} and τ_{nr} are the radiative and nonradiative lifetimes, respectively. $P_{\mathrm{op(int)}}$ and $P_{\mathrm{op(ext)}}$ are the optical power as measured in and out of the device. W_{e}, I and V are the electrical power absorbed, current and voltage of the operating LED. $\hbar\omega$ is the emission energy of the LED

internal quantum efficiency η_{int}	number of photons emitted versus the number of electron–hole pairs generated	$\eta_{\mathrm{int}} = \frac{\tau_{\mathrm{nr}}}{\tau_{\mathrm{nr}} + \tau_{\mathrm{rad}}} = \frac{eP_{\mathrm{op(int)}}}{I\hbar\omega}$
external quantum efficiency η_{ext}	number of photons detected versus numbers of charge injected	$\eta_{\mathrm{ext}} = \frac{eP_{\mathrm{op(ext)}}}{I\hbar\omega}$
power efficiency η_{P}	watts of light detected versus watts of electricity used	$\eta_{\mathrm{P}} = \frac{P_{\mathrm{op(ext)}}}{W_{\mathrm{e}}} = \frac{\eta_{\mathrm{ext}}\hbar\omega}{eV}$

called wall–plug-in efficiency because it is intended that the power should be directly taken by the plug-in. It should be noticed that this general definition can also be used for nonelectrically pumped luminescence mechanisms. In any case, the power efficiency value will always be between 0 and 1.

The mechanism of electrically pumped light generation involves the generation of a photon by the radiative recombination of electrical carriers (one electron and one hole). The capability of a single electron–hole pair to generate a photon is called the internal quantum efficiency, η_{int}. η_{int} is the number of photons generated divided by the number of minority carriers injected into the region where recombination mostly occurs. By definition

$$\eta_{\mathrm{int}} = \frac{\text{(Rate of radiative recombination)}}{\text{(Total rate of recombination (radiative and nonradiative))}},$$

or

$$\eta_{\mathrm{int}} = \frac{1/\tau_{\mathrm{r}}}{(1/\tau_{\mathrm{r}}) + (1/\tau_{\mathrm{nr}})},$$

where τ_{r} is the mean lifetime of a minority carrier before it recombines radiatively and τ_{nr} is the mean lifetime before it recombines without emitting

flux F is the power of light weighted by the human eye responsivity V(λ), as defined in 1924 by CIE [11]. The conversion factor F/P, called the efficacy, has the same spectral shape as the eye responsivity V(λ), with the convention that the efficacy is 683 lm/W at $\lambda = 555$ nm. The *radiant intensity* J is P per unit solid angle and is measured in watts per steradian. The *luminous intensity* I is F per unit solid angle. It is measured in candela, which are lumens per steradian. The *radiance* L is the surface density (with respect to the surface of the source) of radiant intensity J. It is measured in watts per square centimeter per steradian. The *luminance or brightness* B is its photometric counterpart, i.e. the surface density of the luminous intensity I, and it can be measured in candelas per square meters, or nits (nt). Radiance and brightness are important when the source cannot be assumed as a point source.

a photon. If one considers that the number of carriers per unit time electrically injected by a current I is I/e, where e is the electrical charge of one electron, and that the number of photons per unit time carried by the internally generated optical power $P_{\mathrm{op(int)}}$ is $P_{\mathrm{op(int)}}/\hbar\omega$, where $\hbar\omega$ is the emission energy, then $\eta_{\mathrm{int}} = eP_{\mathrm{op(int)}} \,/\, I\hbar\omega$. The *internal* quantum efficiency (IQE), η_{int}, is used to characterize the internal generation of photons. It does not take into account (1) how many electrical carriers injected into the LED do not excite the material to the emitting excited state, and (2) how many generated photons get lost before exiting the device (e.g. by reabsorption processes). In fact, for those photons that are generated, there can still be loss through absorption within the LED material, reflection loss when light passes from a semiconductor to air due to differences in refractive index and total internal reflection of light at angles greater than the critical angle defined by Snell's law. For these reasons one introduces a more general concept to measure the efficiency of a LED, the *external* quantum efficiency (EQE), η_{ext}, which is the ratio of the number of photons actually exiting the LED and the number of electrical carriers entering the LED per unit time. Thus, the following relation holds:

$$\eta_{\mathrm{ext}} = \eta_j \times \eta_{\mathrm{int}} \times \eta_x \tag{1.3}$$

where η_j and η_x are by definition the carrier injection efficiency and the photon extraction efficiencies, respectively [12, 13]. One can give estimates of these last quantities for a LED based on a plane p/n junction.

$$\eta_j = \left(1 + \frac{\mu_{\mathrm{h}} N_{\mathrm{A}}\, L_{\mathrm{e}}}{\mu_{\mathrm{e}} N_{\mathrm{D}} L_{\mathrm{h}}}\right)^{-1} \tag{1.4}$$

where μ_{e} and μ_{h} are the mobilities of electrons and holes, N_{A} and N_{D} are the doping densities on the p-type and n-type side of the junction, and L_{e} and L_{h} are the diffusion lengths of the electrons and holes, respectively. On the other hand, the extraction efficiency can be estimated by considering the fraction of light which is transmitted through a plane interface between a medium of index of refraction n_1 and one of index of refraction n_2 with respect to the total light impinging on the interface at any angle:

$$F_{\mathrm{T}} = \frac{1}{4}\left(\frac{n_2}{n_1}\right)\left[1 - \left(\frac{n_1 - n_2}{n_1 + n_2}\right)^2\right]. \tag{1.5}$$

Finally, by using the EQE one can also derive η_{P}. In fact, $\eta_{\mathrm{P}} = \eta_{\mathrm{ext}}\hbar\omega/eV$, where V is the working voltage of the LED. The power efficiency of a LED should be at least 1% for display applications, whereas for optical interconnects it should be at least 10%, due to the requirement of a small thermal budget on the chip. The operating voltage can be high for display applications, while it should be low (1–5 V) for interconnects. An operating life of more than several tens of thousand hours with no degradation is a must.

1.2 Electroluminescence in Bulk Crystalline Si

A review of LEDs based on Si and Si compatible materials is reported in [14]. Here we discuss those LEDs made from bulk Si; the other approaches to a Si LED are discussed in the following chapters.

1.2.1 Bulk-Si pn Junction LED

EL from bulk crystalline Si can be obtained in the most basic way both by forward or by reverse biasing a pn junction at room temperature (RT), thus injecting minority carriers across the pn junction, and realizing band-to-band radiative recombination of free excess minority carriers after the injection [15]. A pn junction shows a power efficiency of about 10^{-4} at 1.1 μm in forward bias, and of about 10^{-8} in the visible and in reverse bias while in the avalanche breakdown regime [16].

The power quantum efficiency of an ordinary Czochralski (CZ) direct-biased pn junction is low, typically of the order of 10^{-4}% or lower [17]. The most important recombination mechanisms are schematically shown in Fig. 1.4 [13].

EL in bulk Si was first reported by R. J. Haynes and H. B. Briggs in 1952 [19] by direct biasing a Si pn junction at room temperature (RT), and explained by R. J. Haynes and W. C. Westphal in 1956 [18] reporting experiments both at RT and at 77 K (see Fig. 1.5). At 77 K, the emission is not only due to free carriers, but also to exciton recombination [15]. The main feature of Fig. 1.5 at 77 K is the peak at $h\nu_1 = 1.100$ eV (vacuum wavelength 1127 nm). The emission line shape is the product of the Boltzmann occupation factor times the density of states.

For practical applications, however, it is necessary to provide mechanisms which work at RT. By raising the temperature to RT, several changes are observed. The first is a lowering of the peak intensity and of η_{int}. This is due (1) to an increase of nonradiative recombination rates, and (2) to exciton ionization. Since the exciton binding energy is of the order of 15 meV, at RT exciton dissociation is significant (kT is about 25.8 meV at RT). It has been shown that the radiative recombination rate, which is dominated by exciton recombination al low temperature (under 26 K), decreases by raising the temperature. The drop is over an order of magnitude from 100 K to 300 K [20]. The second change is a shift to lower energy because of thermal band-gap shrinkage [17, 18, 21]. The third effect is the broadening of the luminescence band, as a consequence of the rise in the free carrier kinetic energy: the full-width-half-maximum (FWHM) of luminescence is fitted with 1.795 kT. The ratio between the free carrier concentration (n_{c}) and exciton concentration (n_{x}) increases with the temperature according to the mass action law. At RT, various spectral features are observed, whose origin is not yet clear and several models have been presented [22]. They include: interband

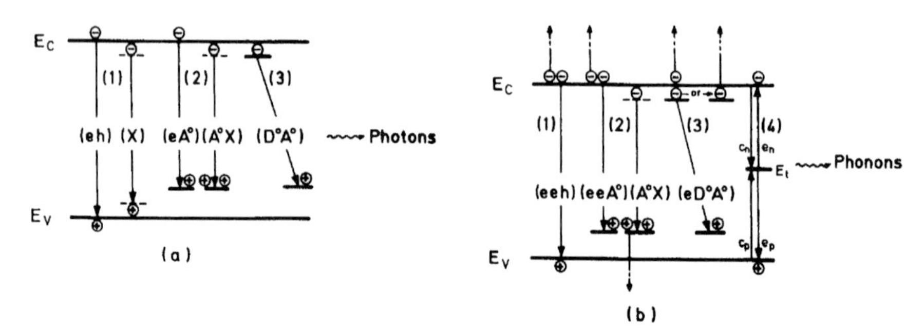

Fig. 1.4. Schematics of selected radiative (**a**) and nonradiative (**b**) recombination mechanisms in Si. (**a**) (1) Intrinsic free electron–hole (eh) and free exciton (X), which both also involve creation or annihilation of a phonon (not shown); (2) extrinsic recombination (in the presence of a neutral acceptor), of a free electron to bound hole (eA^o) and of a bound exciton (A^{o}X) (a similar situation can exist with neutral donors); (3) extrinsic recombination, bound electron (neutral donor) to bound hole (neutral acceptor) (D^{o}A^o). (**b**): Similar mechanisms of type (1), (2), (3) can give rise to nonradiative (Auger) recombination. In this case the released energy is transferred to a free (or bound) carrier rather than generating a photon. (4) is Shockley–Read–Hall recombination. After [13]

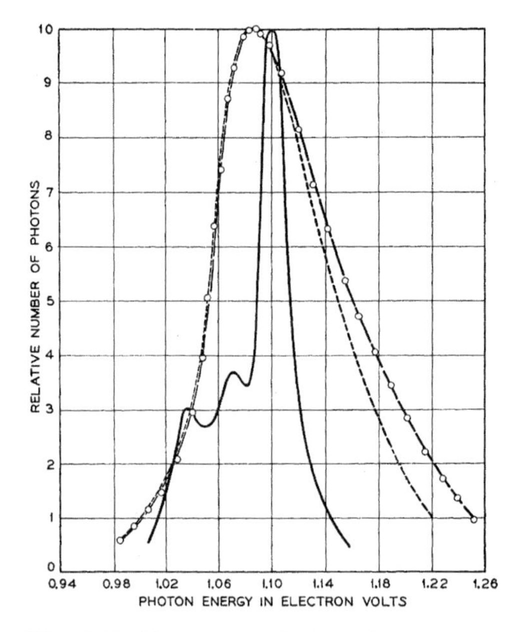

Fig. 1.5. First reported electroluminescence spectrum due to recombination of electrons and holes in Si, as published by J. R. Haynes and W. C. Westphal in 1956. The *solid line* shows the emission at 77 K, whereas the *dashed line* and the *line with circles* show emission at room temperature (the two spectra correspond to slightly different conditions). After [18]

transition, bremsstrahlung, intraband transitions of hole and ionization, and indirect interband recombination under high-field conditions (hot carriers).

1.2.2 Enhanced-Efficiency Bulk-Si pn Junction LED

There exist many strategic schemes to significantly enhance the LED efficiency in bulk-Si based LED. We classify LED as "bulk-Si based LED" if the wide lineshape and the peak position at 1.1 μm of the electroluminescence (EL) spectrum indicate that the emission mechanism is the same as that seen in the preceding section.

A first strategy consists of increasing τ_{nr}. In the ideal limit, the condition $\tau_{\mathrm{nr}} \gg \tau_{\mathrm{r}}$ would imply $\eta_{\mathrm{int}} \simeq 1$. The nonradiative rates can be reduced using (1) high-quality Si substrates, float-zone (FZ) being preferred over CZ, (2) passivation of surfaces by high-quality thermal oxide, in order to reduce surface recombination, (3) small metal areas, and (4) high doping regions limited to contact areas, in order to reduce the SRH recombination in the junction region [17].

A second strategy is to reduce parasitic absorption of photons once they have been generated. For example, the reabsorption can be minimized by keeping the doping level to moderate values, such as $\simeq 1.4 \times 10^{16}$ cm^{-3} [17].

Third, light emission from bulk silicon can be enhanced by suitably texturizing the Si surface. The effect of a texture on the Si/air interface is to increase the absorptivity. Kirchhoff has shown that the spectral emissivity is equal to the absorptivity. Texturized surfaces by themselves can have over one order of magnitude larger emissivity compared to planar surfaces, since they act as light trapping elements. Such texturized surfaces have been used for a long time for Si photovoltaic cells [17].

Combination of all these three strategies has allowed demonstration of bulk-Si LEDs with the highest power efficiency to date, approaching 1% [17]. An example of electroluminescence spectra and of the device structure is reported in Fig. 1.6.

The main drawback of this approach for an integrated light emitting diode is the need of both high purity (no doping) and of texturing which render the device processing incompatible with standard CMOS processing. In addition, no population inversion in bulk Si can be achieved due to fast free carrier absorption [23].

A somehow different approach was reported in [24], see Fig. 1.7.

The idea has been again a reduction of the nonradiative channels by exploiting the strain produced by localized dislocation loops to form energy barriers for carrier flow. The localization increases the recombination rate of injected carriers. Dislocations form potential pockets close to the junction which block the carriers and enhance radiative decay by localizing them in defect-free regions [24]. The size of dislocation loops was in the range of 100 nm, i.e. not enough to cause quantum confinement of the carriers, and

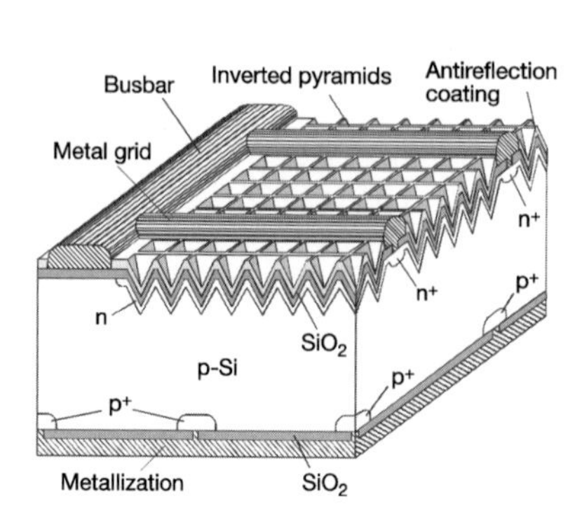

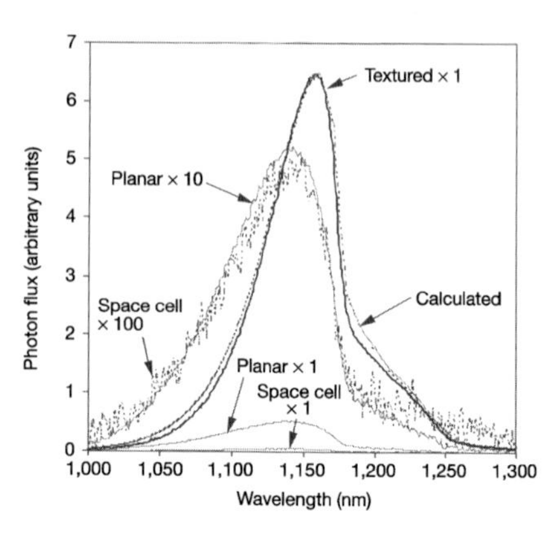

Fig. 1.6. *Left*: Design of the textured Si light emitting device. *Right*: Electroluminescence spectra for textured, planar and baseline space cell diodes under 130 mA bias current at 298 K (diode area 4 cm^2). Calculated values assume a rear reflectance of 96%. After [17]

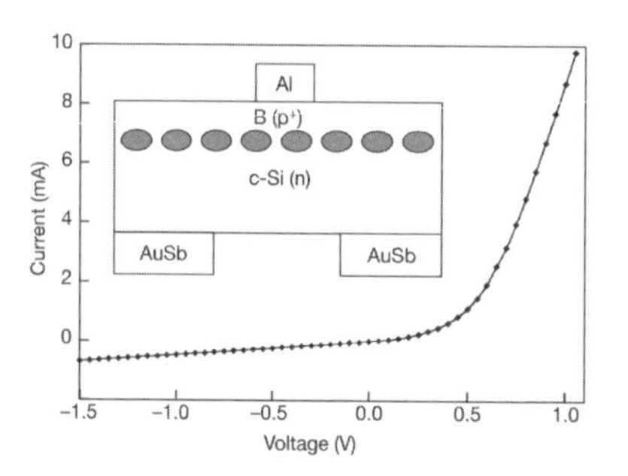

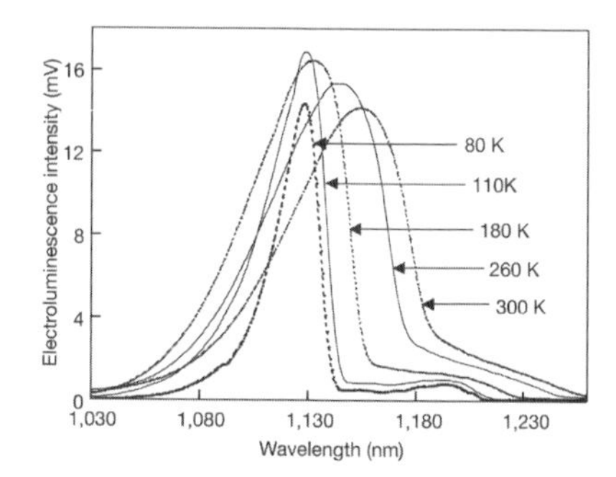

Fig. 1.7. *Left*: The current–voltage characteristic for a device formed by buried dislocation loops measured at room temperature. *Inset*: a schematic of the LED device. The top and bottom ohmic contacts are formed by Al and AuSb respectively. The infrared light is emitted through a window left in the bottom contact. *Right*: Electroluminescence spectra against wavelength at various temperatures. The device was operated at a forward current of 50 mA for all temperatures. After [24]

the loop distances were of the order of 20 nm. The onset of the electroluminescence (EL) at the band edge was observed as the diode turns on under forward bias. No EL was observed under reverse bias.

The electroluminescence spectra at various temperatures are shown in the right part of Fig. 1.7. The EL spectrum does not present significant differences in lineshape or in the peak position compared with bulk Si. The confining potential only acts in preventing free carrier diffusion. A remarkable feature of this device is the high injection efficiency into the confined regions. This is due on one hand to the lack of level quantization. In fact, since the density of states in the active zone is large (comparable to the bulk value), it is not a limiting factor for the free carrier injection, contrary to quantum confined structures. On the other hand, injection is also smooth because there is no wide band-gap material as the confining barrier. Although not explained, this device has the additional and interesting feature of increasing the efficiency with temperature. Power efficiency lower than 0.1% has been reported.

1.2.3 Band-to-Band EL in MOS LED

Recently, field-induced carrier confinement in quasi-2D structures has been observed in metal-oxide-semiconductor (MOS) LED structures [25]. The LED were essentially MOS capacitors with a thin gate oxide (2.7 nm). Both n-type and p-type substrates were used. By biasing the LED in inversion mode (in p-type structures, gate positive with respect to the substrate and vice versa in n-type) two relevant effects were generated: (1) quasi-2D minority carrier layers were created at the SiO_2/substrate interface, and (2) majority carriers were injected by tunneling from the gate. Nondestructive tunnelling was allowed by the thin oxides. The EL lineshape was similar to that observed in a Si pn junction. The peak position was also similar, except for a small low-energy shift (around 100 meV), which was explained by the band bending effect (see Fig. 1.8) [25]. The efficiency was not reported.

Given the EL spectral similarity between these devices and the devices described in previous sections, we think that these devices are essentially based on the same mechanism, despite quasi-2D confinement being claimed for the MOS. However, the report is interesting because it demonstrates that

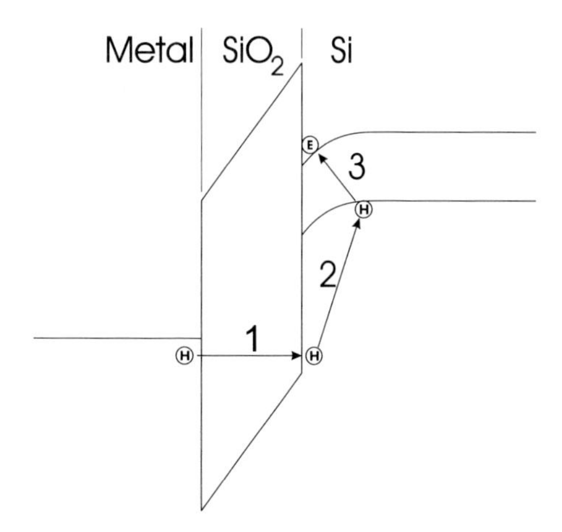

Fig. 1.8. Schematic mechanism of radiative emission in thin-gate Si MOS LEDs. The steps are hole tunneling (1), hole energy loss (2) and electron–hole plasma radiative recombination (3). The energy of emitted photons is lower than the Si band-gap since the difference between electron and hole energies is reduced by band-bending due to the electric field. MOS LEDs are described in [25]

EL at the Si band edge can be excited not only in a pn junction, but also by carrier tunneling through the thin oxide in genuine, plain MOS structures. Additionally, diminished temperature dependence was observed in such MOS LEDs, due to limited availability of impurity related nonradiative recombination centers in the quasi-2D region [26].

1.3 Engineering Luminescent Silicon Structures

The desire for optically active silicon components is so strong that several attempts have been made to circumvent the silicon limitations by improving the spontaneous emission rate and internal quantum efficiency. In order to achieve large internal quantum efficiency two routes are possible: either to strongly reduce the lifetimes of all the nonradiative pathways or to produce much faster radiative lifetimes. Both procedures have been used to engineer luminescent transitions in silicon. Nonradiative rates have been reduced by confining the hole and electron to a small volume so that the probability of finding a nonradiative center strongly decreases and radiative rates have been increased using defect engineering or band gap engineering [27, 28, 29, 30]. All these attempts are described in the next paragraphs. It should be noted that total suppression of the different nonradiative channels is usually not feasible, thus the radiative recombination rate should, anyway, be as large as possible.

1.3.1 Intrinsic and Extrinsic Luminescence

The low temperature photoluminescent (PL) spectra of lightly doped Si are dominated by the optical recombination of free excitons. Due to the indirect nature of Si the radiative recombination of excitons involves phonons. However it is also possible for excitons to relax without phonons in the so-called no-phonon (NP). The presence of localized impurity states can help to overcome the momentum conservation selection rule [31]. Because the electron and the hole are bound to a localized defect state, they are localized in the real space **r**, and, as a consequence of the uncertainty relation $\Delta x \Delta k \sim h/2\pi$, the selection rule relaxes in the wavevector space **k**. Figure 1.9 sketches how the localization process near an impurity delocalizes the wavefunction in reciprocal space [7]. NP lines are usually weak and are quenched by Auger recombinations.

Donors and acceptors can bind free excitons at low temperatures forming bound excitons; thus a typical low temperature PL spectrum in doped Si consists of a bound-exciton NP line at highest energy and phonon replicas at lower energies. The intensity of the NP line depends on the type of impurity (donor or acceptor) and on the exciton binding energy. The NP lines are always weak and the nonradiative Auger recombination dominates the recombination process.

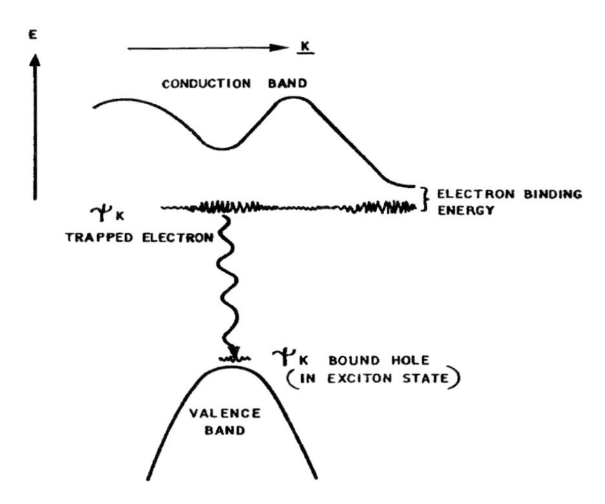

Fig. 1.9. The presence of an impurity leads to spatial localization of the electron and to delocalization of the wavefunction. After [7]

A strategy to make Si luminescent is hence to add localized centers where excitons can be trapped or relax or transfer energy, and recombine radiatively. Among the radiative active impurities, the most interesting, from the point of view of device applications, are the rare earth impurities and the isoelectronic centers. The radiative mechanism is quite different in the two cases. The introduction, for example, of erbium in silicon allows the electronic excitation of an internal $4f$ transition at the impurity through a carrier mediated process with subsequent radiative de-excitation [32, 33, 34, 35]; in the isoelectronic centers an exciton is bound at the impurity trap, often quoted as isoelectronic bound exciton (IBE) [36, 37, 38, 39].

Rare earth doped silicon structures are the core of Chap. 5; here we will discuss only the case of isoelectronic centers. An historical survey and a detailed presentation of recent investigations on radiative isoelectronic impurities in silicon can be found in [39]. An isoelectronic impurity complex must be isovalent with the Si in the lattice, i.e. must be electrically neutral. Examples are thus any other element in the group IV column of the periodic table, such as C, Ge and Sn or any multiple atom complex in such a configuration where dangling bonds are not present. Indeed dangling bonds are very efficient nonradiative recombination centers. Photo- and sometimes electroluminescence has been observed for example in the case of Be-related, S-related, Se-related and O-related complexes at energies of 1.076 [40], 0.968 [41, 42], 0.955 [42], and 0.767 eV [45]. Also carbon complexes created by high-energy irradiation have been considered [46]. The characteristics of the emission processes are: (1) the lifetime is long, of the order of milliseconds, (2) the quantum efficiency decreases rapidly with increasing temperature and at 300 K one has a vanishingly weak luminescence. A major difficulty arises from the necessity of a high impurity concentration that can degrade the electrical properties.

It is worth noticing that claims for stimulated emission observations in intra-impurity transitions, in particular, among shallow donor states (2p $\rightarrow$ 1s), have appeared in the literature [43, 44].

In very dilute systems, in the THz frequency region it is possible to observe emissions due to bound-electron transition among an excited state of the impurity and the ground impurity state. A band diagram showing the particular transition which showed stimulated emission is reported in Fig. 1.10.

Very narrow spectral emission and a light intensity threshold versus pumping power are data reported in support of the stimulated nature of the recombinations. More data are, however, needed to clarify the nature of the emission under investigation.

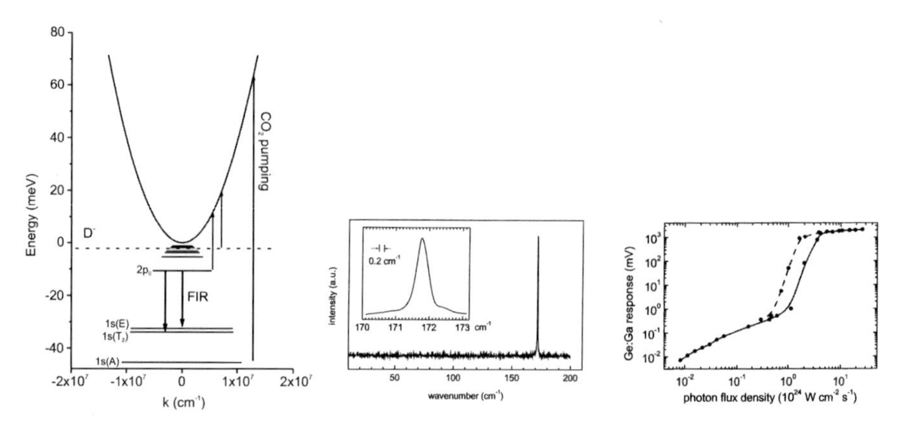

Fig. 1.10. *Left:* Optical transitions in Si:P. The *dashed line* represents the energy level of the D^- center state. After [43]. *Middle* Stimulated emission spectrum from Si:Sb. The emission line is identified with the $2p_0 \rightarrow 1s$ intracenter Sb transition. *Right:* Dependence of the emission on the pump power for 9.6 μm excitation (*dashed line*) or 10.6 μm excitation (*solid line*). After [44]

1.3.2 Alloy-Induced Luminescence

Si-rich group-IV alloys can be used in light emitting systems which would take all the advantages of monolithic integration. In particular, Ge [47], due to its good miscibility, has been intensively investigated. The low temperature excitonic band gap of $Si_{1-x}Ge_x$ alloys is shown in Fig. 1.11, together with the relative luminescence spectra at 4.2 K [48, 49]. The alloys have been produced by liquid-phase (LPE) and vapor-phase epitaxy (VPE).

The band gap varies from the Si free exciton gap of 1.155 eV to the Ge excitonic gap of 0.740 eV. As evident from the figure, around $x = 0.85$,

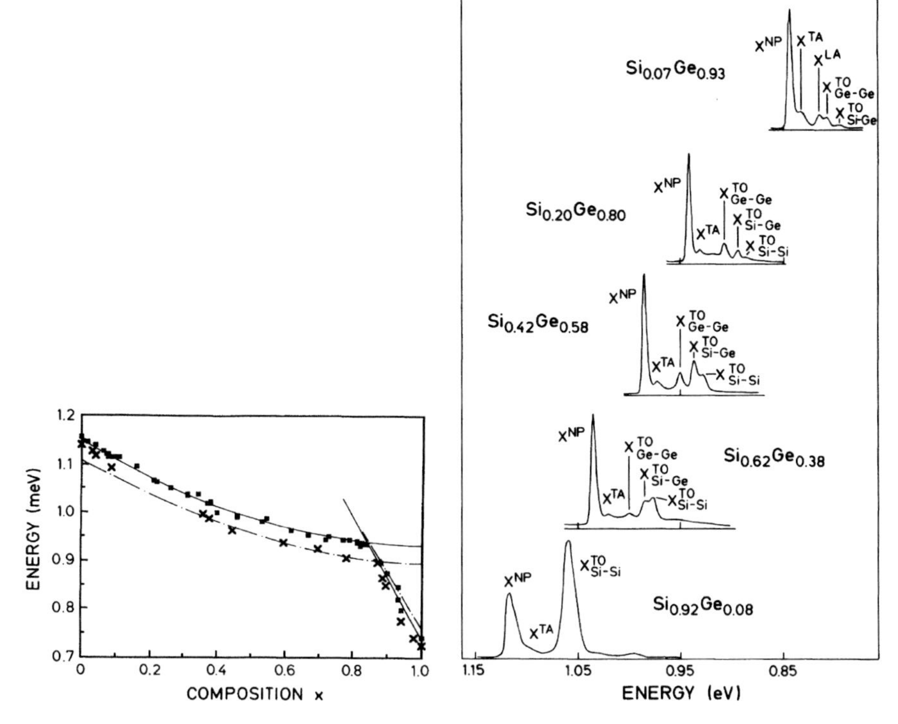

Fig. 1.11. *Left*: Excitonic band gap vs composition x for $Si_{1-x}Ge_x$ alloys. Squares: [48]; crosses: [49]. *Right*: Near band-gap luminescence spectra for several SiGe samples at 4.2 K. The optical transitions are named $X^j{}_i$ where j gives the type of transitions (no-phonon or phonon participation) and i specifies the nature of the phonon. After [48]

the character of the band gap changes suddenly. From $x = 0$ to $x = 0.85$ the minimum of the conduction band is dictated by the Si-like X minima, whereas in the 0.85–1.00 composition range the minimum is related to the Ge-like L minima. These results are reflected in the near-band-gap luminescence spectra, that, by increasing the Si content in the alloy, show that: (1) the PL lines become sharper, (2) the phonon replicas are more structured, (3) the NP line is more intense. This last fact is due to the alloy fluctuations, that act as scattering centers allowing momentum conservation. Unfortunately, luminescence is rapidly quenched at high temperature. Apart from the poor temperature dependence, a problem is related to the presence of dislocations that act negatively on diode characteristics.

Molecular beam epitaxy (MBE) has been widely used for growing SiGe systems on Si substrate in order to obtain dislocation-free alloys. The SiGe alloy is strained due to the difference in lattice parameter (4.2%) between bulk Si and Ge. Strain affects the electronic properties. When the SiGe epilayer is thinner than the critical thickness for strain relaxation, a strained layer is formed which has a lattice constant equal to that of the substrate (pseudomorphic growth). In this case the energy gap of strained $Si_{1-x}Ge_x$ alloys is reduced with respect to the relaxed ones. Both PL and EL have been observed in strained alloys, but only at low temperature [50]. Light emitting diodes and their PL and EL are shown in Fig. 1.12. EL persists only up to about 80 K.

Even if recently the design of these devices has been improved [51], the growth of strained SiGe alloy is limited by its maximum thickness. Because of stability limitations, the critical thickness of the $Si_{1-x}Ge_x$ layer decreases rapidly with increasing Ge concentrations [52]. One way to overcome this problem is the use, instead of a single layer, of a multiple quantum well structure (MQW). PL and EL for SiGe MQW have been observed by many authors (for reviews see [52, 53]) and their temperature dependence has been steadily improved [54, 55, 56, 57, 58]. It is possible to tune the energy gap from 1.3 μm to 1.55 μm. Moreover for these systems also PL and EL at room temperature at 1.3 μm have been observed [58] (see Fig. 1.13). Nevertheless the external quantum efficiency of the LED remains very low, $\simeq 10^{-7}$. SiGe alloys are used as infrared detector devices.

The use of Si alloys both in the form of single layers or MQW is an example of band gap engineering. The goal is to control the band structures through the use of composition and strain. For example by varying the substrate, one

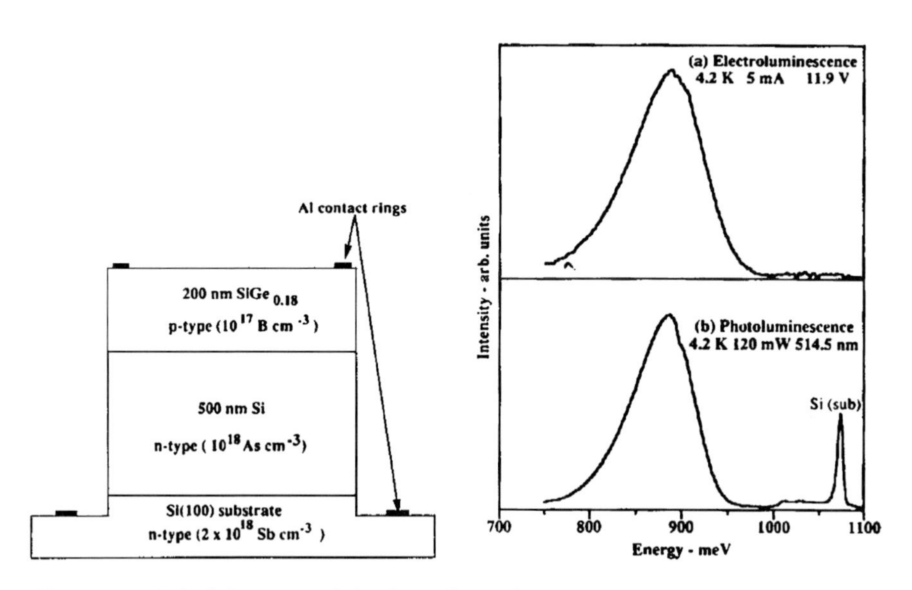

Fig. 1.12. *Left*: Schematic of the $Si_{0.82}Ge_{0.18}$ heterostructure processed into light emitting diodes. *Right*: (**a**) Electroluminescence under a forward bias of 5 mA at 11.9 V and (**b**) photoluminescence at 4.2 K of the $Si_{0.82}Ge_{0.18}$ heterostructure. After [50]

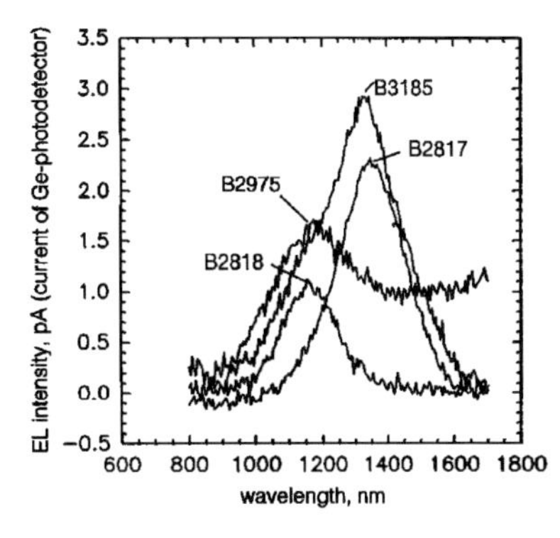

Fig. 1.13. Room temperature electroluminescence spectra of different SiGe MQW diodes. After [58]

might hope to change the nature of the band gap. The ultimate goal is to obtain a direct band gap structure.

1.3.3 Zone Folding: Si-Ge Superlattices

The impressive progress in the epitaxial growth of heterostructures raised the hope of realizing the conversion of the Si indirect band gap into a direct one through Brillouin zone folding. These efforts have been extensively reviewed [47, 59]. The idea was proposed in 1974 [60] and it is sketched in Fig. 1.14.

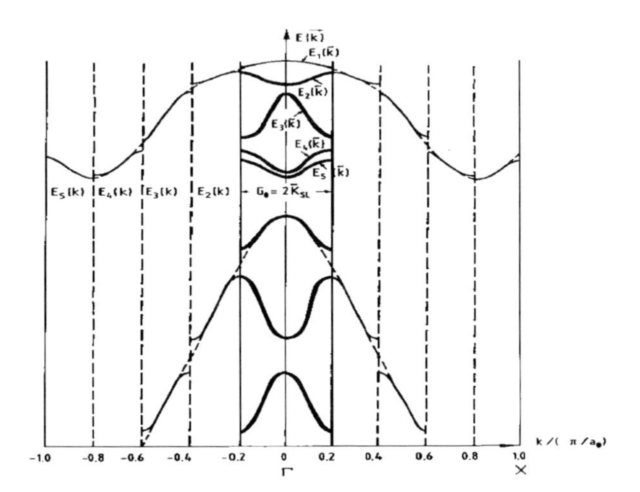

Fig. 1.14. Brillouin zone folding of a Si-like band structure introduced by a superlattice with a period of $L = 5a_0$ (a_0 is the lattice constant of Si). The structure is folded into the first minizone (*bold curves*), which has as extension 1/5th of the original zone. The indirect band gap transforms into a direct band gap at $k = 0$. After [59]

Zone folding is a direct consequence of the superperiod imposed on the lattice of the host crystal when a superlattice is grown. We consider here Si_mGe_n superlattice, where m is the number of Si monolayers and n the number of Ge monolayers. The band structure is folded in minizones within the first Brillouin zone depending on the number of the monolayers. The periodicity of the real lattice is replaced by those of the superlattice. For the right choice of the superlattice periodicity the minimum of the Si conduction band, located in the Δ direction, should be folded back at the center of the Brillouin zone, creating a direct band gap. The suitable candidate is a 10 monolayer superlattice. However, also the strain induced by the buffer layer plays a role with respect to the folded band gap. A superlattice grown directly on Si will still behave, due to the strain, as an indirect semiconductor. The strain will be symmetrized if a Si_mGe_n superlattice is grown on a thin $Si_{1-x}Ge_x$ layer where x equals $n/(n+m)$. Photoluminescence of SiGe superlattices has been reported and studied by many authors [47, 59]. Figure 1.15 reports the results [61] for strain-symmetrized Si_mGe_n superlattices, for the cases $m = 9$ and $n = 6$; $m = 6$ and $n = 4$ and $m = 3$ and $n = 2$.

The PL spectrum for a $Si_{0.6}Ge_{0.4}$ alloy layer is reported for comparison. Taking into account the thicknesses of the superlattices and the thickness of the alloy layer the intensity of the NP line is enhanced by a factor of 150 for

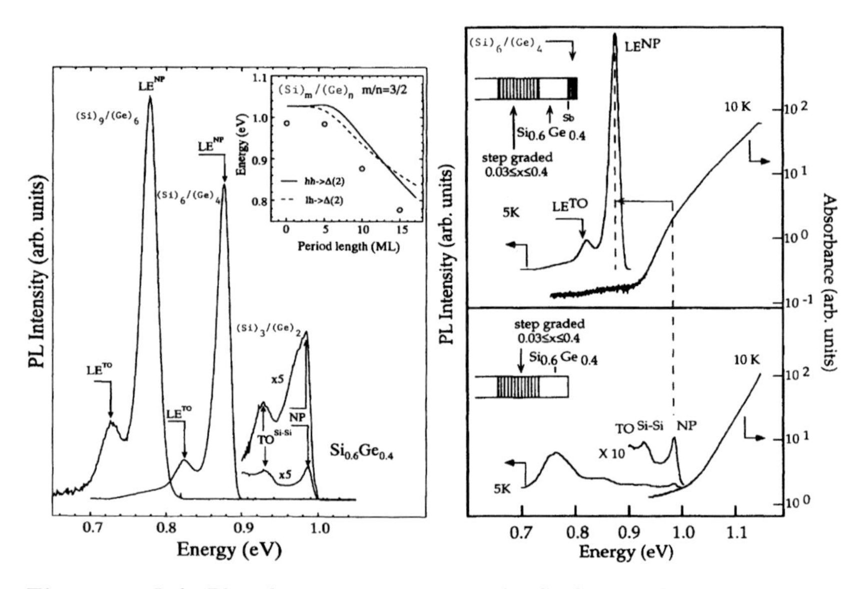

Fig. 1.15. *Left*: Photoluminescence spectra for Si_9Ge_6, Si_6Ge_4, Si_3Ge_2 superlattices and a $Si_{0.6}Ge_{0.4}$ alloy. In the *inset* the PL peak energies of the NP lines are compared with band gaps calculated using an effective mass calculation. *Right*: Photoluminescence and absorption spectra for a Si_6Ge_4 superlattice and a $Si_{0.6}Ge_{0.4}$ alloy. After [61]

the Si_6Ge_4, 90 for Si_9Ge_6 and 5 for Si_3Ge_2 compared to the alloy, in agreement with zone folding arguments. The right part of the figure, where the PL and absorption spectra of the Si_6Ge_4 superlattices are shown together with that of the alloy, is clear evidence that both absorption and PL are due to transitions across the superlattice band gap. The measurements were performed at low temperature, 5–10 K. Increasing the temperature rapidly decreases the intensity of the PL signals. At 29 K no signals from the interband transitions–related PL are observed. The efficiency is limited by the type II band alignment of the superlattices, where the holes remain localized in the Ge-rich layers and the electron in the Si-rich layers. To strength the electron–hole transitions both localizations in reciprocal and real space are important. For example it is possible to embed $Ge_nSi_mGe_n$ structures in cladding layers, forcing the localization of electrons in the $Ge_nSi_mGe_n$ layers. This has been done, using a $Si_{0.85}Ge_{0.15}$ cladding layer [62], and the photoluminescence signal increases up to 185 K and at room temperature is still only 1/2 of the low temperature value as shown in Fig. 1.16.

For LEDs based on this scheme, the electroluminescence signal remains up to room temperature, even if the external quantum efficiency is only 10^{-5} [57].

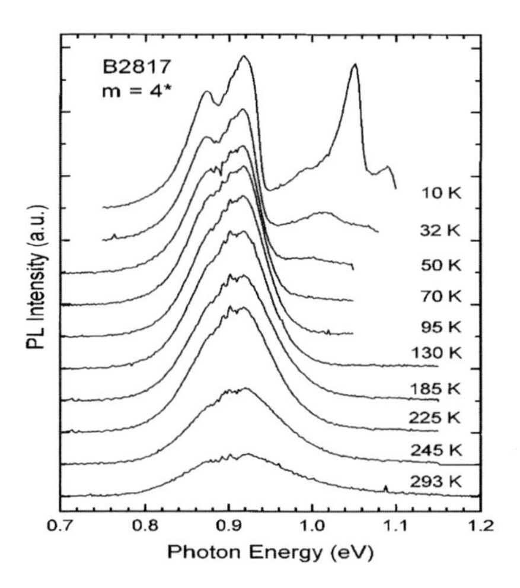

Fig. 1.16. Temperature dependence of the PL of the $Ge_4Si_{20}Ge_4$ structure. After [62]

1.3.4 Silicon–Germanium Quantum Cascade Emitters

Another approach for Si-based light emitting devices is linked to intersubband transitions [63, 64, 65] such as those occurring in quantum cascade lasers which are so effective in III-V based systems [66, 67, 68]. In this case, since the transitions occur within the same band in a unipolar device, they are not limited or influenced by the indirect band gap of silicon. This idea is sketched

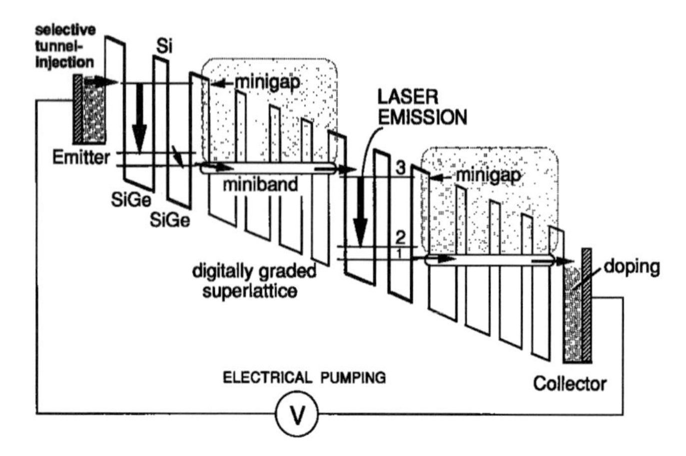

Fig. 1.17. Silicon based intersubband quantum cascade laser. After [63]

in Fig. 1.17 which shows the energy bands of a quantum cascade intersubband laser grown on a Si substrate [63]. The figure shows an electron-injected n-i-n conduction band device, although a hole-injected p-i-p valence band device would be more useful since the valence band offset in SiGe/Si is much larger than the conduction band offset [64]. The carriers make a vertical transition between subbands 3 and 2, then they cascade down the electrically biased staircase. In order to assist population inversion, the lower laser level 2 is rapidly depopulated by phonon emission if, by design, the energy difference between levels 2 and 1 is resonant with phonon energies. Practically one has two identical active regions connected by an injector.

The high control reached in the molecular beam deposition of strain compensated films allows the fabrication of a dislocation–free superlattice by choosing appropriate widths and Ge contents of the SiGe quantum wells. Starting from the possibility of monolithic integration with silicon microelectronics, the Si/SiGe system actually offers numerous advantages over III-V heterostructures for quantum cascade laser application. Due to the covalent bonding, nonpolar phonon scattering is practically ruled out. This is importent because the nonpolar electron–phonon interaction is the dominant nonradiative loss process in III-V quantum cascade lasers. The optical phonon energy in Si is much higher than in GaAs (64 meV compared with 36 meV), providing a larger frequency window within which (nonpolar) optical phonon scattering is suppressed. In Si the thermal conductivity is much larger than that of GaAs, giving better prospects of continuous–wave (CW) operation at non-cryogenic temperatures.

Efficient intersubband electroluminescence in quantum cascade structures has been demonstrated [69, 70, 71, 72, 73, 74]. The levels involved are valence levels, and the radiative transition is between heavy hole states. In the first case [69] the growth was realized pseudomorphically on high-resistivity

Si(100) substrates and intersubband electroluminescence was observed up to 180 K centered at 130 meV. The nonradiative lifetime depends strongly on the quantum well structure design and is comparable to III-V material systems. Varying the well width and the Ge content it is possible to shift the intersubband transition energies as witnessed by the left part of Fig. 1.18 [70]. The observed peak shift is well described by quantum confinement effects; the quantum efficiency estimate is about 10^{-5}. Similar results have been obtained [74], see Fig. 1.18. The absence of the EL peak under reverse bias strongly supports the intersubband origin. Temperature-dependent measurements show nearly identical spectra between 20 and 90 K and a broadening and vanishing of the peak up to 160 K.

It is possible to improve these results by controlling the large accumulation of strain imposed by the use of a Si substrate. This has been done using a $Si_{0.5}Ge_{0.5}$ substrate and growing on it strain compensated $Si_{0.2}Ge_{0.6}$/Si quantum wells. Intersubband transitions have been observed by absorption measurements at 235, 262 and 325 meV. By changing the well width from 3.5 to 2.5 nm, peaks are observed up to room temperature [71]. For similar structures EL has been detected at 80 K [72].

All these approaches address the infrared spectral region, a region not overlapping any of the standard telecommunication windows. Despite some

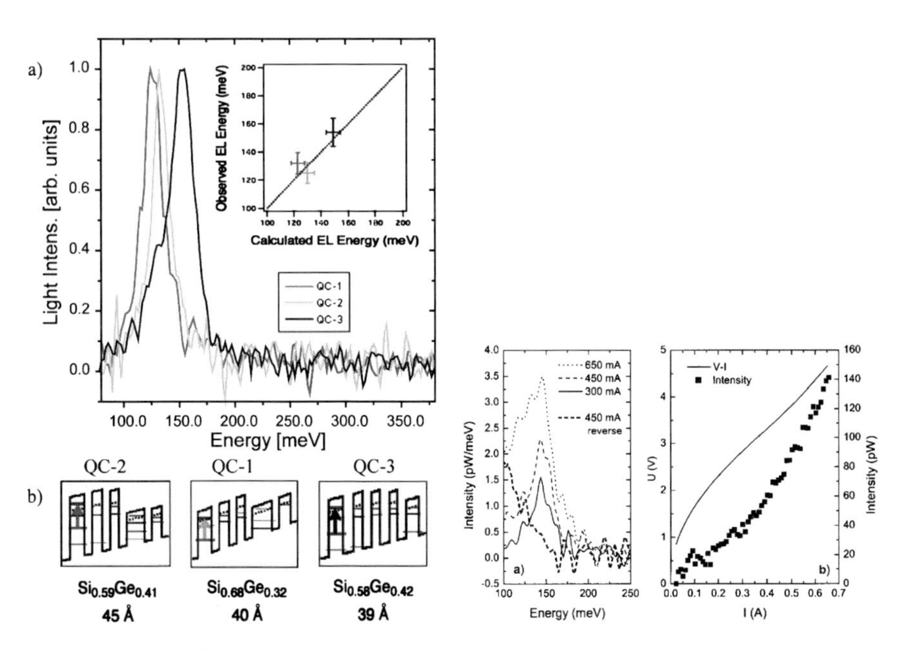

Fig. 1.18. *Left*: (**a**) EL spectra of three Si/SiGe structures, depicted schematically in (**b**). The *inset* in (**a**) compares calculated vs measured intersubband transition energies. After [70]. *Right:* (**a**) Current-dependent EL spectra and a spectrum at reverse bias at 80 K; (**b**) *I-V* curve and integrated EL intensity. After [74]

authors proposing the use of quantum cascade lasers for free-air optical interconnects, such a Si/Ge quantum cascade laser will be of little use for silicon photonics unless many other kinds of devices are developed, such as integrated waveguides or detectors.

Si/SiGe multilayers are also used to achieve THz (6 THz or 40 μm) emission by interband heavy-hole–light-hole transitions in [75]. Intersubband transitions have been observed also at 2.9 THz (103 μm), 8.9 THz (33.7 μm) and 16.2 THz (18.5 μm) in edge emission electroluminescence at 4.2 K, see Fig. 1.19 [76].

Although only spontaneous and not stimulated EL has been observed, the results are very encouraging; a very high THz power of 50 nW has been measured [77], 10^2 times higher than in the case of GaAs/AlGaAs THz quantum cascade structures [78], which, in suitable cavities, yielded THz laser action.

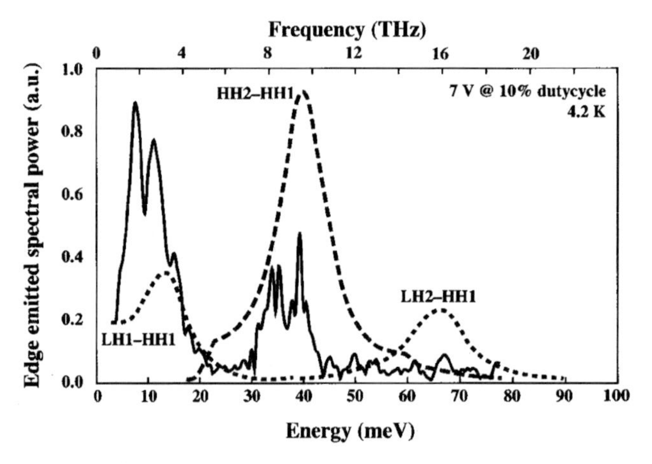

Fig. 1.19. Edge-emission EL spectrum for an applied bias voltage of 7 V at 4.2 K (*solid line*). Calculated edge-emission spectrum for the cascade wafer (*dotted line*) showing the involved transitions. After [76]

1.3.5 Quantum Confinement: Porous Silicon and Silicon Nanostructures

The quantum confinement effect in silicon structures of nanometric size constitutes another approach to engineering a direct transition and the emission of visible light. The larger band gap with respect to bulk Si arises as a consequence of fundamental quantum mechanics where energy levels for a particle inside a potential box will be moved in energy depending on the box size. Valence states will be shifted down and conduction states will be shifted up in energy, so that the effective bandgap will be enlarged (see Sect. 2.1).

Consider, for example, Si and Ge quantum films of different crystallographic orientation. In a simple effective mass approximation the band gap shift due to quantum confinement will be $\Delta E_{\mathrm{g}} \propto 1/m^*$, where m^* is the electron effective mass in the confinement direction. Due to the anisotropic behavior of the electron effective mass for Si and Ge (see Fig. 1.2 and Table 1.1), the electron effective mass for the three main surface orientations will behave quite differently. Table 1.3 and Fig. 1.20 show how the six equivalent valleys of Si and the four equivalent valleys of Ge are affected by quantum confinement [79, 80].

We see that in the case of Si, regarding the Si(111) orientation, all the six valleys having the same effective mass will be equally shifted in energy; in the other directions the six valleys will be split by the confinement in groups of two and four. Since the shift is proportional to the inverse of the effective mass, the valleys with two as degeneracy will be shifted more than the other four in the case of Si(001); the opposite happens for the Si(110) films. The situation is different for Ge owing to the different location of the ellipsoid minima. Here only for the Ge(100) surface will the plane in k space containing the wavevectors for motion parallel to the surface make equal angles with all the four constant-energy surfaces of bulk. For the (110) surface one obtains two lower lying valleys and two higher lying valleys; for the

Table 1.3. Effective masses for three surface orientations, for semiconductors having band structures like those of the conduction band of Si (six {100} ellipsoids of revolution) and of Ge (four {111} ellipsoids of revolution). m_{T} and m_{L} are the transverse and longitudinal effective masses of the electrons in the bulk. n is the degeneracy of each set of ellipses. m^* is the effective mass in the confinement direction. Their values are in units of the free-electron mass. After [79]

		Si			
Surface	(001)		(110)		(111)
Valleys	Lower	Higher	Lower	Higher	All
Degen. n	2	4	4	2	6
mass m^*	m_{L}	m_{T}	$\frac{2m_{\mathrm{T}}m_{\mathrm{L}}}{m_{\mathrm{T}}+m_{\mathrm{L}}}$	m_{T}	$\frac{3m_{\mathrm{T}}m_{\mathrm{L}}}{m_{\mathrm{T}}+2m_{\mathrm{L}}}$
mass m^*	0.916	0.190	0.315	0.190	0.198

		Ge			
Surface	(001)	(110)		(111)	
Valleys	All	Lower	Higher	Lower	Higher
Degen. n	4	2	2	1	3
mass m^*	$\frac{3m_{\mathrm{T}}m_{\mathrm{L}}}{m_{\mathrm{T}}+2m_{\mathrm{L}}}$	$\frac{3m_{\mathrm{T}}m_{\mathrm{L}}}{2m_{\mathrm{T}}+m_{\mathrm{L}}}$	m_{T}	m_{L}	$\frac{9m_{\mathrm{T}}m_{\mathrm{L}}}{m_{\mathrm{T}}+8m_{\mathrm{L}}}$
mass m^*	0.120	0.224	0.082	1.64	0.092

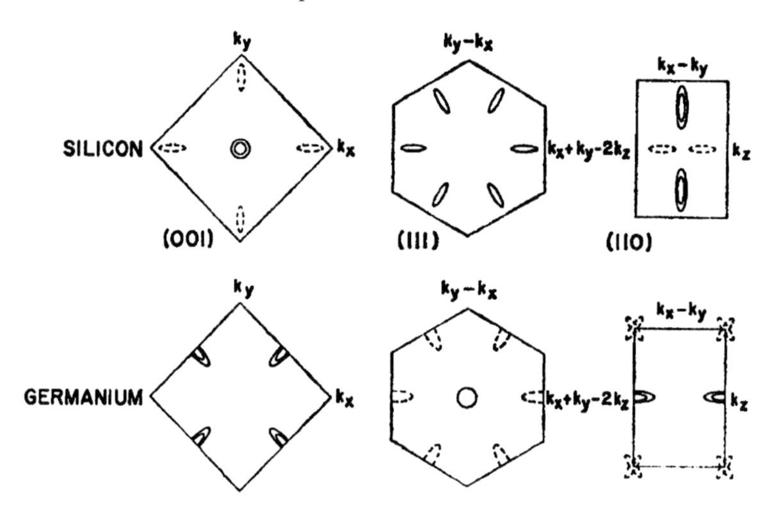

Fig. 1.20. Schematic representation of the constant-energy ellipses and Brillouin zones for the (001), (111), and (100) surfaces of Si and Ge in the effective-mass approximation. Where both *solid* and *dashes curves* appear, the *solid curves* correspond to the electric subband of lower energy. Concentric ellipses are shown to indicate doubly degenerate levels. The radii of the circumscribed circles about the three Brillouin zones are $2\pi/a$, $(4\pi/3a)\sqrt{2}$, and $(\pi/a)\sqrt{3}$ for (001), (111) and (110) respectively. After [79]

(111) surface one lower lying valley and three equivalent higher lying valleys. Looking at Table 1.3 one would expect that the confinement energy shifts for the different films, governed by the lowest-lying valleys, and proportional to the inverse of the perpendicular effective mass, should be in the sequence (111) > (110) > (100) for Si quantum films and (100) > (110) > (111) for Ge quantum films. Moreover looking at Fig. 1.20 we see that in the case of Si(100) and Ge(111) the lowest lying valleys are now located at the center of Brillouin zone; thus one would expect in these cases the presence of a direct band gap, since the maximum of the valence band is also located at the center of the Brillouin zone.

As pointed out before, the confinement will act also on the valence and not only on the conduction band; moreover the effective mass approximation avoids any realistic treatment of the boundary conditions at the surfaces. Thus the effective mass approximation we used before yields only an estimation of the quantum confinement effects. Nevertheless first principle calculations of these effects on Si and Ge quantum films of different orientations saturated by hydrogen show that the underlying trends are correct (see Sect. 2.2.1).

Confinement in a small size has other effects on the recombination mechanisms [81]: carriers become localized and cannot diffuse to defects and thus Shockley–Read–Hall recombination is suppressed. Auger recombination is also absent until two excitons are created within the same nanocrystals. Moreover reducing the size radiative recombination becomes more efficient since electron and hole wavefunctions overlap more and more in space causing faster recombination.

In 1990 Canham demonstrated that Si nanostructures may be useful for optoelectronics due to the room temperature enhanced emission of light with respect to bulk Si for porous silicon [82, 83]. The reason of this is attributed to the low dimensionality of the surviving Si skeleton, see Fig. 1.21.

Research on porous silicon is widely discussed in Chap. 3. After the discovery of bright photoluminescence in porous silicon, research on light emission from Si nanostructure has been very impressive. A survey of the most relevant results is given in Chap. 4. Also Ge and Si/Ge nanostructures have been widely investigated. The interested reader can find exhaustive reviews [84, 85].

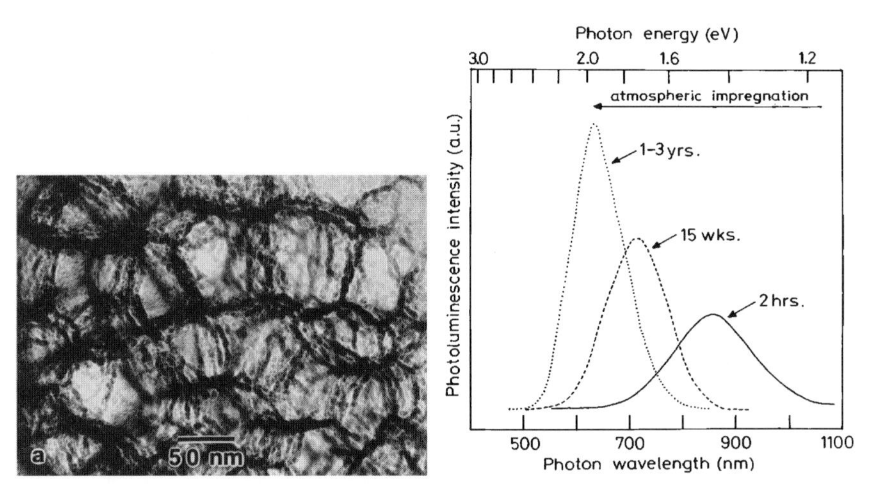

Fig. 1.21. *Left*: Transmission electron microscope image of a highly luminescent, 95% porosity Si sample. Undulating Si columns of 3–5 nm are visible. *Right*: Photoluminescence spectra from a 77% porosity Si sample following storage in air for the indicated times. After [83]

1.3.6 Semiconducting Silicides

Another method suggested to achieve direct band gap behavior in silicon systems is the use of semiconducting silicides. In particular iron disilicide ($FeSi_2$) and diruthenium trisilicide (Ru_2Si_3) have attracted attention in the last decade due to their potential for silicon-based optoelectronic applications [86].

Iron disilicide occurs in two stable phases depending on the temperature. The interesting semiconducting phase is the β-$FeSi_2$ phase. It has been shown theoretically [87, 88] that β-FeSi$_2$ is an indirect semiconductor, with a direct gap only some tens of meV above the indirect one, at about 0.8 eV corresponding to a wavelength of 1.55 μm, which is important for applications in fiber based telecommunications. Moreover the band structure shows a high sensitivity to strain, which can even reverse the role of direct and indirect gaps [89]. Some experimental observations have concluded that the band gap is indirect, while others indicate direct band gap behavior [90]. These differences could be related both to the different growth techniques used and to the use for band gap measurements of optical absorption or photoluminescence. Indeed even absorption and PL measurements taken on the same sample sometimes, show different results. It is interesting to note that: (1) light emission has not been observed in bulk β-FeSi$_2$, but only in thin films or precipitates, where stress is likely to play an important role, and (2) in a series of experiments Grimaldi and coworkers have demonstrated that light emission does not occur in highly strained small precipitates but only in large precipitates, apparently not strained, with extremely long radiative lifetimes [91, 92].

Interest has greatly increased after the demonstration of a working and robust LED based on β-FeSi$_2$ [93]. The device was fabricated by forming ion beam synthesized iron disilicide into a pn junction epitaxial layer grown by MBE. Successively also a LED formed completely by ion implantation (FI) has been fabricated [94]. Both devices are shown schematically in Fig. 1.22.

The electroluminescence spectra, measured at 80 K, are shown in Fig. 1.23. The intensities from the devices are similar. The EL reduces with increasing temperature but could still be seen at room temperature [93].

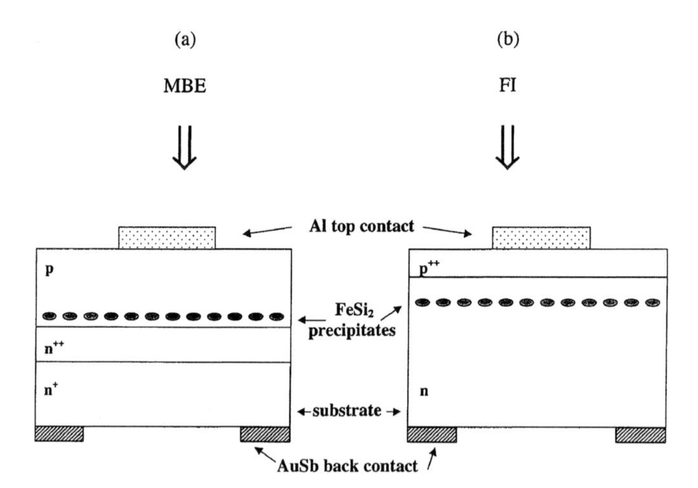

Fig. 1.22. Schematic of LEDs made by (**a**) synthesis into an existing MBE grown pn junction and (**b**) a fully implanted (FI) LED. After [94]

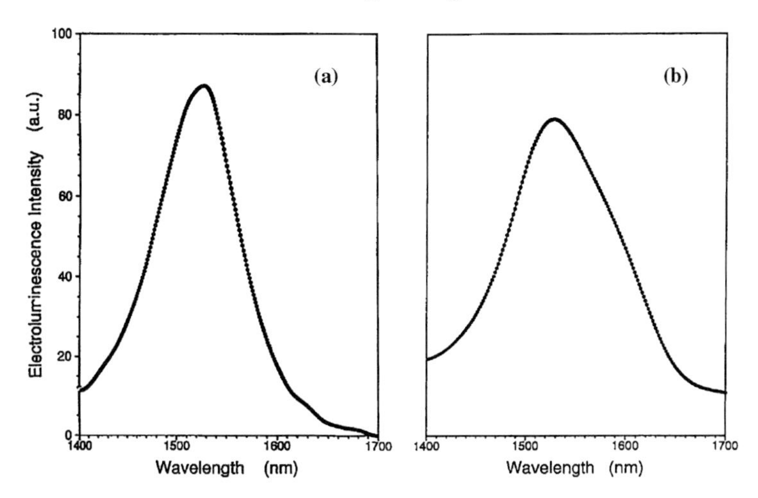

Fig. 1.23. Electroluminescence spectra, measured at 80 K, from (**a**) the MBE LED and (**b**) the FI LED. After [94]

The debate about the nature of the band gap and the origin of the EL in β-$FeSi_2$ is still open. Two very recent works reach opposite conclusions. In the first [95] a detailed analysis of the band gap behavior is done using optical absorption, PL and photoreflectance measurements. The direct gap revealed at 0.815 eV by the photoreflectance spectra agrees well with the PL results if temperature difference and trap recombination effects are taken into account; moreover also the measured optical absorption results, if corrected through the effects of stress, are in good agreement. For the authors, these results provide very convincing evidence for intrinsic light emission. In the second work [96] continuous wave and time dependent PL measurements in iron disilicide precipitate layers have been employed to achieve important information on the physics of the recombination mechanisms. The long decay times of 10 μs at 10 K and the low quantum efficiency found have been interpreted in terms of an indirect transition. No evidence in the time resolved spectra for a contribution that can be solely related to the silicide has been found.

Ruthenium occupies in the periodic table the place immediately below iron, thus one expects a similar chemistry in this case. In particular orthorhombic Ru_2Si_3 represents a promising compound since *ab initio* theoretical calculations indicate that Ru_2Si_3 shows a direct transitions at the Γ point [97, 98, 99], even if with low oscillator strength [100]. Absorption measurements on Ru_2Si_3 single crystals yield a direct band gap at 0.84 eV [101]. In samples fabricated by ion beam synthesis [102] also a direct band gap at 0.90 eV has been observed [103]. In the case of Ru_2Si_3 films grown by MBE on silicon substrate, the measurements do not allow an evaluation of the direct band gap because of the very low absorption coefficient [100, 104].

Unfortunately, no photoluminescence has been observed in the near-infrared region [100]. Further studies on this new material are needed.

1.3.7 Hybrid Approaches

One way to overcome the inability of silicon to emit light is to integrate III-V semiconductors onto a Si wafer, combining the optoelectronic properties of the first with the electronic ones of the second [27, 28, 30, 105, 106, 107, 108]. This can be an effective way, and technologically is the more viable today. Here the main problem is related to the growth of III-V semiconductors on silicon, to the different properties of the two materials, where the atoms of one are used as dopants for the other, and to the economical feasibility of the two different processes. Two main approaches have been developed and used: hetero-growth and bonding.

In the first case a III-V epilayer is directly grown on a Si wafer. III-V emitters directly grown on silicon have serious problems due to the lattice mismatch between the two semiconductors (4% in the case of GaAs/Si, 8% in the case of InP/Si) that induces a very high defect density. Nevertheless modulator [109, 110], LED [111, 112, 113], AlGaAs/GaAs laser [114, 115], GaInAsP/InP laser [116], InGaAs/InGaAsP MQW laser [117] and vertical cavity surface emitting laser (VCSEL) [118] have been successfully integrated to silicon. However difficulties remain related to the thermal mismatch of the involved semiconductors (note that the thermal mismatch of InP on Si is half the value of GaAs on Si) and to the necessity to prevent any of the silicon oxide in the circuit from being exposed to Ga or In which makes the oxide conducting. Very interesting results have been recently achieved for the growth of GaN on Si and of InAs quantum dots on Si [119].

In the other hybrid approach, a III-V device is grown on a lattice-matched substrate and subsequently transferred to a silicon wafer or processed chip. Various techniques have been demonstrated and used [106, 107, 108]. In the epitaxial lift-off procedure [120], a selective etching process is used to physically separate the epitaxial layer from the substrate and, then, the epitaxial layer is attached by van der Waals bonding to Si. This has been used in order to transfer LED arrays [121, 122] and VCSEL [123] on silicon integrated circuits. It is also possible to suppress the lift-off step and perform a direct bonding between III-V wafers and Si wafers. This technique is termed wafer-bonding or wafer-fusion and is very versatile. It has been used for integrating VCSEL to Si [124, 125]. Other used hybrid techniques are polyimide bonding [126], fluidic self-assembly [127], DNA-assisted self-assembly and flip-chip bonding [106]. In the lift-off procedures the sacrificial substrate is removed before the photonic devices are attached to the host silicon; in the flip-chip bonding the substrate can be removed after the devices are bonded. This allows one to integrate large VCSEL arrays to Si [128] and to CMOS VLSI chips [106].

In summary tremendous progress has been made, in the last few years, in the integration of III-V with Si. Major concerns are still materials quality, reliability, packaging, and costs. However this approach is up-to-now the only viable technological scheme to add optical functionality to a silicon chip.

References

1. J.R. Chelikowsky, M.L. Cohen: Phys. Rev. B **14**, 556 (1976)
2. M.L. Cohen, J.R. Chelikowsky: *Electronic Structure and Optical Properties of Semiconductors*, 2nd edn., Springer Ser. Solid-State Sci. Vol. 75 (Springer, Berlin, Heidelberg 1989)
3. D.E. Aspnes, A.A. Studna: Phys. Rev. B **27**, 985 (1983)
4. J.H. Davies: *The Physics of Low-Dimensional Semiconductors* (Cambridge University Press 1998)
5. M. Levinshtein, S. Rumyantsev, M. Shur: *Handbook Series on Semiconductors Parameters: Volume 1* (World Scientific, Singapore 1996)
6. L.T. Canham: 'Progress towards understanding and exploiting the luminescent properties of highly porous silicon'. In: *Optical Properties of Low Dimensional Silicon Structure* ed. by D.C. Bensahel, L.T. Canham, S. Ossicini, NATO ASI Series, Vol. 244 (Kluwer Academic Publishers, Dordrecht 1993) pp. 81–94
7. B. Hamilton: 'Si-based light-emitting materials'. In: *Silicon-Based Microphotonics: from Basics to Applications.* ed. by O. Bisi, S.U. Campisano, L. Pavesi, F. Priolo (IOS Press, Amsterdam 1999) pp. 21–46
8. P.T. Landsberg: *Recombination in Semiconductors* (Cambridge University Press 1991)
9. M. Lax: Phys. Rev. **119**, 1502 (1960)
10. J. Dzievor, W. Schmidt: Appl. Phys. Lett. **31**, 346 (1977)
11. CIE Proceedings, Commission Internationale de l'Eclairage (CIE) (Cambridge University Press, Cambridge, UK 1924)
12. C.J. Nuese, J.I. Pankove: 'Light-Emitting Diodes – LEDs'. In: *Display Devices,* ed. by J.I. Pankove (Springer, Berlin 1980)
13. M.H. Pilkuhn, W. Shairer: 'Light Emitting Diodes' In: *Device Physics,* series Handbook on Semiconductors vol. 4, ed. by C. Hilsum (North-Holland, Amsterdam 1993)
14. Z. Gaburro, L. Pavesi: 'Light Emitting Diodes for Si Integrated Circuits' In: *Handbook of Luminescence, Display Materials, and Nanocomposites,* Volume 1, ed. by H.S. Nalwa, L.S. Rohwer (American Scientific Publishers 2003)
15. P.Y. Yu, M. Cardona: *Fundamentals of Semiconductors* (Springer-Verlag 1996)
16. J. Kramer, P. Seitz, E.F. Steigmeier, H. Auderset, B. Delley, H. Baltes: Sensors and Actuators A **37–38**, 527 (1993)
17. M.A. Green, J. Zhao, A. Wang, P.J. Reece, M. Gal: Nature **412**, 805 (2001)
18. J.R. Haynes, W.C. Westphal: Phys. Rev. **101**, 1676 (1956)
19. J.R. Haynes, H.B. Briggs: Phys. Rev. **86**, 647 (1952)
20. H. Schlangelotto, H. Maeder, W. Gerlach: Phys. Stat. Sol. (a) **21**, 357 (1974)
21. V. Alex, S. Finkbeiner, J. Weber: J. Appl. Phys. **79**, 6943 (1996)

22. N. Akil, S.E. Kerns, D.V. Kerns Jr., A. Hoffmann, J.-P. Charle: Appl. Phys. Lett. **73**, 871 (1998)
23. W.P. Dumke: Phys. Rev. **127**, 1559 (1962)
24. W.L. Ng et al.: Nature **410**, 192 (2001)
25. C.W. Liu, M.H. Lee, Miin-Jang Chen, I.C. Lin, Ching-Fuh Lin: Appl. Phys. Lett. **76**, 1516 (2000)
26. Ching-Fuh Lin, Miin-Jang Chen, Eih-Zhe Liang, W.T. Liu, C.W. Liu: Appl. Phys. Lett. **78**, 261 (2000)
27. S.S. Iyer, Y.-H. Xie: Science **260**, 40 (1993)
28. S.S. Iyer, Y.-H. Xie: 'Light emission in silicon'. In: *Porous Silicon*, ed. by Z.C. Feng, R. Tsu (World Scientific, Singapore 1994) pp. 53–76
29. L.C. Kimerling, K.D. Kolenbrander, J. Michel, J. Palm: 'Light emission from silicon'. Solid State Phys. **50**, 333 (1997)
30. D.J. Lockwood: 'Light emission in silicon'. In: *Light Emission in Silicon From Physics to Devices*. Semiconductor and Semimetals Vol. 49, ed. by D.J. Lockwood (Academic Press, San Diego 1998) pp. 1–36
31. G. Davies: Physics Reports **176**, 83 (1989)
32. H. Ennen, G. Pomrenke, A. Axmann, K. Eisele, W. Haydl, J. Schneider: Appl. Phys. Lett. **46**, 381 (1985)
33. A. Polman: J. Appl. Phys. **82**, 1 (1997)
34. J. Michel, L.V.C. Assali, M.T. Morse, L.C. Kimerling: 'Erbium in Silicon'. In: *Light Emission in Silicon From Physics to Devices*. Semiconductor and Semimetals Vol. 49, ed. by D.J. Lockwood (Academic Press, San Diego 1998) pp. 111–156
35. F. Priolo: 'Light emission of Er^{3+} in crystalline silicon'. In: *Silicon-Based Microphotonics: from Basics to Applications*. ed. by O. Bisi, S.U. Campisano, L. Pavesi, F. Priolo (IOS Press, Amsterdam 1999) pp. 279–346
36. R. Sauer, J. Weber: 'Photoluminescence of defect complexes in silicon'. In: *Defect Complexes in Semiconductor Structures*. ed. by J. Giber (Springer, New York 1983) pp. 79–94
37. S.T. Pantelides: *Deep Centers in Semiconductors* (Gordon and Breach, New York 1986)
38. D.G. Hall: 'The role of silicon in optoelectronics'. In: *Proceedings of the Materials Research Society: Silicon Based Optoelectronic Materials*. (MRS, San Francisco 1987) pp. 367–371
39. T.G. Brown, D.G. Hall: 'Radiative isolectronic impurities in silicon and silicon-germanium alloys and superlattices'. In: *Light Emission in Silicon From Physics to Devices*. Semiconductor and Semimetals Vol. 49, ed. by D.J. Lockwood (Academic Press, San Diego 1998) pp. 77–110
40. M.L.W. Thewalt, S.P. Watkins, U.O. Ziemelis, E.C. Lightowlers, M.O. Henry: Sol. State Comm. **44**, 573 (1982)
41. T.G. Brown, D.G. Hall: Appl. Phys. Lett. **49**, 245 (1986)
42. P.L. Bradfield, T.G. Brown, D.G. Hall: Phys. Rev. B **38**, 3533 (1988)
43. S.G. Pavlov et al.: Phys. Rev. Lett. **84**, 5220 (2000)
44. S.G. Pavlov et al.: J. Appl. Phys. **92**, 5632 (2002)
45. G. Bohnert, K. Weronek, A. Hangleiter: Phys. Rev. B **48**, 14973 (1993)
46. L.T. Canham, K.G. Barraclough, D.J. Robbins: Appl. Phys. Lett. **51**, 1509 (1987)

47. G. Abstreiter: 'Band gaps and light emission in Si/SiGe atomic layer structures' In: *Light Emission in Silicon From Physics to Devices.* Semiconductor and Semimetals Vol. 49, ed. by D.J. Lockwood (Academic Press, San Diego 1998) pp. 37–76
48. J. Weber, M.I. Alonso: Phys. Rev. B **40**, 5683 (1989)
49. R. Braunstein, A.R. Moore, F. Herman: Phys. Rev. **109**, 695 (1958)
50. N.L. Rowell, J.P. Noel, D.C. Houghton, M. Buchanan: Appl. Phys. Lett. **58**, 957 (1990)
51. L. Vescan, T. Stoica: J. of Lumin. **80**, 485 (1999)
52. T.P. Pearsall: Progr. Quant. Opt. **18**, 97 (1994)
53. S. Fukatsu: Solid-State Electronics **4–6**, 817 (1994)
54. D.J. Robbins, P. Calcott, W.Y. Leong: Appl. Phys. Lett. **59**, 1350 (1991)
55. Q. Mi, X. Xiao, J.C. Sturm, L.C. Lenchyshyn, M.L.W. Thewalt: Appl. Phys. Lett. **60**, 3177 (1992)
56. S. Fukatsu, N. Usami, Y. Shiraki, Y. Nishida, K. Nagkawa: Appl. Phys. Lett. **63**, 967 (1993)
57. M. Gail et al.: Solid State Phenom. **47–48**, 473 (1996)
58. H. Presting, T. Zinke, A. Splett, H. Kibbel, M. Jaros: Appl. Phys. Lett. **69**, 2376 (1996)
59. H. Presting et al.: Semicond. Sci. Technol. **7**, 1127 (1992)
60. U. Gnutzmann, K. Clausecker: Appl. Phys. **3**, 9 (1974)
61. U. Menczigar et al.: Phys. Rev. B **47**, 4099 (1993)
62. M. Gail et al.: Appl. Phys. Lett. **66**, 2978 (1995)
63. R.A. Soref: J. Vac. Sci. Technol. A **14**, 913 (1996)
64. R.A. Soref: Thin Solid Films **294**, 325 (1997)
65. R.A. Soref, L. Friedman, G. Sun: Superlatt. Microstruct. **23**, 427 (1998)
66. J. Faist et al.: Science **264**, 553 (1994)
67. R.Q. Yang et al.: Appl. Phys. Lett. **70**, 2013 (1997)
68. C. Sirtori et al.: Appl. Phys. Lett. **73**, 3486 (1998)
69. G. Dehlinger et al.: Science **290**, 2277 (2000)
70. G. Dehlinger et al.: Mat. Sci. Engin. B **89**, 30 (2002)
71. L. Diehl et al.: Appl. Phys. Lett. **80**, 3274 (2002)
72. L. Diehl et al.: 'Strain compensated Si/SiGe quantum cascade emitters grown on SiGe pseudosubtrates'. In: *Towards a Silicon Laser.* NATO Science Series II, Mathematics, Physics and Chemistry vol. 93, ed. by L. Pavesi, S.V. Gaponenko, L. Dal Negro (Kluwer Academic Publishers, Dordrecht 2003) p. 325–330
73. R.A. Kaindl et al.: Phys. Rev. Lett. **86**, 1122 (2001)
74. I. Bormann et al.: Appl. Phys. Lett. **80**, 2260 (2002)
75. S.A. Lynch et al.: Appl. Phys. Lett. **81**, 1543 (2002)
76. S.A. Lynch et al.: Mat. Sci. Engin. **B89**, 10 (2002)
77. R.W. Kelsall et al.: 'Terahertz emission from silicon-germanium quantum cascades'. In: *Towards a Silicon Laser.* NATO Science Series II, Mathematics, Physics and Chemistry vol. 93, ed. by L. Pavesi, S. V. Gaponenko, L. Dal Negro, (Kluwer Academic Publishers, Dordrecht 2003) pp. 367–382
78. R. Kohler et al.: Appl. Phys. Lett. **80**, 1867 (2002)
79. F. Stern, W.E. Howard: Phys. Rev. **163**, 816 (1967)
80. T. Ando, A.B. Fowler, F. Stern: Rev. Mod. Phys. **54**, 437 (1982)

81. J. Linnros: 'Silicon nanostructures'. In: *Silicon-Based Microphotonics: from Basics to Applications.* ed. by O. Bisi, S.U. Campisano, L. Pavesi, F. Priolo (IOS Press, Amsterdam 1999) pp. 47–85
82. L.T. Canham: Appl. Phys. Lett. **57**, 1046 (1990)
83. A.G. Cullis, L.T. Canham, P.D.J. Calcott: J. Appl. Phys. **82**, 909 (1997)
84. Y. Kanemitsu: 'Silicon and germanium nanoparticles'. In: *Light Emission in Silicon From Physics to Devices.* Semiconductor and Semimetals Vol. 49, ed. by D.J. Lockwood (Academic Press, San Diego 1998) pp. 157–205
85. K. Brunner: Rep. Prog. Phys. **65**, 27 (2002)
86. V.E. Borisenko (ed.): *Semiconducting Silicides* (Springer, Berlin, Heidelberg 2000)
87. N.E. Christensen: Phys. Rev. B **42**, 7148 (1990)
88. A.B. Filonov et al.: J. Appl. Phys. **79**, 7708 (1996)
89. L. Miglio, V. Meregalli, O. Jepsen: Appl. Phys. Lett. **75**, 385 (1999)
90. K.J. Reeson et al.: Microel. Eng. **50**, 223 (2000)
91. M.G. Grimaldi, S. Coffa, C. Spinella, F. Marabelli, M. Galli, L. Miglio, V. Meregalli: J. Lumin. **80**, 467 (1999)
92. C. Spinella, S. Coffa, C. Bongiorno, S. Pannitteri, M.G. Grimaldi: Appl. Phys. Lett. **76**, 173 (2000)
93. D. Leong, M.A. Harry, K.J. Reeson, K.P. Homewood: Nature **387**, 686 (1997)
94. K.P. Homewood et al.: Thin Solid Films **381**, 188 (2001)
95. A.G. Birdwell, S. Collins, R. Glosser, D.N. Leong, K.P. Homewood: J. Appl. Phys. **91**, 1219 (2002)
96. B. Schuller, R. Carius, S. Lenk, S. Mantl: Microel. Eng. **60**, 205 (2002)
97. W. Henrion, M. Rebien, V.N. Antonov, O. Jepsen, H. Lange: Thin Solid Films **313–314**, 218 (1998)
98. W. Wolf, G. Bihlmayer, S. Blügel: Phys. Rev. B **55**, 6918 (1997)
99. A.B. Filonov et al.: Phys. Rev. B **60**, 16494 (1999)
100. V.L. Shaposhnikovov et al.: Opt. Mater. **17**, 339 (2001)
101. W. Henrion, M. Rebien, A.G. Birdwell, V.N. Antonov, O. Jepsen: Thin Solid Films **364**, 171 (2000)
102. J.S. Sharpe et al.: Appl. Phys. Lett. **75**, 1282 (1999)
103. J.S. Sharpe et al.: Nucl. Instr. and Meth. B **161–163**, 937 (2000)
104. D. Lenssen, D. Guggi, H.L. Bay, S. Mantl: Thin Solid Films **368**, 15 (2000)
105. R.A. Soref: Proc. IEEE **81**, 1687 (1993)
106. A.V. Krishnamoorthy, K.W. Goossen: IEEE J. Select. Topics Quantum Electron. **4**, 899 (1998)
107. M. van Rossum: 'Compound semiconductors on silicon'. In: *Silicon-Based Microphotonics: From Basics to Applications.* ed. by O. Bisi, S.U. Campisano, L. Pavesi, F. Priolo (IOS Press, Amsterdam 1999) pp. 327–346
108. D.A.B. Miller: IEEE J. Select. Topics Quantum Electr. **6**, 1312 (2000)
109. K.W. Goossen et al.: IEEE Photon. Technol. Lett. **1**, 304 (1989)
110. L.A. D'Asaro et al.: IEEE J. Quantum Electron. **29**, 670 (1993)
111. A. Hashimoto et al.: IEDM Techn. Digest. 658 (1985)
112. H.K. Choi, G.W. Tumer, T.H. Windhorn, B.Y. Tsaur: IEEE Electr. Device Lett. **7**, 500 (1986)
113. K.V. Shenoy, C.G. Fonstad Jr., A.C. Grot, D. Psaltis: IEEE J. Photon. Technol. Lett. **7**, 508 (1995)
114. R. Fischer et al.: Appl. Phys. Lett. **48**, 1360 (1986)

115. H.Z. Chen, J. Paslaski, A. Yariv, H. Morkoc: Appl. Phys. Lett. **52**, 605 (1988)
116. M. Razeghi et al.: Appl. Phys. Lett. **53**, 725 (1988)
117. M. Sugo, H. Mori, Y. Sakai, Y. Itoh: Appl. Phys. Lett. **60**, 472 (1992)
118. S. Matsuo et al.: IEEE J. Photon. Technol. Lett. **7**, 1165 (1995)
119. R. Leon: 'Self-assembled quantum dots: potential for silicon optoelectronics'. In: *Silicon Monolithic Integrated Circuits in RF Systems.* (Digest of Papers 2001) pp. 79–87. For a recent reviews on III-V on silicon see the special issue of Phys. Stat. Sol. (a) **194 n. 2** edited by A. Hoffmann and A. Rizzi.
120. E. Yablonovitch et al.: Appl. Phys. Lett. **56**, 2419 (1990)
121. I. Pollentier et al.: IEEE Electron. Lett. **26**, 193 (1990)
122. B.D. Dingle et al.: Appl. Phys. Lett. **62**, 2760 (1993)
123. D.L. Mathine, R. Droopad, G.N. Maracas: IEEE J. Photon. Technol. Lett. **9**, 869 (1997)
124. H. Wada, T. Takamori, T. Kamijoh: IEEE J. Photon. Technol. Lett. **8**, 181 (1997)
125. N.M. Margalit et al.: IEEE Electron. Lett. **32**, 1675 (1996)
126. S. Matsuo et al.: IEEE Electron. Lett. **33**, 1148 (1997)
127. H.J.J. Yeh, J.S. Smith: IEEE J. Photon. Technol. Lett. **6**, 706 (1994)
128. Y. Ohiso et al.: IEEE J. Photon. Technol. Lett. **8**, 1115 (1996)

2 Electron States and Optical Properties in Confined Silicon Structures

At the heart of solid state physics lies the Bloch theorem, which states that the eigenstates of the one-electron Hamiltonian can be described as plane waves times a function with the periodicity of the lattice:

$$\Psi_{n,\boldsymbol{k}}(\boldsymbol{r}) = \mathrm{e}^{\mathrm{i}\boldsymbol{k}\times\boldsymbol{r}} u_{n,\boldsymbol{k}}(\boldsymbol{r}) \tag{2.1}$$

where $u_{n,\boldsymbol{k}}$ is the Bloch function and n and $\boldsymbol{k}$ are the band index and the wavevector. From the solution of the Hamiltonian, the band structure, the density of states and, consequently, the optical properties are determined. Hence the physical properties of a solid are closely related to the three-dimensional periodicity of the lattice. For this reason, a change in the dimensionality of the system has profound consequences on all its properties.

Low-dimensional semiconductor systems can be grouped in three different classes: two-dimensional (2D) quantum wells or quantum slabs, one-dimensional (1D) quantum wires and zero-dimensional (0D) quantum dots [1]. In the first case, semiconductor heterojunctions produce potential wells whose widths are equal to the thickness of the low energy band-gap material and whose depths depend on the energy band-gap difference among the various materials. Assuming a quantum well of thickness L_z formed by a very thin slab of a material sandwiched by infinitely high energy barriers, the problem simplifies. Let z be the direction normal to the interfaces. The electrons are free to move in the xy plane and constrained in the z direction. The electronic system becomes quantum confined and electrons behave as two-dimensional particles. Their energy spectrum modifies becoming discrete. The plane wave along z in (2.1) has to be replaced by standing waves. Through another lateral confinement (say in the y direction) one obtains a quantum wire, where the electrons are free to move only in the x direction and also along y the wavefunction becomes a quantized wave. Finally, in quantum dots the translational motion is completely suppressed and no more plane waves appear in (2.1). Therefore, passing from three to two, one, and zero dimensions the density of states changes drastically from $N_3(E) \sim E^{1/2}$, to $N_2(E) \sim \Theta(E)$, to $N_1(E) \sim 1/\sqrt{E}$, to a delta-like dependence for the zero-dimensional case [2, 3]. For 2D systems, the formation of standing waves for the motion in the confinement directions implies that the allowed wavevectors k_z are quantized and satisfy the standing wave conditions

$$k_z = \frac{n\pi}{L_z}, \quad n = 1, 2, 3, \ldots \tag{2.2}$$

Consequently, the ground state energy is

$$\Delta E = \frac{\hbar^2 \pi^2}{2m^* L_z^2} \ . \tag{2.3}$$

In particular for the energy of the band gap ϵ_{g}^*, one has

$$\epsilon_{\mathrm{g}}^* = E_{\mathrm{g}} + \frac{1}{L_z^2} \left(\frac{1}{m_{\mathrm{e}}^*} + \frac{1}{m_{\mathrm{h}}^*} \right) \frac{\hbar^2 \pi^2}{2} \tag{2.4}$$

where m_{e}^* and m_{h}^* are the effective masses for electron and hole, respectively, and E_{g} is the bulk band gap. In (2.4) excitonic effects are neglected. Thus, the band gap of quantum wells increases as $1/L^2$ with decreasing thickness L. The term in (2.3) can be considered as additive for all the confinement directions. Hence the band-gap enlargement will be larger for quantum wires and quantum dots. Since, for most semiconductors, the hole effective mass is roughly the double of the electron mass, one would expect that the conduction band will shift by essentially the double of the valence band.

It is clear that such a simplified picture neglects any surface/interface related effects and relies heavily on the effective mass approximation. More sophisticated methods are necessary to elucidate the relationship between confinement and changes in the electronic and optical properties of low-dimensional Si structures. For this reason in the following, a review of the computational methods and results is presented.

2.1 Calculational Methods of Electronic and Optical Properties: an Overview

A reliable description of the electronic and optical properties of confined silicon structures is a formidable task. It is not only necessary to study the structural relaxations that occur at microscopic levels, but also to give a reasonable description of localized and excited electronic states. Moreover, it is necessary to include properly in the calculations correlation effects, which are expected to be especially important in small systems. Consequently, several methods have been used to study the optoelectronic properties of Si quantum wells, wires and dots. They essentially belong to five classes: effective mass approximation (EMA), Hartree–Fock (HF) based methods, empirical tight-binding (ETB), empirical pseudopotential (EPS) and, finally, first-principle techniques. A number of review articles are devoted to this subject [4, 5, 6, 7, 8, 9, 10]. In the following we present the ingredients of these methods, with particular emphasis on the sophisticated *ab initio* techniques.

The EMA has been extensively used [11, 12, 13, 14]. The simplest version of EMA assumes a single parabolic band extremum and an infinite potential well, neglecting the band nonparabolicity, the coupling to other band extrema and the properties of the confining materials. In more sophisticated versions multiband coupling is included [12], the nonparabolicity band dispersion of Si is taken into account via an energy dependent mass, and a finite potential barrier is considered [11]. The main shortcoming of the EMA is the impossibility of treating deviations from the perfect crystalline conditions other than idealized confinement, for example changes in symmetry.

In the HF approximation, the many-body function has the form of a Slater determinant constructed from single-particle functions that obey the Pauli exclusion principle, so that the exchange energy of the many-body system of electrons is properly described. Due to the lack of correlation effects, however, the gap energy of bulk Si is overestimated. Correlation effects can be included using configuration interaction (CI). In fact, in HF theory unoccupied eigenvalues carry no precise physical meaning, so that the HOMO–LUMO (Highest Occupied–Lowest Unoccupied Molecular Orbital) eigenvalue difference cannot be taken as a gap energy. CI method includes correlation by calculating the total energy of a pure excited configuration. This is associated with a Slater determinantal function, with a hole in the formerly occupied states and an electron in one of the formerly unoccupied states, or two holes and two electrons, and so on. Full correlation can then be achieved allowing interaction between the ground and all such excited configurations. Advantages of the HF scheme are that (1) calculations can be performed through self-consistency in the charge density, (2) charge rearrangements can be taken into account which allows one to determine stable geometry, and (3) the study of excited electronic states is possible. The main problems are related to (1) the frozen-core approximation (only valence electrons are treated self-consistently), (2) the use of a minimum basis set, and (3) the use of parametrized electronic and nuclear energies with parameters that must be transferable from one environment to another. Different approximations to the HF Hamiltonian exist. These distinguish one technique from another; within the same technique different parametrizations are present [8].

The idea behind the empirical tight-binding (ETB) method is the treatment of the Hamiltonian matrix elements as adjustable parameters. Usually, the parameters are fitted on the band structure, the lattice parameter, the elastic constants and the cohesive energy of bulk Si. The fit is either performed on experimental data or on HF and *ab initio* results. Then the Hamiltonian is applied to low-dimensional calculations with fixed matrix elements. The main advantage of ETB is the limited computational effort needed, i.e. much larger systems can be studied with respect to first-principle methods. As what concerns Si nanostructures, the accuracy of the method depends on the fitting procedure: the third nearest neighbor TB method (TNN-TB) [15] seems to work better than the nearest neighbor TB technique (NN-TB) [16].

Moreover the reliability of results strongly depends on the fit quality: models that predict a flat Si conduction band yields too small nanostructure band gaps [17].

The EPS has been applied for Si nanostructures [5, 18] through the use of the one-particle Schrödinger equation:

$$\left\{-\tfrac{1}{2}\nabla^2 + V(\boldsymbol{r})\right\} \Psi_i(\boldsymbol{r}) = E_i \Psi_i(\boldsymbol{r}) \,. \tag{2.5}$$

The mean-field potential $V(\boldsymbol{r})$ is constructed as a superposition of atom-centered potentials

$$V(\boldsymbol{r}) = \sum^{\text{atom}} v(\boldsymbol{r} - \boldsymbol{R}^{\text{atom}}) \,, \tag{2.6}$$

$\boldsymbol{R}^{\text{atom}}$ represents atomic position vectors. The wavefunctions Ψ_i are expanded in terms of plane waves

$$\Psi_i(\boldsymbol{r}) = \sum_{q} a_i(\boldsymbol{q}) \mathrm{e}^{\mathrm{i}\boldsymbol{q}\times\boldsymbol{r}} \,. \tag{2.7}$$

$a_i(\boldsymbol{q})$ are the variationally determined expansion coefficients at the reciprocal lattice vector $\boldsymbol{q}$ of the supercell. The potentials $v(\boldsymbol{r})$ can be fitted to a series of experimental data and detailed first-principle calculations. Differently from the ETB approaches, it is possible to compare the obtained potential $V(\boldsymbol{r})$ with *ab initio* results. A major advantage of the method is that the energies of excited states are comparable with experiments. This method can be used to calculate electronic and optical properties of systems containing up to 10^3–10^4 atoms.

The most reliable first principle scheme is developed within the density functional theory (DFT) [19]. This method describes the electronic properties of the ground state: the particle density $n(\boldsymbol{r})$ and the total energy of the electron system are obtained by a suitable set of one-particle equations.

The eigenvalues of the so-called one-particle Kohn–Sham equations [20] are generally interpreted as the quasiparticle energies of the system in analogy with the equations of the Hartree–Fock method. This approach is not completely adequate if the eigenvalues are seen as one-particle energies from which the excitation energies are obtained as differences between empty and filled states. In fact, the semiconductor band gap is usually underestimated [21, 22, 23]. However DFT is a powerful tool to describe ground state properties of solids and, in particular, the electron charge density distribution, the total energy, the equilibrium structure and the lattice parameter, the formation of a defect in the bulk, as well as the energy changes induced by the displacements of atoms from their equilibrium positions. The eigenvalues of the one-electron equations provide a description of the band structure, even if some care must be taken in identifying them with the true electron and hole quasiparticle energies.

By means of the slab method it is possible to treat surfaces and low-dimensional systems using the same computing scheme as for bulk calculations. The essence of the method is to reproduce the periodicity in the problem by introducing an artificial crystal which is formed by periodic repetition of an atomic slab and void. A supercell is thus constructed by intercalating a number of atomic planes with a vacuum region, which is a few interplanar distances thick. In this way the periodicity is restored and all the computational methods valid for bulk systems are still useful for the low-dimensional system. The only difficulties are to accurately describe the vacuum in all the confinement directions.

The general aim of a theory of the electronic properties is to solve the Schrödinger equation

$$\left\{-\frac{\hbar^2}{2m}\sum_{i=1}^{N}\nabla_i^2+\sum_{i=1}^{N}V(\boldsymbol{r}_i)+\frac{1}{2}\sum_{i,j=1}^{N}\frac{e^2}{|\boldsymbol{r}_i-\boldsymbol{r}_j|}-E\right\}\psi(\boldsymbol{r}_1,\ldots,\boldsymbol{r}_N)=0 \tag{2.8}$$

for an interacting many-electron system in an external potential $V(\boldsymbol{r})$. As a consequence of the many-body character of the Hamiltonian a single-particle potential and single-particle states cannot be written. In DFT, the total energy is written as a functional of the electron density $n(\boldsymbol{r})$,

$$E[n]=\int V(\boldsymbol{r})n(\boldsymbol{r})\mathrm{d}\boldsymbol{r}+\frac{e^2}{2}\int\mathrm{d}\boldsymbol{r}\int\mathrm{d}\boldsymbol{r}'\frac{n(\boldsymbol{r})n(\boldsymbol{r}')}{|\boldsymbol{r}-\boldsymbol{r}'|}+T_0[n]+E_{\mathrm{xc}}[n] \tag{2.9}$$

where $V(\boldsymbol{r})$ is the external potential experienced by the electrons due to the nuclei (in the case of an all-electron calculation) or to the ions (in a pseudopotential calculation). In the latter case $V(\boldsymbol{r})$ is a superposition of single ionic pseudopotentials which can be nonlocal operators. The quantity $T_0[n]$ is the kinetic energy of a system of noninteracting electrons with density $n(\boldsymbol{r})$ and $E_{\mathrm{xc}}[n]$ is the exchange-correlation energy. Also these terms are assumed to be a universal function of the density. The energy $E[n]$ achieves its minimum value in the ground state. To minimize it one can choose a form

$$n(\boldsymbol{r})=\sum_{i}^{\mathrm{occupied}}|\psi_i(\boldsymbol{r})|^2\,. \tag{2.10}$$

The problem is then transformed into the solution of a system of N one-particle equations [20]:

$$\left\{-\frac{\hbar^2}{2m}\nabla^2+V(\boldsymbol{r})+e^2\int\mathrm{d}\boldsymbol{r}'\frac{n(\boldsymbol{r})}{|\boldsymbol{r}-\boldsymbol{r}'|}+V_{\mathrm{xc}}(\boldsymbol{r})\right\}\psi_i(\boldsymbol{r})=\epsilon_i^{\mathrm{DF}}\psi_i(\boldsymbol{r}) \tag{2.11}$$

where the exchange-correlation potential

$$V_{\mathrm{xc}}(\boldsymbol{r}) = \frac{\delta E_{\mathrm{xc}}}{\delta n} \tag{2.12}$$

is the functional derivative of the exchange-correlation energy. The last two terms in the curly brackets of (2.11) depend on the set of solutions via (2.10) and are determined through an iterative procedure, which evaluates at each step a total effective potential

$$V_{\mathrm{eff}}(\boldsymbol{r}) = V(\boldsymbol{r}) + \int \mathrm{d}\boldsymbol{r}' \frac{e^2}{|\, \boldsymbol{r} - \boldsymbol{r}' \,|} + V_{\mathrm{xc}}(\boldsymbol{r}) \tag{2.13}$$

computed from the density that is constructed with the eigenfunctions. At each step of the procedure, a new output potential is calculated which must be compared with the input potential in order to verify whether self-consistency is achieved. Regarding the physical meaning of the one-particle eigenvalues ϵ_i^{DF} in (2.11), we notice that they are Lagrange parameters and consequently they have no formal justification as one-particle energies. Rather they must be considered as ingredients to calculate the total energy of the system. The great advantage of this result is that, if the exact $E_{\mathrm{xc}}[n]$ functional was known, (2.10), (2.11) and (2.12) would give the exact density and total energy of the system under consideration. Clearly the limitation is related to the fact that we cannot give an exact functional for the exchange and correlation energy.

The exchange-correlation potential is commonly assumed to be a local function of the density $n(\boldsymbol{r})$. This is the local density approximation (LDA), which is correct if the density varies slowly in space. The form of the local functional dependence of E_{xc} is chosen from the computed exchange-correlation density for a homogeneous electron systems $\epsilon_{\mathrm{xc}}[n(\boldsymbol{r})]$,

$$E_{\mathrm{xc}}^{\mathrm{LDA}} = \int \mathrm{d}\boldsymbol{r} \epsilon_{\mathrm{xc}}[n(\boldsymbol{r})] n(\boldsymbol{r}) \tag{2.14}$$

$$V_{\mathrm{xc}}^{\mathrm{LDA}} = \epsilon_{\mathrm{xc}}[n(\boldsymbol{r})] + n(\boldsymbol{r}) \frac{\mathrm{d}\epsilon_{\mathrm{xc}}}{\mathrm{d}n} \ . \tag{2.15}$$

It has already been pointed out that the band gap of semiconductors is underestimated in a well-performed LDA calculation, although the total energy is described very well. In the last decades several methods have been developed to overcome the shortcomings of LDA. Among these methods we remember the generalized gradient approximation (GGA) [24], the screened exchange LDA [25], the time dependent density functional method and the more rigorous self-energy approach.

In describing the band structure we have assumed that (2.11) could give (through ϵ_i^{DF}) the excitation energies of the system, i.e. the energy of an added particle (for empty states) and of a created hole (for filled states). This property is true if we write, instead of (2.11):

$$\left\{-\frac{\hbar^2}{2m}\nabla^2 + V(\boldsymbol{r}) + e^2 \int \mathrm{d}\boldsymbol{r}' \frac{n(\boldsymbol{r})}{|\boldsymbol{r}-\boldsymbol{r}'|}\right\}\psi_i(\boldsymbol{r}) + \int \mathrm{d}\boldsymbol{r}' \Sigma(\boldsymbol{r},\boldsymbol{r}',E)\psi_i(\boldsymbol{r}') = \epsilon_i \psi_i(\boldsymbol{r}) \quad (2.16)$$

where $\Sigma(\boldsymbol{r},\boldsymbol{r}',E)$ is the self-energy and not the exchange-correlation potential. The eigenvalues ϵ_i are in principle complex: their real parts are here the quasiparticle energies while the imaginary parts are the inverse lifetimes. The evaluation of the real parts of ϵ_i can be done in the GW approximation starting from the wavefunctions obtained in the LDA calculation [26]. This method used for bulk semiconductors gives good values for the energies of the bandgaps [21, 22, 23]. An estimate of these effects can be derived by examining the gap of bulk Si. The LDA value, obtained by solving (2.11), is underestimated by about 50–60% in a fully converged calculation. The GW approximation has been applied also to low-dimensional systems [27, 28, 29, 30, 31, 32]. Since the computational cost increases dramatically on going from LDA to the GW method, and since the errors in the gaps are predictable and can be easily corrected by using the the so-called scissor operator, i.e. by rigidly up-shifting all conduction bands, DFT-LDA is very often used to determine the band structure and optical properties even in the case of bulk systems. Its success in mimicking quasiparticle states can be ascribed to the similarity of its exchange-correlation potential to the exchange-correlation self-energy, whose expectation values differ by only 5–10%.

The calculation of the optical spectra should in principle include a very large class of different effects. Limiting ourselves to the independent-electron approximation, the imaginary part ϵ_2 of the dielectric function is obtained as [33]:

$$\epsilon_2^{\alpha}(\omega) = \frac{4\pi^2 e^2}{m^2\omega^2} \sum_{v,c,k} \frac{2}{V} |\langle\psi_{c,k}|p_\alpha|\psi_{v,k}\rangle|^2 \delta[E_c(k) - E_v(k) - \hbar\omega] \quad (2.17)$$

where $\alpha = (x, y, z)$, E_v and E_c denote the energies of the valence $\psi_{v,k}$ and conduction $\psi_{c,k}$ band states at a k point, and V is the supercell volume. The major problem in the evaluation of $\epsilon_2(\omega)$ is the computation of the optical matrix elements.

From the calculated ϵ_2 the real part ϵ_1 is obtained by a Kramers–Kronig transformation:

$$\epsilon_1(\omega) = \frac{2}{\pi} P \int_0^\infty \mathrm{d}\omega' \frac{\omega'\epsilon_2(\omega')}{\omega'^2 - \omega^2} \,. \quad (2.18)$$

The absorption coefficient

$$\alpha(\omega) = \frac{\omega}{nc}\epsilon_2(\omega) \quad (2.19)$$

is directly related to ϵ_2, thus the imaginary part of the dielectric function contains all the necessary information about the absorption properties of the material. The knowledge of both ϵ_2 and ϵ_1 allows the determination of the real part of the refractive index,

$$n(\omega) = \left\{ \tfrac{1}{2} [(\epsilon_1^2(\omega) + \epsilon_2^2(\omega))^{\frac{1}{2}} + \epsilon_1(\omega)] \right\}^{\frac{1}{2}} , \tag{2.20}$$

of the extinction coefficient,

$$K(\omega) = \left\{ \tfrac{1}{2} [(\epsilon_1^2(\omega) + \epsilon_2^2(\omega))^{\frac{1}{2}} - \epsilon_1(\omega)] \right\}^{\frac{1}{2}} , \tag{2.21}$$

and of the normal incidence reflectance

$$R(\omega) = \frac{(n(\omega) - 1)^2 + K^2(\omega)}{(n(\omega) + 1)^2 + K^2(\omega)} , \tag{2.22}$$

which are the quantities usually determined in experiments.

Optical spectra calculated for bulk semiconductors within the one-particle approximation yield only qualitative agreement with the experimental spectra, no matter which method, semi-empirical or *ab initio*, LDA or GW, is used. In Fig. 2.1 we show the calculated absorption spectra of bulk Si within LDA and GW [34].

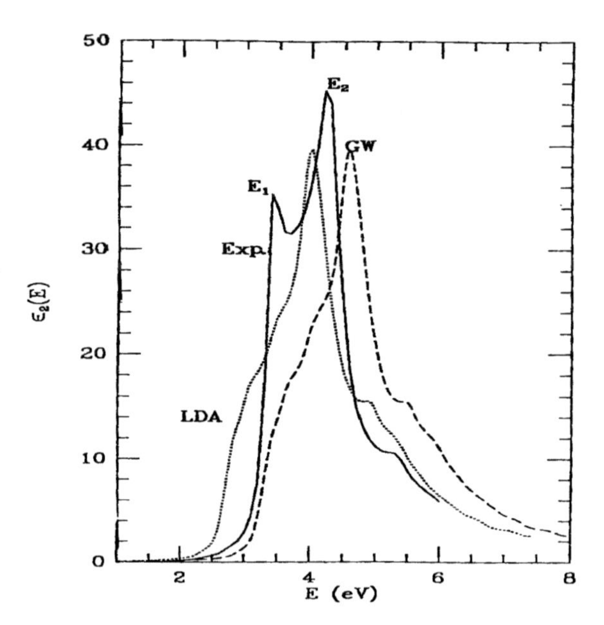

Fig. 2.1. Absorption spectra of bulk Si. *Full line*: experiment [35]; *dotted line*: DFT-LDA calculation; *dashed line:* GW calculation. After [34]

Whereas, the LDA result for the energy onset is red shifted with respect to experiment, the GW result is slightly blue shifted. Moreover in both cases we note a severe underestimation of the E_1 experimental peak. This is due to use in both cases of the one-particle approximation. An optical process involves two particles, the hole and the electron that are simultaneously created. The energy of the absorbed photon can be quite different from the bare sum of the hole and electron energies. It is often necessary to taken into account their mutual interaction. This interaction can not only produce absorption under the gap, due to bound exciton states, but can also induce appreciable distortions of the spectral lineshape above the continuous-absorption edge. Excitonic effects must be included in the calculations. This has been rarely done from first principles. Some efforts have been done by including the electron–hole interaction in a model dielectric function [36], or by using time-dependent LDA [37] which is a formalism that allows the electronic screening and correlation effects to be incorporated in the LDA method, or through the Bethe–Salpeter (BS) equation for the two-particle Green's function [29, 38]. The absorption spectra of bulk Si calculated within the BS approach are shown in Fig. 2.2 [38].

The peak positions and the relative intensities of the main structures are now in good agreement with experiment. The remaining slight overestimate is of the order of what has been predicted to be the contribution of dynamical

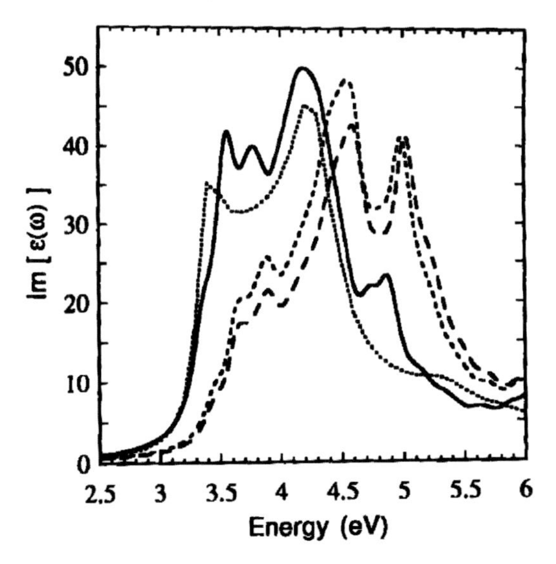

Fig. 2.2. Absorption spectra of bulk Si. Result obtained with the inclusion of the random phase approximation (RPA) and quasiparticle shifts (*short-dashed line*); result obtained with the inclusion of local field effects (*long-dashed line*), result obtained with both electron–hole attraction and local field effects (*solid line*). Experimental result (*dotted line*). After [38]

effects [39]. These types of calculation, even in the "simple" case of bulk Si, are really computationally demanding. However, it has been demonstrated by using the GW approximation and the BS equation that the self-energy correction and the Coulomb corrections almost exactly cancel for Si nanocrystals with radius larger than 0.6 nm [40]. Consequently the lowest excitonic energy is correctly predicted by LDA for a low-dimensional Si structure when the bulk correction for the energy band-gap is considered.

Another important issue is related to the investigation of the stable atomic structure of the confined systems. Structural properties can be studied within the simple valence-force schemes based on the Keating model [41] with tight-binding molecular dynamics [42] or with DFT-LDA [43]. Here the electron–ion interaction contribution to the total potential is described in the pseudopotential approach. For any fixed geometry, the total energy per unit cell, the one-electron energies and wavefunctions, and the forces on the atoms are calculated. The atoms are then moved along the force directions and the total energy is computed again until the stable structure is found. The most powerful and efficient theoretical technique to investigate stable structures is, however, the dynamical approach provided by the Car–Parrinello method (CP) [44] based on the DFT. It allows us in principle to obtain reconstructed configurations starting from ideal bulk-terminated geometries with no assumption of a given symmetry. Within the CP method it is possible to minimize the energy simultaneously with respect to the ionic and electronic degrees of freedom. Moreover, one can realize finite temperature molecular dynamics simulation searching for a global minimum in complicated multiminima potential energy surfaces. However the use of sufficiently large supercells that allows one to describe a system without imposing any symmetry constraints or artificial periodicity requires a large number of atoms per cell and, consequently, large computer resources.

2.2 Silicon Quantum Wells

All the calculations performed on confined Si structures give a similar picture concerning the quantum confinement effect: a widening of the band gap from the near-infrared wavelength region to and beyond the visible region. Furthermore an enhancement of the dipole matrix elements responsible of the radiative transitions is found. Figure 2.3 shows the predicted band gaps versus size as obtained from DFT-LDA calculations for hydrogen terminated quantum wells, wires and dots. They are compared to the results obtained by EPS and ETB.

Apart from dimensionality effects, it is important to understand both the role of the chemical species which passivate the dangling bonds at the surfaces and the effects of the symmetry of the confinement directions. Thus we discuss the computed electronic and optical properties of Si quantum films, where H atoms are used to terminate dangling bonds at the surfaces,

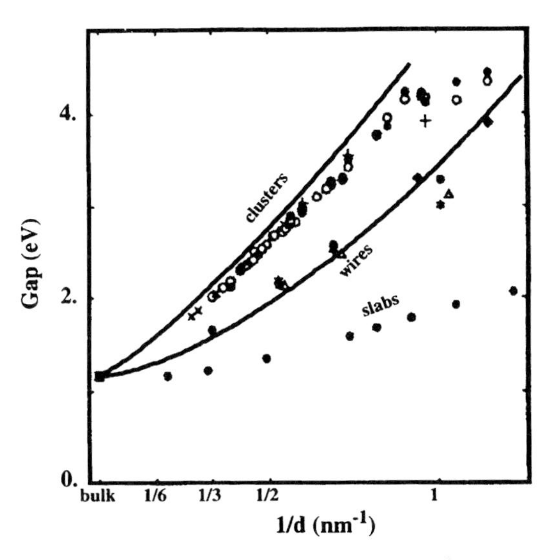

Fig. 2.3. Energy gap versus confinement parameter $1/d$ for H-terminated Si dots, wires and wells ($d = aN/8$ for wells, $d = a(0.5N/\pi)^{1/2}$ for wires, $d = a(0.75N\pi)^{1/3}$ for dots, where a = 0.54 nm and N is the number of Si atoms in the unit cell). LDA: (• and ∘) [45], (+) [46], (*filled diamonds*) [47], (Δ) [48], (*) [49]. EPS: (x) [50]. ETB: (*straight lines*) [16], [51]. After [6]

and whose surfaces are oriented along the (100), (110) and (111) directions. Moreover we present the calculated optoelectronic properties of Si quantum wells with CaF_2(111) or SiO_2(001) barriers to compare with experimental data; see Sect. 4.1.

2.2.1 Electronic Properties

The size dependence of the band gap for Si(001) oriented quantum wells, whose surfaces were free or covered by hydrogen, has been determined within EPS [50]. The size dependencies of the valence band maximum (VBM) and of the conduction band minimum (CBM) are shown in Fig. 2.4 and compared to the EMA results. EMA leads to both quantitative and qualitative errors. Not only is the band gap opening greatly overestimated for thicknesses below 2 nm, but also the nonmonotonic oscillations in the valence band energies and the appearance of valence band states whose energies are independent of the thicknesses for layers with an even number of Si planes are missed.

Another important point is the direct or indirect nature of the band gap in Si quantum wells. To answer this question, first-principle calculations for Si and Ge free-standing films, which are constructed with the same lattice periodicity and the same interatomic distance as in the bulk, have been performed [52, 53, 54]. Their surfaces are oriented along (100), (110), and (111) directions and saturated with H. For each orientation, two effective thicknesses

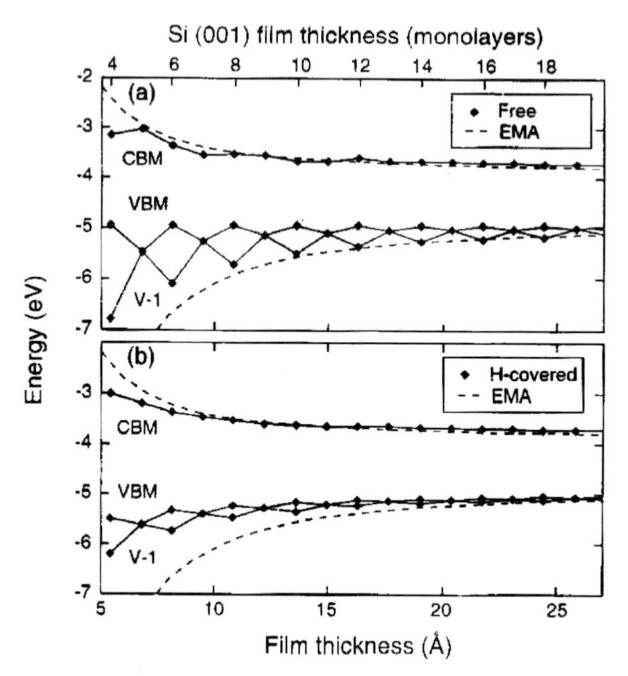

Fig. 2.4. Calculated neargap energy levels of (001) oriented H-free (*upper part*) and H-covered (*lower part*) Si quantum films. All states shown are bulk-like, i.e. surface states are omitted. Note the oscillations in the VBM and in the next-highest occupied valence state (V-1), absent in EMA. After [5]

have been chosen for the analysis, namely approximately 0.65 and 1.0 nm. The Si-H and Ge-H bond lengths have been taken to be 0.1481 and 0.1525 nm, respectively, corresponding to those in SiH_4 and GeH_4 molecules. Due to the LDA, the gaps are underestimated by 0.66 eV and 0.76 eV in the case of Si and Ge, respectively. The energy bands calculated along high symmetry directions of the two-dimensional Brillouin zone are shown in Fig. 2.5 for both Si and Ge quantum wells which are about 1.0 nm thick. The valence band dispersion is very similar for both Si and Ge in one orientation due to the similarity in the bulk valence band structure (see Figs. 1.1 and 1.2). It is also important to point out the absence of surface related states in the band-gap as a consequence of the H passivation of the dangling bonds [55].

Looking at the (100)-oriented films, we note that the band gap appears to be direct at Γ for both Si and Ge. The lowest indirect transitions occur only a few meV higher than the direct ones. A quite different character of the band gap is evident for the (110) films. In the case of Si it is still direct at Γ, whereas for Ge it is indirect and lies in the Γ–X_1 direction. Certain differences in the conduction band dispersion can be seen for the (111) films. For Si the folding of the bulk energy bands results in an indirect band gap along the Γ–M direction. It should be remarked here that the Γ–M direction

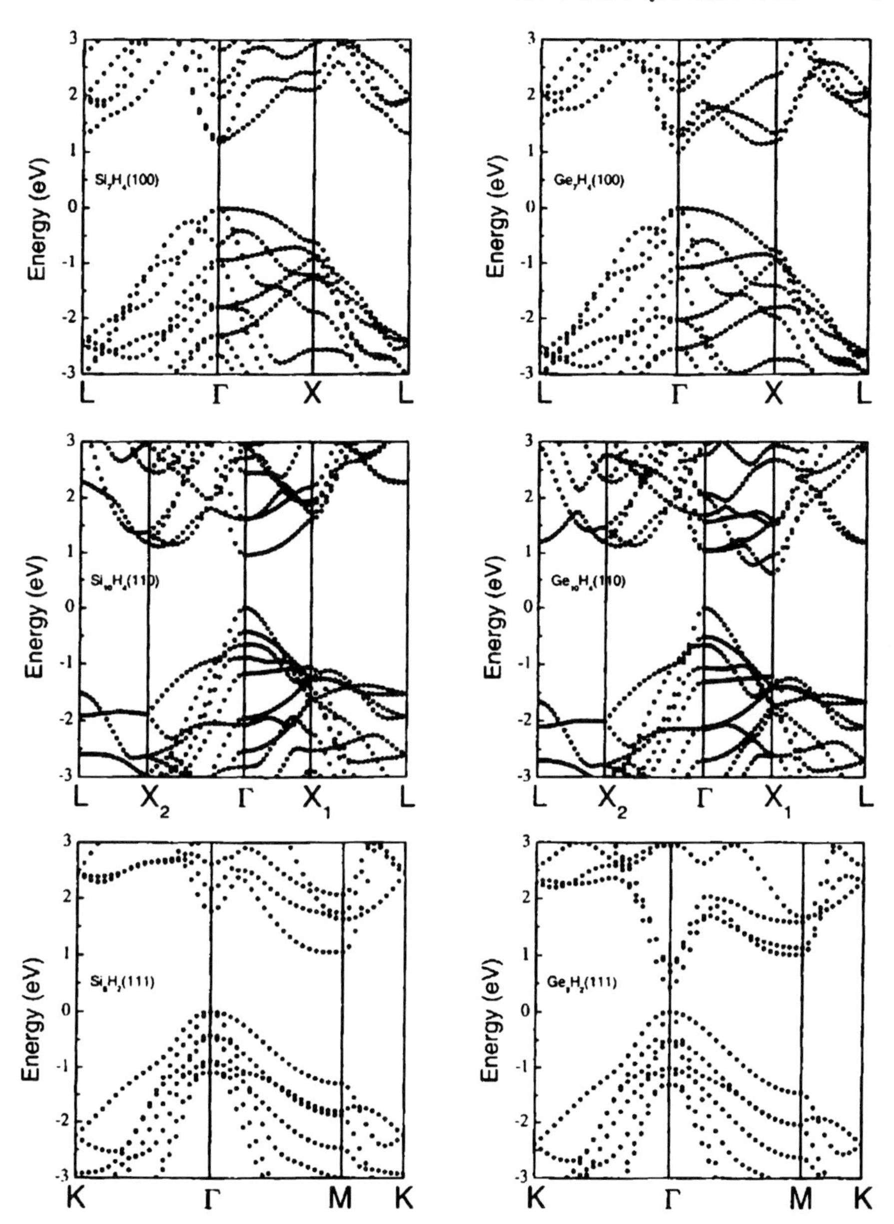

Fig. 2.5. Electronic band structures of Si and Ge quantum films of about 1.0 nm thickness for different orientation: $Si_7H_4(100)$, $Si_{10}H_4(110)$, $Si_6H_2(111)$, $Ge_7H_4(100)$, $Ge_{10}H_4(110)$, $Ge_6H_2(111)$. The numbers refer to the amount of semiconductor and H layers in the different structures. After [52, 53, 54]

of the 2D hexagonal Brillouin zone (BZ) corresponds to the Γ–X direction of the 3D face-centered cubic BZ where the bulk Si conduction band minimum occurs. For Ge films, however, there is a well-resolved conduction band minimum at the Γ point thus indicating direct band gap character. The different behavior (direct versus indirect band gap) of Si and Ge with respect to the film orientation can be explained with the folding of the conduction bands. Whereas the six equivalent ellipsoidal conduction band minima of bulk Si occur in the $\langle 100 \rangle$ directions near to the zone boundary, in bulk Ge there are eight symmetry-related ellipsoids with long axis along the $\langle 111 \rangle$ directions and centered at the midpoints of the hexagonal zone faces. Thus, when the confinement direction is the $\langle 111 \rangle$, the conduction band minimum is folded onto the Γ point for Ge and not for Si, while the opposite is true when the confinement direction is $\langle 001 \rangle$. Also the different confinement energy shift with respect to the orientation of the layer, which increases for Si on going from (110) to (111) to (100) and for Ge from (111) to (110) to (100), can be roughly interpreted in term of the different highly anisotropic behaviors of the effective masses for bulk Si and Ge (see Sect. 1.3.1 and Table 2.1).

Surface passivation affects these results. The substitution of the H-Si bonds on one of the two surfaces of a confined Si(111) film with OH-Si bonds causes a marked lowering of the band gap and the appearance of a direct band gap. This is a consequence of the strongly attractive potential of the O atom which has a very low electronegativity [32].

CaF_2/Si systems can be epitaxially grown because of a room temperature lattice mismatch of only 0.06%. This feature makes the system particularly

Table 2.1. Calculated direct, at Γ, and indirect energy gaps (in eV) for Si and Ge quantum films covered by H for three different orientations, each with two different thicknesses

Si(100)			Ge(100)		
d(nm)	E_{direct}(eV)	E_{indirect}(eV)	d(nm)	E_{direct}(eV)	E_{indirect}(eV)
0.71	1.56	1.91	0.74	1.38	1.47
0.99	1.17	1.33	1.02	0.99	1.15
Si(110)			Ge(110)		
d(nm)	E_{direct}(eV)	E_{indirect}(eV)	d(nm)	E_{direct}(eV)	E_{indirect}(eV)
0.63	1.41	1.63	0.65	1.50	1.03
1.01	0.96	1.11	1.05	1.03	0.61
Si(111)			Ge(111)		
d(nm)	E_{direct}(eV)	E_{indirect}(eV)	d, (nm)	E_{direct}(eV)	E_{indirect}(eV)
0.69	1.93	1.32	0.71	0.68	1.38
1.00	1.77	1.05	1.04	0.44	1.01

appealing for the experimental realization of nanostructures (see Sect. 4.1). The CaF_2-Si(111) system is also interesting from a theoretical point of view, because it allows first principle investigation of the interaction between a polar insulator with ionic bonding (CaF_2) and a homopolar semiconductor with covalent bonding (Si). Detailed calculations [56] for Si quantum wells of different thickness embedded in a CaF_2 matrix have been performed within DFT-LDA in the linear muffin-tin orbital (LMTO) method [57]. Figure 2.6 shows the electronic band structure of a quantum well formed by two double layers (DL) of Si atoms embedded in CaF_2 (thickness of 8.7 Å).

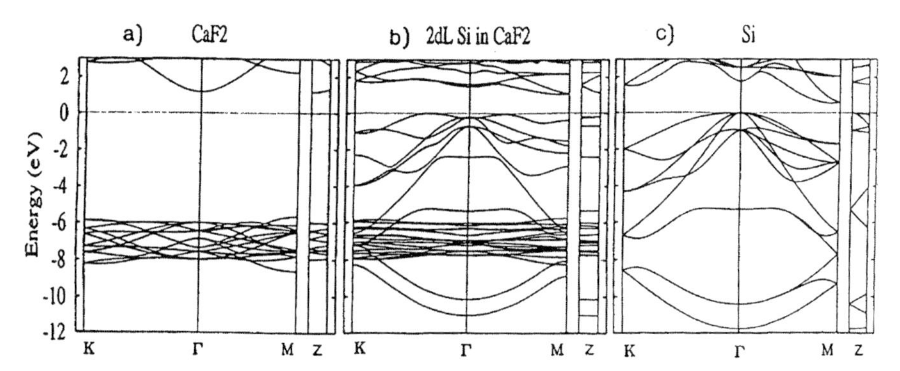

Fig. 2.6. Band structure projected along two symmetry directions (Γ-M and Γ-K) of the hexagonal 2D Brillouin zone and along the direction (Z) perpendicular to the surface Brillouin zone. Energies are referred to the valence band maximum. (**a**) bulk CaF_2, (**b**) two DL Si embedded in CaF_2, and (**c**) bulk Si. After [58]

It can be seen that the major changes occur in the energy region around the gap, while deeper in the valence band one can recognize the almost unaffected Si and CaF_2 band structures. There are three main effects related to layering to be noticed: (1) the band gap increases; taking into account the LDA underestimation of the gap, this is now in the optical region; (2) the top of the valence band is no longer at Γ; and (3) the first conduction band states are very flat in comparison with bulk Si. As a result, the conduction band and valence band dispersions run almost parallel over a large part of the Brillouin zone and display a gap at finite in-plane wavevectors, which cause an enhancement of the transition rate. Even for a thicker slab the bulk-like situation is not recovered: in fact an interface state appears just below the conduction band at Γ, and a state emerges from the valence band also at Γ. These two interface states are the bonding–antibonding states resulting from the Ca-Si bond at the interfaces. These results show that both confinement and hybridization play a role in the opening of the gap in this system. In particular the Si valence band is largely affected by confinement effects, whereas the conduction band is dominated by hybridization effects [56, 58]. Remarkably, by alternating Si slabs of different thicknesses (four Si

DL and two Si DL), one can achieve a quasidirect band gap in CaF_2-Si-CaF_2 superlattices[56].

The last system we consider in this section is Si/SiO_2. Here we present results related to the first *ab initio* calculation of the electronic and optical properties of Si-SiO_2(001) superlattices or multiple quantum wells, where the thickness of the Si slabs has been varied in the nanometer range [59]. In the DFT-LDA calculation, supercells formed by a variable number of Si elementary cells, separated by SiO_2 layers, have been considered. The SiO_2 was described by the β-cristobalite structure which leads to a simple Si/SiO_2 interface with a minimum number of dangling bonds at the interface: one interface Si atom with two dangling bonds. It is possible to fully passivate the system through an extra double-bonded O atom that saturates the Si dangling bonds. Recent experimental studies evidence β-cristobalite formation at the first interface between Si and SiO_2 [60]. Figure 2.7 reports the calculated fundamental energy gap as a function of the Si layer thickness, d.

Instead of the particle-in-a-box $1/d^2$ dependence, a weaker d-dependence of the energy gap is found [61]. A reason for this is the presence of interface states, which are less affected by the dimensionality. Indeed, the computed electronic band structure of Si-SiO_2 superlattices shows a state near the top of the valence band caused by the interaction between the interface Si and its double-bonded O atom. If the extra oxygen atom is removed, the system is still a semiconductor with a new state, a defect state, at the top of the valence band that reduces the band gap by 0.12 eV. The interesting fact about this state is that its energy distance to the bottom of the conduction band is almost unaffected by the Si thickness whereas the distance with respect to the top of the valence band depends on d. This result has a profound influence on the optical properties.

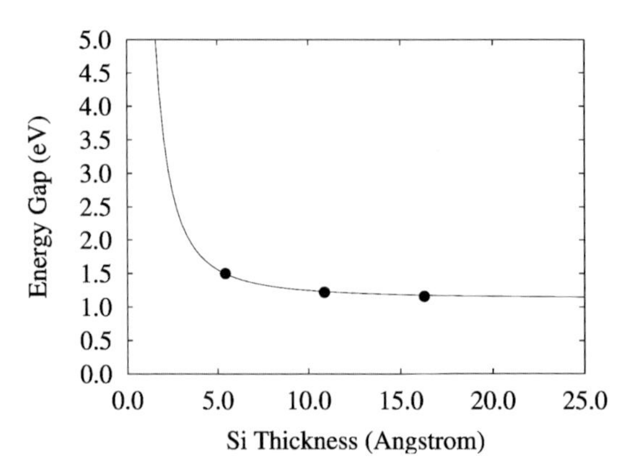

Fig. 2.7. The fundamental energy gap of Si-SiO_2 multiple quantum wells. The points are the computational results while the *line* is a fit to the *points*. After [61]

2.2.2 Optical Properties

The optical properties are studied by computing the dielectric function ϵ. The imaginary part ϵ_2 (2.17) of free-standing Si and Ge quantum wells is shown in Fig. 2.8; the relative band structures were shown in Fig. 2.5.

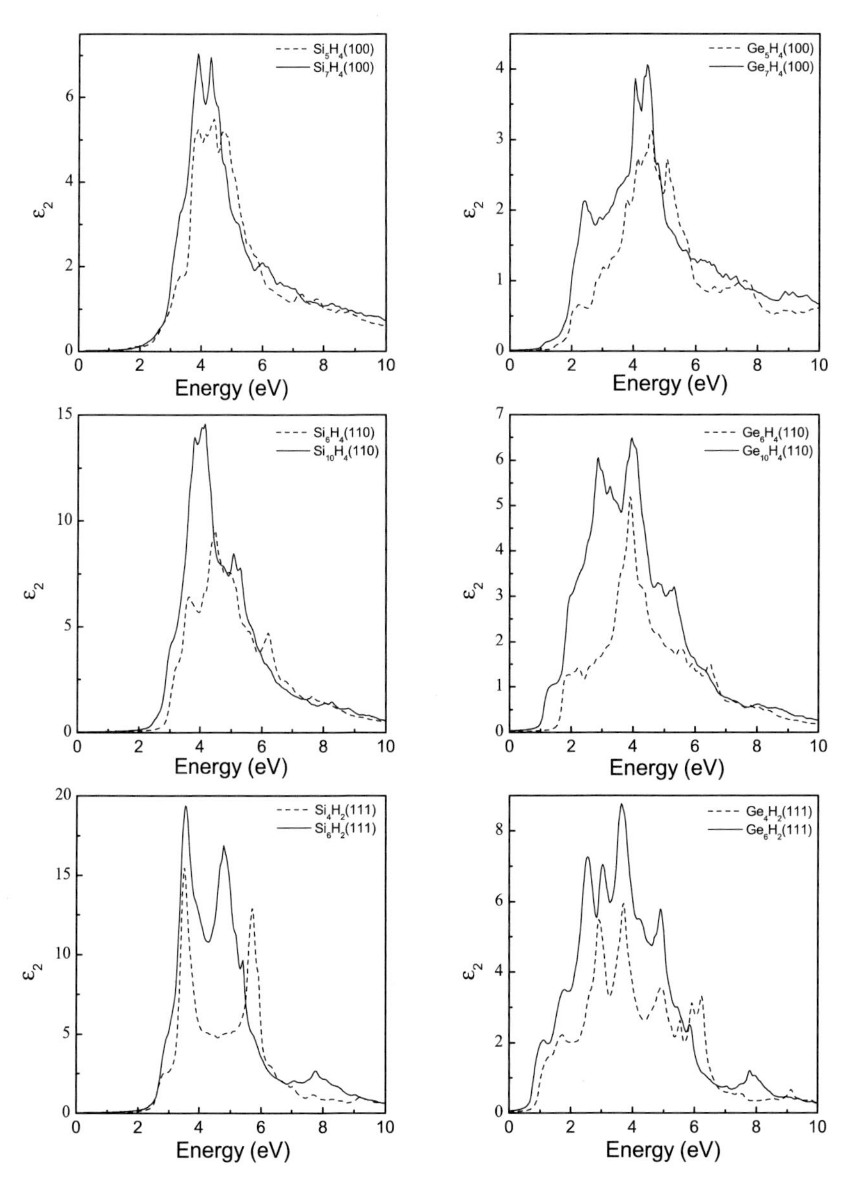

Fig. 2.8. Calculated imaginary part of the dielectric function of Si and Ge quantum films for different orientations and thicknesses. The curves are dressed with a broadening of 0.1 eV. After [53, 54]

The ϵ_2 of both Si and Ge films has a major peak around 4 eV. This peak corresponds to the E_2 transition of the bulk material (see Fig. 2.1, for a comparison). For both systems a reduction of the peak intensity with respect to the bulk value can be observed as a consequence of the quantum confinement. However, for (100) and (110) surface oriented films the peak is also high-energy shifted, whereas for (111) films it moves to lower photon energies. A different behavior was observed for extremely thin layers of porous Si [62].

If we concentrate on the low energy region, we note the presence of new features that are absent for bulk Si and Ge. We remember that Si(100), Si(110), Ge(100) and Ge(111) quantum wells have a direct band gap. A deeper insight is gained by analyzing the strength of the optical transitions computed as the squared optical matrix elements for the first two direct transitions between valence and conduction bands at Γ (Table 2.2). We note that the presence of a direct band-gap does not ensure large optical matrix elements for the optical transitions of free-standing quantum wells. The optical transition strength reflects also the character of the involved states, i.e. how the zone folding acts on the confinement. As a comparison, in the direct band-gap semiconductor GaAs the first transition at Γ has a squared optical matrix element $\langle P^2 \rangle$ of about 0.31 a.u. For Si and Ge, the values are much lower. They grow rapidly by reducing the film thickness. For a single Si layer covered by H, values comparable with those of GaAs have been found [32].

Starting from $\epsilon_2(\omega)$ it is possible, through (2.15), to obtain $\epsilon_1(\omega)$, the real part of the dielectric function. Its value at zero energy is the static dielectric constant. The calculated values of $\epsilon_1(\omega)$ for the Si and Ge films considered are reported in Table 2.3 and compared with the corresponding bulk values. Once more the confinement effect is clear: $\epsilon_1(\omega)$ is reduced in thin films. Moreover, for the same film thickness, it appears to be higher for the Si structures than for the Ge ones, despite the fact that the bulk Ge value has a higher value than Si.

The $\epsilon_2(\omega)$ for Si-CaF_2 multiquantum wells shows new features in the low-energy region (see Fig. 2.9 [63]).

The low-energy region is dominated by the band edges and, in particular, by the interface states whose character affects the last occupied and last unoccupied states for very thin Si slabs. The low-energy peaks in $\epsilon_2(\omega)$ are caused by significant matrix elements between the top valence states, which are p-like Si derived states, and the bottom conduction ones, which are mixed p-like Si and s-like and d-like Ca derived states. Their oscillator strengths are reported in Table 2.4 and compared to those of bulk Si. It is worthwhile to note that the oscillator strengths for the matrix elements between these states increase very rapidly as the thickness of the Si well decreases. For well thicknesses less than 2 nm they are of the same order of magnitude as the direct transition at Γ in bulk Si and only one order of magnitude smaller than that of GaAs. The reason of the rapid increase of the oscillator strength with

Table 2.2. Calculated squared optical matrix elements $\langle P^2 \rangle$ (in atomic units) for significant direct transitions E_{optic} at the Γ point for Si and Ge quantum films covered by H for three different orientations, each with two different thicknesses d. The three highest valence subbands and the three lowest conduction subbands are termed v1, v2, v3 and c1, c2, c3, respectively

Si(100)			Ge(100)		
d(nm)	E_{optic}(eV)	$\langle P^2 \rangle$(a.u.)	d(nm)	E_{optic}(eV)	$\langle P^2 \rangle$(a.u.)
0.71	v1-c1 1.56	0.009	0.74	v1-c1 1.38	0.03
	v1-c2 1.71	0.006		v1-c2 1.78	0.04
0.99	v1-c1 1.17	0.00004	1.02	v1-c1 0.99	0.06
	v1-c2 1.23	0.0006		v1-c2 1.28	0.0004
Si(110)			Ge(110)		
d(nm)	E_{optic}(eV)	$\langle P^2 \rangle$(a.u.)	d(nm)	E_{optic}(eV)	$\langle P^2 \rangle$(a.u.)
0.63	v1-c1 1.41	0.00018	0.65	v1-c1 1.50	0.001
	v3-c1 2.52	0.0004		v1-c2 1.73	0.34
1.01	v1-c1 0.96	0.00001	1.05	v1-c1 1.03	0.06
	v1-c2 1.23	0.00004		v1-c2 1.06	0.26
Si(111)			Ge(111)		
d(nm)	E_{optic}(eV)	$\langle P^2 \rangle$(a.u.)	d(nm)	E_{optic}(eV)	$\langle P^2 \rangle$(a.u.)
0.69	v1-c1 1.93	0.00	0.71	v1-c1 0.68	0.00
	v1-c2 2.48	0.25		v1-c2 0.91	0.32
1.00	v1-c1 1.77	0.0022	1.04	v1-c1 0.44	0.002
	v1-c2 2.17	0.00		v1-c2 0.72	0.32

Table 2.3. Calculated values of the static dielectric constant $\epsilon_1(\omega)$ for Si and Ge quantum films in comparison with those of bulk

			Si			
bulk		(100)		(110)		(111)
13.82	Si_5H_4	3.25	Si_6H_4	4.30	Si_4H_2	4.93
	Si_7H_4	3.6	$Si_{10}H_4$	5.63	Si_6H_2	7.26
			Ge			
bulk		(100)		(110)		(111)
14.80	Ge_5H_4	2.57	Ge_6H_4	2.94	Ge_4H_2	4.49
	Ge_7H_4	3.12	$Ge_{10}H_4$	4.83	Ge_6H_2	6.62

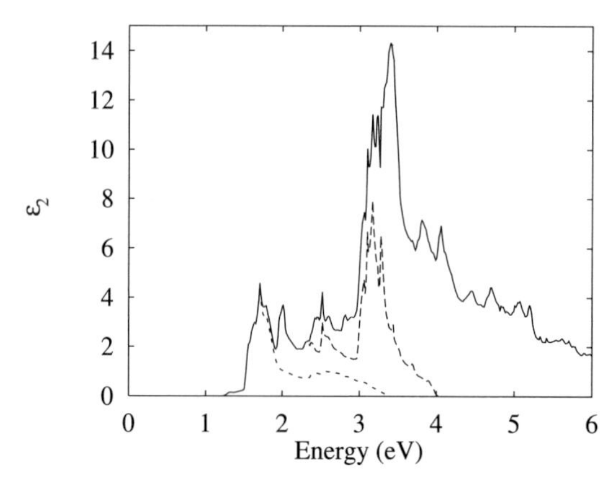

Fig. 2.9. ϵ_2 of two DL Si-CaF_2 multiquantum wells. *Solid line:* total ϵ_2; *short-dashed line:* contribution from the transition between the last occupied state and the first unoccupied one; *long-dashed line:* the same considering the last two occupied states and the first two empty states. After [63]

decreasing well thickness is the different localization of the states involved in the transition. The intensity of the oscillator strength depends not only on the localization in reciprocal space, but also on the localization in real space. For very thin wells both the valence and conduction band states are strongly localized at the interface.

The confinement results in a reduction of the dielectric constant, of the refractive index and of the reflectivity (Table 2.4). A similar lowering of the optical constants has been found for Si quantum wells embedded in SiO_2 [59].

The influence of an O defect on the low energy ϵ_2 of Si/SiO_2 quantum wells is shown in Fig. 2.10. The lineshape of ϵ_2 at the energy onset is asymmetric and can be fitted by two Gaussians: the first is due to defect interface

Table 2.4. Calculated energies ΔE_{cv} and oscillator strengths f_{cv} for significant direct transition at Γ for the different Si-CaF_2 multiquantum wells. The values are compared with that of bulk Si. The LDA results have been rigidly shifted with a scissor operator of 0.60 eV. The table contains also the results for the dielectric constant $\epsilon_1(0)$, the refractive index $n(0)$, and the reflectivity $R(0)$. After [63]

Lattice	bulk Si	7 DL	4 DL	2 DL	4-2 DL
Size (nm)	∞	2.44	14.9	8.7	14.9-8.7
Gap (eV)	1.16	1.16	1.23	1.84	1.44
ΔE_{cv}	3.19	2.04	2.33	2.41	2.17
f_{cv}	2.53	0.004	0.42	0.66	2.33
$\epsilon_1(0)$	10.53	9.26	7.98	5.32	
$n(0)$	3.24	3.04	2.82	2.31	
$R(0)$	0.28	0.26	0.23	0.13	

state transitions, the second to bulk-like transitions. In Fig. 2.10 these Gaussian bands are labeled with an **I** for the interface band and with a **Q** for the bulk-like band. They are shown for three different Si slab thicknesses. A shift to higher energies is evident for **Q** peak when the thickness of the Si slab decreases. This behavior is due to quantum confinement. The **I** peak, instead, is almost unaffected by the dimension of the Si slab and does not shift. This is in agreement with experimental results for the behavior of the photoluminescence of crystalline Si/SiO_2 single quantum wells [64]. Remarkably also in the case of the Si-SiO_2 system the transition probability rapidly increases when the thickness of the Si slab is of the order of a few nanometers.

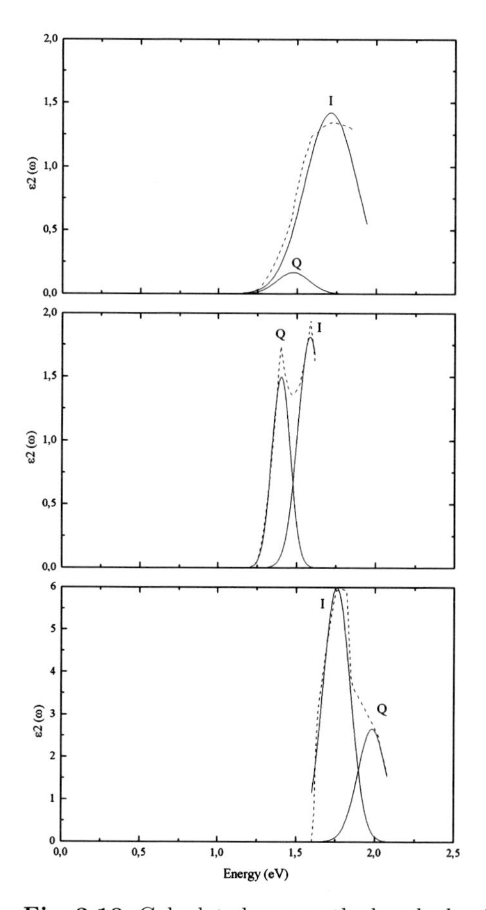

Fig. 2.10. Calculated ϵ_2 near the band edge (*dashed line*) and its Gaussian fit (*solid line*) for $Si_{[3]}SiO_2$, $Si_{[2]}SiO_2$, $Si_{[1]}SiO_2$ multiquantum wells. $[n]$ refers to the number of Si unit cell in the calculation. The letter **I** refers to the interface Gaussian band, while **Q** indicates the bulk-like Gaussian band. A self-energy correction through a scissor operator has been considered. After [59]

Finally, owing to the strong asymmetry between the directions parallel and perpendicular to the plane of the Si well, all the calculated ϵ_2 show a marked anisotropic behavior.

2.3 Silicon Quantum Wires

Freshly etched porous silicon shows the structure of a crystalline silicon skeleton with a connected undulating-wire morphology (see Chap. 3). Thus the investigation of the electronic and optical properties of Si quantum wires was one of the main computational efforts of the last decade [9, 65, 66]. Here we will discuss some selected results devoted to the investigations of how the electronic states and optical properties of Si quantum wires are influenced by quantum confinement effects, symmetry considerations and different passivation regimes.

2.3.1 Electronic Properties

Most of the investigated wire structures consist of a single, infinite quantum wire of rectangular cross-section with its axis along the $\langle 001 \rangle$ direction and its surfaces oriented along the (110) and $(1\bar{1}0)$ planes of bulk Si. The two sides of the cross-section are delimited by zig-zag chains of M and N Si atoms; this structure is usually termed an $M \times N$ wire. The Si atoms at the surfaces are passivated by H. Along the wire there are Si atoms with no H-bond, i.e. Si atoms in the core of the wire which feel a pure Si crystalline environment, surface Si atoms single bonded to an H atom and surface Si atoms double bonded to H atoms. Total energy calculations show that the Si-Si bond length in the wire remains practically unchanged with respect to the bulk Si; there is only a small reorganization of the H-Si distance at the surfaces [49, 67]. Thus these wire structures, albeit so small, represent rather ideal crystalline Si wires. Figure 2.11 shows the DFT-LDA computed band structures for a fully H-saturated 5×4 Si quantum wire of section 7.68×5.76 Å^2.

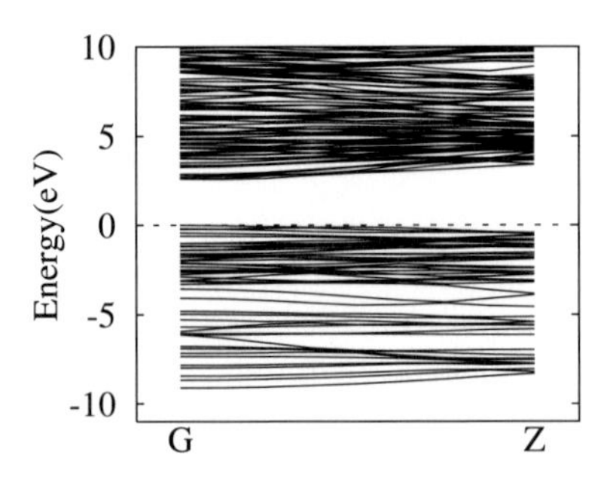

Fig. 2.11. One-dimensional band structure of the H fully passivated 5×4 Si quantum wire. After [67]

Along the growth direction of the wire the translational symmetry holds and a direct band gap appears at Γ for $\boldsymbol{k} = 0$ [67]. For wires having a different orientation, like $\langle 111 \rangle$, the band gap is still direct. This shows that the second direction of confinement contributes to folding at Γ the Si conduction band minimum independently on the wire orientation which contrasts to the quantum well case [69]. Table 2.5 collects the results of the calculated Si wire band gaps for different wire orientations and sections. The great majority of the different results shows remarkable agreement. Structural relaxations result in an opening of the gap of only $\sim$ 0.1–0.2 eV. Once more the computed band gaps reduce with increasing size. Differently from the simple EMA prediction, the dependence of the gap on the wire dimensions is d^{-n} with $n \sim 1.5$. Excitonic corrections to the band gaps can be large as $\sim$ 200 (100) meV for wires of $\sim$ 1 (2) nm mean width [70, 71].

An important point still under discussion is the nature of the quantum wire band gap: is it really a direct one or is it pseudodirect? A careful study demonstrates that the valence band maximum (VBM) originates mainly from the folding of two coupled bulk valence bands at an off-Γ k point [68]. This shows that the VBM does not correspond to a surface state even if the wave function is localized near the surface. The conduction band minimum (CBM) comes mainly from the two lowest bulk conduction bands which are located at a different off-Γ k point than the VBM. The fact that the VBM and CBM wire states unfold into bulk states of different $\boldsymbol{k}$ wavevectors proves that the

Table 2.5. Collected results for calculated energy band gaps of Si quantum wires with respect to wire orientation and wire mean width. LDA means *ab initio* DFT-LDA calculations, LDA* means LDA calculations where also structural relaxation has been considered, and EMP means empirical or semi-empirical methods

Orien.	Width(Å)	Gap(eV)	Method	Orien.	Width(Å)	Gap(eV)	Method
(100)	4.80	4.07	LDA [72]	(111)	5.51	3.63	LDA* [69]
(100)	6.72	3.38	LDA [72]	(100)	6.72	3.53	LDA* [48]
(100)	6.72	3.40	LDA* [67]	(100)	7.6	3.13	LDA* [69]
(100)	7.68	3.29	LDA* [73]	(100)	7.68	3.29	EMP [74]
(100)	7.68	3.52	EMP [75]	(100)	7.68	3.35	LDA [76]
(100)	7.68	3.33	LDA [4]	(100)	7.68	2.44	EMP [77]
(110)	7.68	2.24	EMP [77]	(111)	7.68	2.04	EMP [77]
(100)	8.64	3.03	LDA* [67]	(100)	8.64	3.08	LDA* [73]
(100)	9.60	2.89	EMP [68]	(100)	10.56	2.88	LDA* [48]
(100)	11.4	2.64	LDA* [49]	(100)	11.52	2.5	EMP [74]
(100)	11.52	2.73	LDA [76]	(100)	11.52	2.69	LDA [4]
(100)	13.44	2.32	EMP [68]	(100)	14.40	2.50	LDA* [48]
(100)	15.36	2.1	EMP [74]	(100)	15.36	2.12	EMP [75]
(100)	15.36	2.28	LDA [68]	(100)	15.6	2.23	LDA* [49]
(100)	17.28	1.98	EMP [68]	(100)	21.12	1.78	EMP [68]
(100)	23.04	1.63	EMP [75]	(100)	30.72	1.40	EMP [75]

Si quantum wire band gap is pseudodirect and not direct. This fact influences the optical properties, too.

Additional information on the character of states can be provided by the density of states and site-projected density of states. Detailed analysis of these quantities [7] shows that the bonding Si-H states are located well inside the valence band and the antibonding ones inside the conduction band. Thus the main contribution to the near band gap states of fully passivated wires originates from the crystalline core region. These states are very similar to those of bulk Si. When H atoms are removed, these bonding Si-H states become fully localized dangling-bond states within the band gap.

The role of oxygen on the electronic properties of Si quantum wires has been investigated through the substitution of some of the H atoms with O-H complexes. The total density of states (TDOS) of a 5×4 Si quantum wire modified by the presence of four OH complexes is compared with the fully H passivated TDOS in Fig. 2.12. The TDOS are aligned in energy, the energy zero fixed by the top of the valence band of bulk Si. This alignment is performed by looking at the energy of the Si $2p$ core level located at the center of the wires, that is, in a crystalline environment [78]. The results provide clear evidence of the main OH related effects: (1) the band gap decreases in energy, (2) this decrease is strongly related to the O-H concentration, and (3) it is mostly due to the lowering of conduction band states at the CBM.

Another interesting point is related to the effects of the interaction between the wires, which was calculated by reducing the vacuum between the wires [67]. A strong reduction of the band gap is found. The presence of localized states, which originate from the valence band, leads to a lowering of

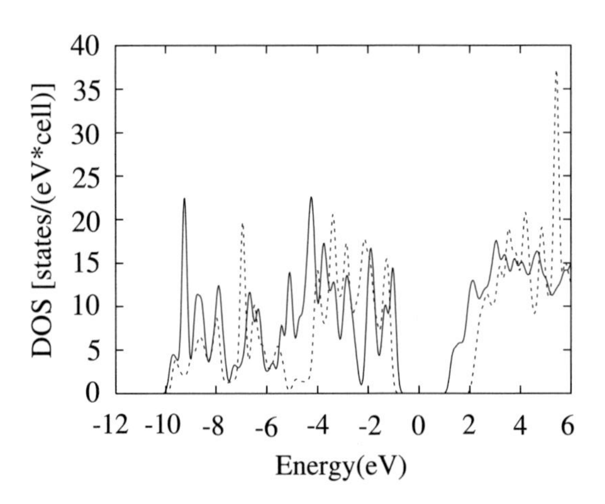

Fig. 2.12. Total density of states of a 3×4 fully H-passivated wire (*dashed line*) and of the 3×4 wire where four OH groups are present at the surface (*solid line*), after a Gaussian broadening of 0.1 eV. The alignment procedure is discussed in the text. The energies are referred to the bulk Si valence band maximum. After [78]

the DOS in the region where the Si-H bonding states were present. In the presence of wire-to-wire interaction, these states are strongly localized at the surface of the Si wire and their tails extend in the vacuum region forming interwire bonded states.

Few investigations have been devoted to undulating Si quantum wires [79, 80, 81, 82]. All the works underline the importance of localized states which arise from width fluctuations in the wire. Similar results have been obtained within a LDA calculation for Si wires, where the rectangular cross-section varies along the growth axis of the wire [83] thus simulating the presence of restrictions and bulges. These calculations show that localized states are not necessarily caused by extrinsic mechanisms such as surface trapping.

2.3.2 Optical Properties

The analysis of the optical properties of Si quantum wires has been concentrated on the investigation of the imaginary part of the dielectric function, the behavior of the dielectric constant and the calculation of radiative lifetimes of the most important transitions. The results confirm that quantum confinement causes new radiative transitions to appear in the optical region, the reduction of the dielectric constant and the decrease of the radiative lifetimes down to μs and ns.

A typical result for ϵ_2 in the 1–4 eV region is shown in Fig. 2.13 [49]. Given the highly anisotropic nature of the wire system, the figure considers separately the component of ϵ_2 along the direction parallel to the axis of the wire and the component in the orthogonal plane. The first quantity is

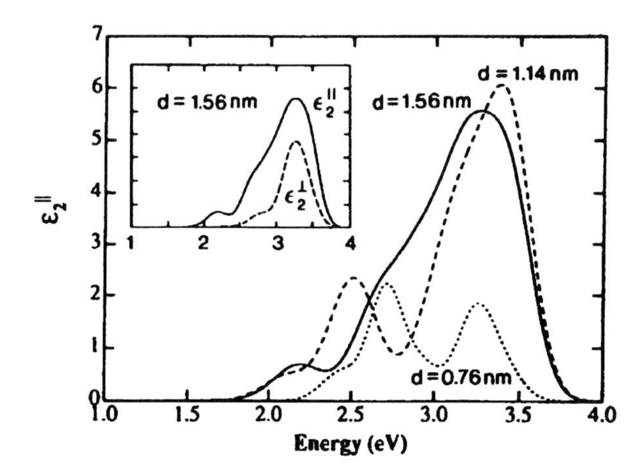

Fig. 2.13. ϵ_2 polarized in the direction of the wire for three different values of the wire side length. *Inset*: the same for the largest wire (*solid line*) compared to the component in the orthogonal wire (*dashed line*). After [49]

reported for three different wire sizes. In the inset ϵ_2 for two polarizations and for the largest wire is reported. Owing to the confinement effects in the plane orthogonal to the wire, the perpendicular component shows a pronounced blue shift. Moreover the parallel component has a side peak at low energies which is absent in the orthogonal one and which shifts to the blue as the wire dimension is reduced. While the position of the high-energy ϵ_2 peak (at about 3.5 eV) remains fixed with the size, the position of the low energy side peak strongly depends on the size. This reflects the fact that the main peak is essentially due to bulk-like transitions, while the side peak is a wire-related feature.

The real part of the static dielectric constant $\epsilon_1(\omega = 0)$ is reduced from the bulk Si value (11.4) to 4.2 and 3.3 for Si quantum wires of mean width of 6.72 and 8.64 Å, respectively [67, 78].

Few theoretical works have investigated the influence of defects at the the wire surfaces, e.g. dangling bonds which are formed by removal of H [67, 68] or OH impurities which are substitutional of H [78, 84]. The presence of dangling bonds causes drastic changes in the optical properties. The ϵ_2 of partially passivated 3×4 wires is reported in Fig. 2.14.

Two H atoms have been removed over the 14 present in the unit cell, leading to a H vacancy concentration which is 14% of the total Si surface bonds. The first structure in ϵ_2 (solid line), below 1 eV, corresponds to dangling-bond-to-dangling bond transitions. These are intense transitions and dominate the spectrum. After a gap, one finds the dangling bond-to-band state transitions, while the band-to-band transitions, characteristic of the fully passivated wire (dotted line) start only over 3 eV. It is evident that the ϵ_2 spectrum is entirely changed. It is interesting to note that the dangling bond related bands do not show a degree of polarization. A good passivation of the Si nanostructures is necessary to improve photoluminescence properties.

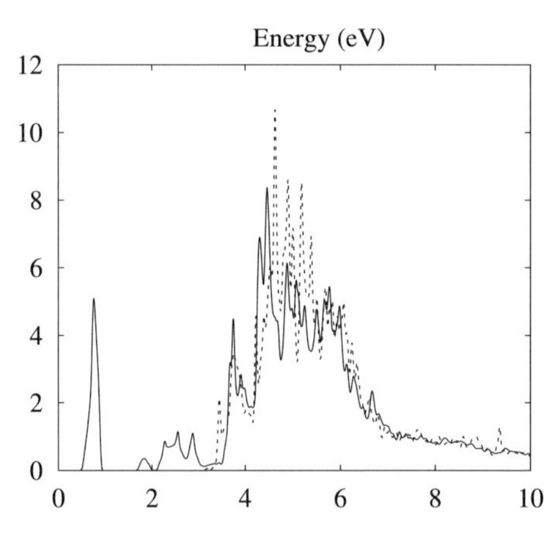

Fig. 2.14. ϵ_2 for a 3×4 Si wire. *Solid line:* partially passivated wire; *dotted line:* fully passivated line. After [84]

If one replaces H atoms with OH complexes, not only does the energy gap shrink, but also the intensity of the lower energy side peaks is reduced [79].

The radiative lifetimes τ are directly related to the inverse of the oscillator strengths, thus very fast transitions correspond to very strong dipole matrix elements. The calculated τ values for Si quantum wires show remarkable properties. In fact, within a given wire, several low-energy transitions (only 0.1–0.2 eV above the lowest ones) exist with widely different radiative lifetimes that can differ by orders of magnitude and which depend also on the symmetry of the states involved in the transition. Moreover for very small wires (width less than 1 nm) these lifetimes can be of the order of nanoseconds [9, 48, 68, 73, 75].

2.4 Silicon Quantum Dots

There has been a lot of interest, in the last few years, in the properties of semiconductor quantum dots [10]. In particular, considerable effort has been spent in investigating the structural, electronic and optical properties of Si nanocrystals [5, 6, 85].

2.4.1 Electronic Properties

One of the most challenging problems concerning Si nanocrystals is the accurate determination of their energy gap. A large number of calculations using different methods has been devoted to this aim. Figure 2.15 collects the results for the band gaps relative to three different shapes of Si nanocrystals covered by H and performed using EPM [5]. These results are compared with

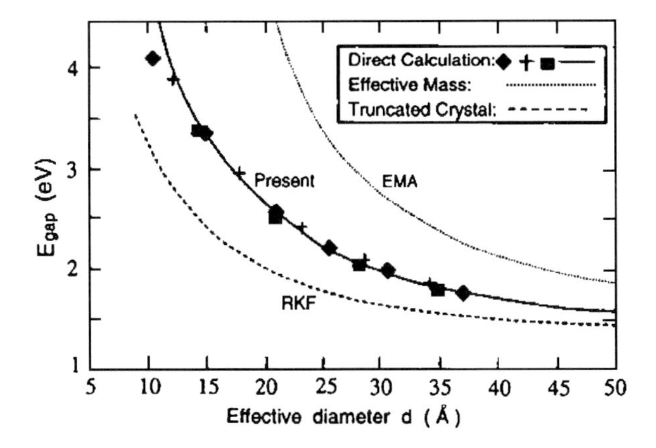

Fig. 2.15. Band gap (without Coulomb corrections) for three quantum dot shapes versus the effective diameter. (*diamond*:) spheres, (+): rectangular boxes, (*squares*:) cubic boxes. EMA [12] and RKF [86] results are also shown. After [6]

EMA [12] and a model calculation (RKF) [86]. All the EPM data are interpolated by a single curve even though they corresponds to nanocrystals of different shapes. EMA and RKF results are far from this curve.

This is another indication that reliable predictions for the energy gap of Si nanocrystals are obtained only when a method is able to provide a good description of the Si bulk band structure. The energy gap dependence of the data in Fig. 2.15 can be fitted by the equation:

$$E_{\mathrm{g}}(d) = 1.167 + \frac{88.34}{d^{1.37}}(\mathrm{eV}). \tag{2.23}$$

An ETB calculation gives similar behavior with practically the same exponent [16, 51].

For a given nanocrystal size, the theoretical data are often contradictory, see Table 2.6. The transition energies can differ by more than 1.5 eV from one method to another.

Even accounting for the difference between methods, this discrepancy is large, and is due to the fact that the studied structures are at the limit between pure molecules and bulk solids, where structural relaxations, localizations, delocalizations, and correlation effects have a great influence. It is thus important to use methods that describe properly both small molecules and bulk solids.

A procedure to adapt semi-empirical HF techniques designed to describe accurately the geometrical features of molecules to the study of complex semiconductor structures, comprising compatible use of a full Bloch-periodic

Table 2.6. Energy gap of H-saturated Si nanocrystals calculated through different methods as a function of the number of Si atoms N in the nanocrystal. The different sigla stand for TB with inclusion of third neighbors [87], ETB with inclusion of nearest neighbors only [88], LDA-LCAO with a LCAO basis set [89], LDA with a constant shift of +0.6 eV in the gap energy [90], MNDO-PM3 semi-empirical HF with standard parametrization [91], and MNDO/ZINDO HF scheme [92]. Adapted from [8]

N	Energy gap (eV)					
	TB	ETB	LDA-LCAO	LDA-PW	MNDO-PM3	M/ZINDO
10			5.10	4.62	3.43	
17	4.29		5.05		3.10	
29			4.95	3.32	2.76	
35		2.95	4.90		2.74	3.25
44						2.98
66			3.98	2.95	2.62	
87	3.10		4.05		3.10	
123			3.60	2.45	2.45	
239		1.68	3.50			

supercell model and a cluster model with adequate treatment of surface dangling bonds, was introduced in [8, 93, 94, 95]. The main results of these calculations, which allow the study of both ground and excited states, are related to the importance of structural relaxations, and the role of nanocrystal surfaces will be discuss in some detail in Sect. 2.4.2. Here we emphasize that even for nanocrystals with as few as 29 Si atoms, in a tetrahedral arrangement, the vibrational spectrum reproduces quite well the phonon dispersion of the bulk.

In the last few years, the most advanced *ab initio* methods have been employed to study the electronic structure of Si hydrogenated nanocrystals [29, 37]. The calculated variation of the optical gaps as a function of the nanocrystal size is shown in Fig. 2.16 and are compared with the HF MINDO/ZINDO results [92]. The comparison demonstrates the importance of structural relaxation and correlations effects.

We have already pointed out that it has been recently shown that the self-energy correction with respect to the bulk result and the Coulomb correction almost exactly cancel each other for Si nanocrystals [40]. Consequently the lowest excitonic energy is correctly predicted by LDA with the bulk correction for the gap.

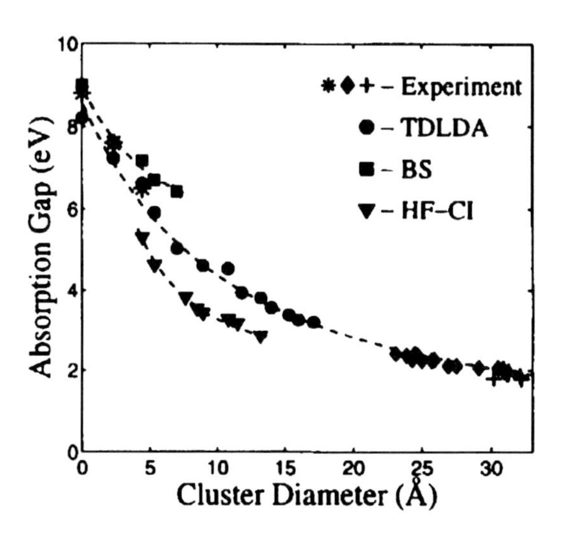

Fig. 2.16. Absorption gap vs. Si cluster diameter. Theoretical values are referred to TDLDA [37], BS [29], HF methods [92]. Experimental results are related to Si nanoparticles [89, 96, 97, 98]. After [37]

2.4.2 Optical Properties

A consistent description for the optical processes in Si nanocrystals is given by the HF MNDO/ZINDO results [8, 92]. The optical absorption can be explained in term of quantum confinement in the crystalline Si region, even if structural relaxations are important in determining the gap energy values. The absorption occurs at the "crystalline" core of the particles and is exciton-like in character, i.e. involves an almost pure one-electron excitation. Optical emission, instead, can occur through different channels, depending on the chemical state of the surface.

For perfectly hydrogenated regions, there is a fast channel, Stokes shifted in energy from the absorption by only $\sim$ 0.2–0.3 eV, which is also tied to the core of the nanocrystal. There is also a slow channel, strongly red shifted and pinned in energy (see Fig. 2.17). It originates from the coupling of the electronic excitation to the surface modes, with generation of a metastable Si-H-Si bridge-like defect.

The nanocrystal in the excited states undergoes a strong symmetry-lowering spontaneous distortion, extremely localized on the two neighboring surface atoms, each bonded to just one H atom. This localized defect occurs both for spherical and cubic Si nanocrystals. Initial oxidation of the surface, as Si-OH units or as Si-O-Si backbonds, does not affect the energy of the first absorption transition, or the decay through the bridge defect. Further oxidation hardens the nanocrystal surface and decreases the number of effective sites for optical decay. Incorporation of a Si=O silanone unit perturbs both absorption and emission properties, introducing a first transition in the red-orange, localized over the defect.

The existence of localized states to account for the large Stokes shift between absorption and luminescence has also been postulated by Allan

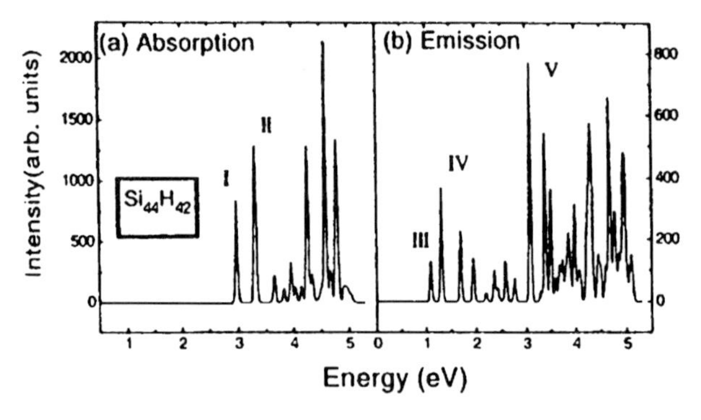

Fig. 2.17. Optical spectra calculated for a $Si_{44}H_{42}$ cluster in its ground (**a**) and excited (**b**) state configuration corresponding to absorption and emission spectra, respectively. Note the huge Stokes shift. The *lines* are Gaussian broadened by 0.01 eV. After [92]

et al. [99]. From total energy calculations, the existence of a self-trapped exciton at some surface bonds of silicon nanocrystals has been demonstrated. The self-trapped exciton recombines with a luminescence energy almost independent of the nanocrystal size which explains the huge Stokes shift. The stabilization of the self-trapped exciton is due to dimer bond passivation by H atoms at the surface of the Si nanocrystals as witnessed by Fig. 2.18. Both results [92, 99] point to the fact that small Si nanocrystals are intrinsically subjected to significant distortion when excited.

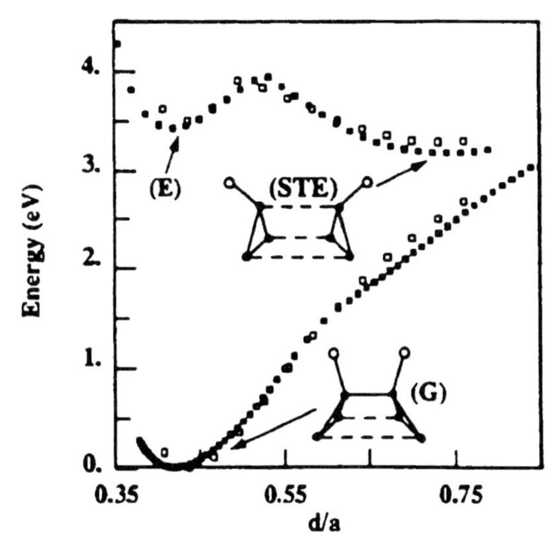

Fig. 2.18. Total energy (*full symbols:* TB; *empty symbols:* LDA) of a Si_{29} hydrogenated nanocrystal in the ground and excited state as a function of the dimer interatomic distance d (a = 0.54 nm). The *arrows* indicate the energy minima. Schematic side views of cluster surface dimer in the ground state (G) and in the self-trapped state (STE) are also shown. After [99]

Other work emphasizes the relevance of Si=O surface defects [100]. It is assumed that, when a Si nanocrystals is passivated by H atoms, recombination is via free exciton states for all sizes. However, if the nanocrystal is passivated by O, a stable electronic state is formed at the Si=O bond. For O passivated nanocrystals, three different recombination mechanisms are suggested (see Fig. 2.19).

In zone I (large nanocrystals) recombination is via free excitons. The PL energy increases with the confinement. In Zone II, recombination involves a trapped electron and a free hole. As the size decreases, the PL emission energy still increases, but not as fast as predicted by simple quantum confinement, since the trapped electron state energy is almost size independent. In Zone III the recombination is via trapped excitons. The PL energy remains

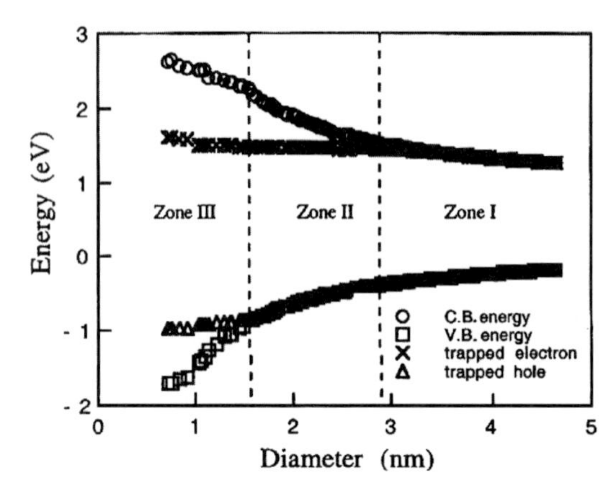

Fig. 2.19. Electronic states in Si nanocrystals as a function of cluster size and surface passivation. The trapped electron state is a p-state localized on the Si atom of the Si=O bond and the trapped hole state is a p-state localized on the O atom. After [100]

constant, thus there is a large PL redshift when the nanocrystal surface is exposed to oxygen.

Several theoretical works have pointed out, recently, that the chemistry of the surface (in particular the presence of double bonded oxygen) can have a substantial impact on the structural, electronic and optical properties of Si quantum dots [101, 102, 103, 104, 105]. Usually one starts the calculations from fully H-passivated Si nanoclusters and then H is substituted by O, considering two types of Si-O bonds: the Si-O-Si backbond and the Si=O double bond. The presence of O atoms in backbond positions produces a huge variation of the surface structure, whereas Si=O bonds cause only small local distortions. On the contrary, whereas the Si-O-Si bond does not affect too much the cluster energy gap value, the Si=O bond originates a huge redshift of the fully H-covered related band gap (see Fig. 2.20a). If multiple oxidation is considered, i.e. new Si=O bonds were added, the gap tends to decrease. The reduction however is not linear with the number of Si=O bonds. For all the clusters sizes, in fact, the strongest redshift is achieved with the first Si=O bond; a second Si=O bond produces a further reduction but weaker than the first and the more Si=O bonds are added the smaller their contribution to the gap lowering (Fig. 2.20b) [105].

Thus a sort of saturation limit seems to be reached; consequently these results reproduce well the experimental findings of oxidation of H-covered Si nanoparticles that show a saturation limit for the PL energies almost independent on the size [100].

Very recently also the first *ab initio* calculation of the structural, electronic and optical properties of Si nanocrystals embedded in SiO_2 has been published [105]. To design the simplest model for a Si nanocluster embed-

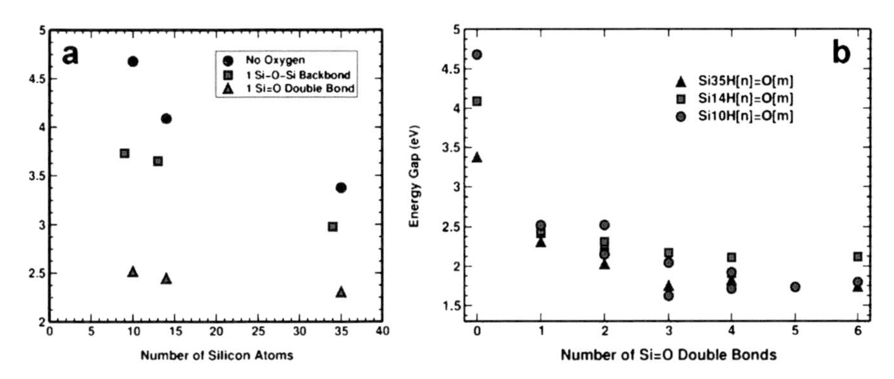

Fig. 2.20. (**a**) Electronic energy gap as a function of the Si atom for the different bond configurations. (**b**) Electronic energy gap as a function of the number of Si = O bond double bonds at the cluster surface. *Circle*: $Si_{10}H_n = O_m$, *square*: $Si_{14}H_n = O_m$, and *triangle*: $Si_{35}H_n = O_m$. After [105]

ded in a SiO_2 matrix, a cubic supercell of SiO_2 β-cristobalite (BC) has been built up repeating the BC unit cell twice along each axis; the diamond-like geometry allows the drawing of sharp Si-SiO_2 interfaces. Then from the SiO_2 structure, 12 O atoms have been extracted and the Si atoms left with dangling bonds have been linked together, forming a supercell of 64 Si and 116 O atoms with 10 Si atoms bonded together in a small nanocluster. The tetrahedral coordination is ensured by the BC symmetry but the Si-Si bond length is highly strained with respect to the bulk Si case (33%). There are no defects at the interface and the O atoms at the nanocluster surface are single bonded with Si. This initial structure has been optimized by total energy minimization runs, leaving us free to relax all the atoms positions and the cell parameters. The study of the structural properties shows that in the relaxed structure, the Si nanocluster in SiO_2 still has a crystalline-like geometry with a Si-Si bond length of 2.67 Å (only 14% strained with respect to bulk Si). This compression towards the bulk geometry causes a complex deformation of the SiO_2 matrix around the nanocrystal, both in bond lengths and angles. Nevertheless the deformation doesn't affect all the SiO_2 matrix. It is actually possible to still find a good BC crystalline structure at a distance from the nanocluster atoms of 0.8–0.9 nm. The nanocluster is therefore surrounded by a cap-shell of stressed SiO_2 with a thickness of about 1 nm which progressively goes towards a pure crystalline BC. Figure 2.21a shows the result for the calculated band structure in the relaxed geometry.

The system is still a semiconductor with a reduced energy gap, in the visible range, with respect to the oxide value. The strong reduction with respect to the bulk SiO_2 energy gap is originated by the presence, at the valence and conduction band edges, of confined, flat of states that are only due to the presence of the Si nanocrystals as confirmed by the analysis of the orbital spatial distribution; see part (b) of Fig. 2.21. These features are

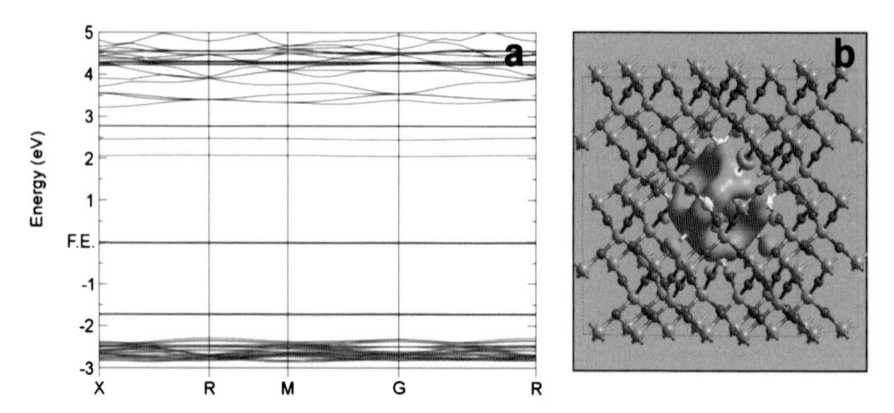

Fig. 2.21. (**a**) Band structure for the relaxed Si nanocluster in the SiO_2 matrix. F.E. is the Fermi energy of the system. (**b**) HOMO square modulus isosurface at 25% of maximum amplitude for the relaxed Si nanocluster in the SiO_2 matrix. *Gray*: O atoms; *pale gray*: Si atoms in SiO_2; *white*: Si atoms in Si nanocluster. After [105]

entirely new and may be the origin of the observed photoluminescence in the red optical region for Si nanocrystals immersed in a SiO_2 matrix [61]. Deep inside the valence and conduction bands the more k-dispersed states related to the SiO_2 matrix are still present. Moreover the result concerning the role of both the Si nanocrystal and the interface Si-O region with respect to the absorption process is in close agreement with the X-ray absorption fine structure measurements that indicate the presence of an intermediate region between the Si nanocrystal and the SiO_2 matrix that participates in the light emission processes [61, 106].

Finally, all the calculations for Si nanocrystals yield a reduction of the static dielectric constant which depends on the nanocrystal size. The reduction is significant for nanocrystal size smaller than 2 nm [18, 107, 108].

References

1. A.D. Yoffe: Adv. Phys. **42**, 173 (1993)
2. M.J. Kelly, R.J. Nicholas: Rep. Prog. Phys. **48**, 1702 (1985)
3. M.J. Kelly: 'Fundamentals of low dimensional physics'. In: *The Physics and Fabrication of Microstructures and Microdevices.* ed. by M.J. Kelly, C. Weisbuch (Springer, Berlin, Heidelberg 1986) pp. 174–196
4. G.C. John, V.A. Singh: Phys. Rep. **263**, 1 (1995)
5. A. Zunger, L.-W. Wang: Appl. Surface Sci. **102**, 350 (1996)
6. M. Lannoo, G. Allan, C. Delerue: 'Theory of optical properties and recombination processes in porous silicon'. In: *Structural and Optical Properties of Porous Silicon Nanostructures.* ed. by G. Amato, C. Delerue, H.-J. von Bardeleben (Gordon and Breach Science Publishers, Amsterdam 1997) pp. 187–220

7. S. Ossicini, E. Degoli: 'Electron states in confined silicon systems'. In: *Silicon-Based Microphotonics: from Basics to Applications.* ed. by O. Bisi, S.U. Campisano, L. Pavesi, F. Priolo (IOS Press, Amsterdam 1999) pp. 191–260
8. M.J. Caldas: Phys. Stat. Sol. (b) **217**, 641 (2000)
9. O. Bisi, S. Ossicini, L. Pavesi: Surface Sci. Rep. **38**, 5 (2000)
10. A.D. Yoffe: Adv. Phys. **50**, 1 (2001)
11. M.S. Hybertsen: 'Mechanism for light emission from nanoscale silicon'. In: *Porous Silicon Science and Technology.* ed. by J.-C. Vial, J. Derrien (Springer, Berlin, Heidelberg 1995) pp. 67–90
12. T. Takagahara, K. Takeda: Phys. Rev. B **46**, 15578 (1992)
13. M. Voos, Ph. Uzan, C. Delalande, G. Bastard, A. Halimaoui: Appl. Phys. Lett. **61**, 1213 (1992)
14. G. Fishman, I. Mihalcescu, R. Romenstain: Phys. Rev. B **48**, 1464 (1993)
15. S.Y. Ren, J.D. Dow: Phys. Rev. B **45**, 6492 (1992)
16. J.P. Proot, C. Delerue, G. Allan: Appl. Phys. Lett. **61**, 1948 (1992)
17. N.A. Hill, K.B. Whaley: Phys. Rev. Lett. **75**, 1130 (1995)
18. L.-W. Wang, A. Zunger: J. Chem. Phys. **100**, 2394 (1994)
19. P. Hohenberg, W. Kohn: Phys. Rev. **136B**, 864 (1964)
20. W. Kohn, L.J. Sham: Phys. Rev. **140A**, 1133 (1965)
21. M.S. Hybertsen, S.G. Louie: Phys. Rev. B **34**, 5390 (1986)
22. R.W. Godby, M. Schlüter, L.J. Sham: Phys. Rev. Lett. **56**, 2415 (1986)
23. R.W. Godby, M. Schlüter, L.J. Sham: Phys. Rev. B **37**, 10159 (1988)
24. M. Fuchs, M. Bochstedte, E. Pehlke, M. Scheffler: Phys. Rev. B **57**, 2134 (1998)
25. G.E. Engel: Phys. Rev. Lett. **78**, 3350 (1997)
26. L. Hedin: Phys. Rev. **139**, 796 (1965)
27. M.S. Hybertsen, S.G. Louie: Phys. Rev. Lett. **58**, 1551 (1987)
28. G. Onida, L. Reining, R.W. Godby, R. Del Sole, W. Andreoni: Phys. Rev. Lett. **75**, 818 (1995)
29. M. Rohlfing, S. Louie: Phys. Rev. Lett. **80**, 3320 (1998)
30. O. Pulci, G. Onida, R. Del Sole, L. Reining: Phys. Rev. Lett. **81**, 5374 (1998)
31. G. Onida, R. Del Sole, M. Palummo, O. Pulci, L. Reining: Phys. Stat. Sol. (a) **170**, 365 (1998)
32. C.G. Van de Walle, J. E Northrup: Phys. Rev. Lett. **70**, 1116 (1993)
33. F. Bassani, G. Pastori Parravicini: *Electronic and Optical Transition in Solids* (Pergamon Press, London 1975)
34. R. Del Sole: 'Reflectance spectroscopy theory'. In: *Photonic Probes of Surfaces.* ed. by P. Halevi (Elsevier, Amsterdam 1995) pp. 131–152
35. D.E. Aspnes, A.A. Studna: Phys. Rev. B **27**, 985 (1983)
36. L.X. Benedict, E.L. Shirley, R.B. Bohn: Phys. Rev. Lett. **80**, 4514 (1998)
37. I. Vasiliev, S. Ogut, J.R. Chelikowsky: Phys. Rev. Lett. **86**, 1813 (2001)
38. S. Albrecht, L. Reining, R. Del Sole, G. Onida: Phys. Rev. Lett. **80**, 4510 (1998)
39. B. Adolph, V.I. Gavrilenko, K. Tenelsen, F. Bechstedt, R. Del Sole: Phys. Rev. B **53**, 9797 (1996)
40. C. Delerue, M. Lannoo, G. Allan: Phys. Rev. Lett. **84**, 2457 (2000)
41. N. Keating: Phys. Rev. **145**, 637 (1966)
42. M. Tang, L. Colombo, J. Zhu, T.D. de la Rubia: Phys. Rev. B **55**, 14279 (1997)

43. J. Ihm, A. Zunger, M.L. Cohen: J. Phys. C **12**, 4409 (1979)
44. R. Car, M. Parrinello: Phys. Rev. Lett. **55**, 2471 (1985)
45. B. Delley, E.F. Steigmeier: Appl. Phys. Lett. **67**, 2370 (1995)
46. M. Hirao, T. Uda, Y. Murayama: Mater. Res. Soc. Symp. Proc. **283**, 425 (1993)
47. J.W. Mintmire: J. Vac. Sci. Technol. A **11**, 1733 (1993)
48. A.J. Read, R.J. Needs, K.J. Nash, L.T. Canham, P.D.J. Calcott, A. Qteish: Phys. Rev. Lett. **69**, 1232 (1992)
49. F. Buda, J. Kohanoff, M. Parrinello: Phys. Rev. Lett. **69**, 1272 (1992)
50. L.-W. Wang, A. Zunger: J. Phys. Chem. **98**, 2158 (1994)
51. C. Delerue, G. Allan, M. Lannoo: Phys. Rev. B **48**, 11024 (1993)
52. A.N. Kholod, A. Saúl, J. Fuhr, V.E. Borisenko, F. Arnaud D'Avitaya: Phys. Rev. B **62**, 12949 (2000)
53. A.N. Kholod, S. Ossicini, V.E. Borisenko, F. Arnaud, D'Avitaya: Phys. Rev. B **65**, 1115315 (2002)
54. A.N. Kholod, S. Ossicini, V.E. Borisenko, F. Arnaud, D'Avitaya: Surface Sci. **527**, 30 (2003)
55. S. Ossicini, A. Fasolino, F. Bernardini: phys. stat. sol (b) **190**, 117 (1995)
56. S. Ossicini, A. Fasolino, F. Bernardini: Phys. Rev. Lett. **72**, 1044 (1994)
57. O.K. Andersen: Phys. Rev. B **12**, 3060 (1975)
58. S. Ossicini, A. Fasolino, F. Bernardini: 'The electronic properties of low-dimensional Si structures'. In: *Optical Properties of Low Dimensional Silicon Structure* ed. by D.C. Bensahel, L.T. Canham, S. Ossicini, NATO ASI Series, Vol. 244 (Kluwer Academic Publishers, Dordrecht 1993) pp. 219–228
59. E. Degoli, S. Ossicini: Surface Sci. **470**, 32 (2000)
60. N. Ikarashi, K. Watanabe: Jpn. J. Appl. Phys. **39**, 1278 (2000)
61. G. Vijaya Prakash, N. Daldosso, E. Degoli, F. Iacona, M. Cazzanelli, Z. Gaburro, G. Puker, P. Dalba, F. Rocca, E. Ceretta Moreira, G. Franzó, D. Pacifici, F. Priolo, C. Arcangeli, A.B. Filonov, S. Ossicini, L. Pavesi: J. Nanosci. Nanotech. **1**, 159 (2001)
62. M. Ben-Chorin, B. Averboukh, D. Kovalev, G. Polisski, F. Koch: Phys. Rev. Lett. **77**, 763 (1996)
63. E. Degoli, S. Ossicini: Phys. Rev. B **57**, 14776 (1998)
64. Y. Kanemitsu, S. Okamoto: Phys. Rev. B **56**, R15561 (1997)
65. S. Ossicini: 'Porous silicon modelled as idealised quantum wires'. In: *Properties of Porous Silicon.* ed. by L.T. Canham (IEE INSPEC, London 1997) pp. 207–211
66. S. Ossicini, O. Bisi: 'Electronic and optical properties of silicon quantum wires at different passivation regimes'. In: *Structural and Optical Properties of Porous Silicon Nanostructures.* ed. by G. Amato, C. Delerue, H.-J. von Bardeleben (Gordon and Breach Science Publishers, Amsterdam 1997) pp. 187–226
67. S. Ossicini, C.M. Bertoni, M. Biagini, A. Lugli, G. Roma, O. Bisi: Thin Solid Films **276**, 154 (1997)
68. C.-Y. Yeh, S.B. Zhang, A. Zunger: Phys. Rev. B **50**, 14405 (1994)
69. A.M. Saitta, F. Buda, G. Fiumara, P.V. Giaquinta: Phys. Rev. B **53**, 1446 (1996)
70. T. Ohno, K. Shiraishi, T. Ogawa: Phys. Rev. Lett. **69**, 2400 (1992)
71. C.Y. Yeh, S.B. Zhang, A. Zunger: Appl. Phys. Lett. **63**, 3445 (1993)

72. L. Dorigoni, O. Bisi, S. Ossicini, F. Bernardini: Phys. Rev. B **53**, 4557 (1996)
73. M.S. Hybertsen, M. Needels: Phys. Rev. B **48**, 4608 (1993)
74. H.M. Polatoglou: J. Lumin. **57**, 117 (1993)
75. G.D. Sanders, Y.C. Yang: Phys. Rev. B **45**, 6492 (1992)
76. S.-G. Lee, B.-H. Cheong, K.-H. Lo, K.J. Chang: Phys. Rev. B **51**, 1762 (1995)
77. A.B. Filonov, G.V. Petrov, V.A. Novikov, V.E. Borisenko: Appl. Phys. Lett. **67**, 1090 (1995)
78. S. Ossicini, O. Bisi: Solid State Phenomena **54**, 127 (1997)
79. K.J. Nash: 'Porous silicon modelled as undulating quantum wires'. In: *Properties of Porous Silicon*. ed. by L.T. Canham (IEE INSPEC, London 1997) pp. 216–220
80. T.C. Tigelis, J.P. Xanthaxis, J.L. Vomvoridis: phys. stat. sol. (a) **65**, 125 (1998)
81. D. Ninno, G. Iadonisi, F. Buonocore: Solid State Commun. **112**, 521 (1999)
82. D. Ninno, F. Buonocore, G. Cantele, G. Iadonisi: phys. stat. sol. (a) **182**, 285 (2000)
83. E. Degoli, M. Luppi, S. Ossicini: phys. stat. sol. (a) **182**, 301 (2000)
84. S. Ossicini: phys. stat. sol. (a) **170**, 377 (1998)
85. C. Delerue, M. Lannoo, G. Allan: 'Porous silicon modelled as idealised quantum dots'. In: *Properties of Porous Silicon*. ed. by L.T. Canham (IEE INSPEC, London 1997) pp. 212–215
86. M.V. Rama Krishna, R.A. Friesner: Phys. Rev. Lett. **67**, 629 (1991)
87. L. Vervoort, A. Saúl, F. Bassani, F.A. D'Avitaya: Thin Solid Films **297**, 163 (1997)
88. T. Huaxiang, Y. Ling, X. Xide: Phys. Rev. B **48**, 11204 (1993)
89. B. Delley, E.F. Steigmeier: Phys. Rev. B **47**, 1397 (1993)
90. M. Hirao, T. Uda: Surface Sci. **306**, 87 (1994)
91. R. Kumar, Y. Kitoh, K. Shegematsu, K. Hara: Jpn. J. Appl. Phys. **33**, 909 (1994)
92. R.J. Baierle, M.J. Caldas, E. Molinari, S. Ossicini: Solid State Commun. **102**, 545 (1997)
93. R.J. Baierle, M.J. Caldas, E. Molinari, S. Ossicini: Braz. J. Phys. **26**, 631 (1996)
94. R.J. Baierle, M.J. Caldas, E. Molinari, S. Ossicini: Mater. Sci. Forum **258–263**, 11 (1998)
95. R.J. Baierle, M.J. Caldas: J. Mod. Phys. B **13**, 2733 (1999)
96. U. Itoh, Y. Toyoshima, H. Onuki, N. Washida, T. Ibuki: J. Chem. Phys. **85**, 4867 (1986)
97. S. Furukawa, T. Miyasato: Phys. Rev. B **38**, 5726 (1988)
98. D.J. Lockwood, A. Wang, B. Bryskiewicz: Solid State Commun. **89**, 587 (1994)
99. G. Allan, C. Delerue, M. Lannoo: Phys. Rev. Lett. **78**, 3161 (1997)
100. M. Wolkin, J. Jorne, P.M. Fauchet, G. Allan, C. Delerue: Phys. Rev. Lett. **82**, 197 (1999)
101. A. Puzder, A.J. Williamson, J.C. Grossman, G. Galli: Phys. Rev. Lett. **88**, 097401 (2002)
102. I. Vasiliev, J.R. Chelikowsky, R.M. Martin: Phys. Rev. B **65**, R121302 (2002)
103. I. Vasiliev, S. Ogut, J.R. Chelikowsky: Phys. Rev. B **65**, 115416 (2002)
104. A.B. Filonov, S. Ossicini, F. Bassani, F. Arnaud D'Avitaya: Phys. Rev. B. **65**, 195317 (2002)

105. M. Luppi, S. Ossicini: SPIE **4808**, 73 (2002); phys. stat. sol. (a) **197**, 251 (2003)
106. N. Daldosso et al.: Physica E **16**, 321 (2003), Phys. Rev. B **68**, 085327 (2003)
107. R. Tsu, D. Babic: Appl. Phys. Lett. **64**, 1806 (1994)
108. M. Lannoo, C. Delerue, G. Allan: Phys. Rev. Lett. **74**, 3415 (1995)

3 Porous Silicon

In 1990 it was proposed that Si when rendered porous can be an efficient light emitting system, as efficient as many III-V semiconductor systems [1]. The reason for this was attributed to the low dimensionality of the surviving Si skeleton. In addition other effects were recognized: (1) the large photon extraction efficiency due to the decrease in the effective refractive index and (2) the confinement of the carriers in regions free of recombination centers. After this very surprising report, a huge number of papers was published trying to understand the optical properties of porous silicon (PS) and the physical mechanism at the heart of the large luminescence efficiency. Soon it was realized that PS is a very fragile and highly reactive material so that many of its properties are age dependent and unstable. Over the years, sizeable progress was made both in terms of improvement of its stability and in term of efficiency of the device. Microelectronics compatibility of PS was demonstrated and integration of driving circuits with a light emitting element was performed [2]. At the same time microcavities with improved light emission properties were demonstrated [3]. At the end of 2000 another paper shocked the silicon community: the report on light amplification or optical gain in silicon nanocrystals [4]. This paper raises big hopes of the development of a silicon laser even though it is not yet clear whether PS can be used as an active material.

In this chapter we will try to review the most important results about the exploitation and the understanding of the properties of PS. The chapter is divided into six parts. The first introduces how porous silicon is made. The second is aimed to demonstrate that the structure of PS is a collection of nanometric objects that contribute to the light emission process. The relevant role of the surface chemistry is underlined. The third and fourth parts discuss the essential features of the absorption and luminescence of PS in order to discriminate between the models of the recombination mechanisms. The fifth presents recent progress in the field of electrical properties of PS. The section six is an overview of the application of PS in light emitting diodes.

In the last few years various conference proceedings [5], review articles [6] and edited books [7] have been published on PS.

3.1 How Porous Silicon is Made

PS is formed by the electrochemical etching of Si in an HF solution. Following the electrochemical reaction at the Si surface partial dissolution of Si settles in. Let us concentrate on the various factors which rule this process.

3.1.1 Current–Voltage Characteristics

When a potential is applied to Si with respect to an electrode placed in an HF-rich solution, a measurable current flows through the system. For any current to pass the Si/electrolyte interface, a change from electronic to ionic current must occur. This means that a specific chemical redox reaction must occur at the Si/electrolyte interface. Application of a potential then induces a precise chemical reaction, the nature of which is fundamental to the formation of PS.

Figure 3.1 shows the "typical" i-V curves for n- and p-type doped Si in aqueous HF[8]. The i-V curves show some similarities to the normal Schottky diode behavior expected for a semiconductor/electrolyte interface, but some important differences occur. For instance, while the sign of majority carriers changes between n- and p-type, the chemical reactions at the interface remain the same.

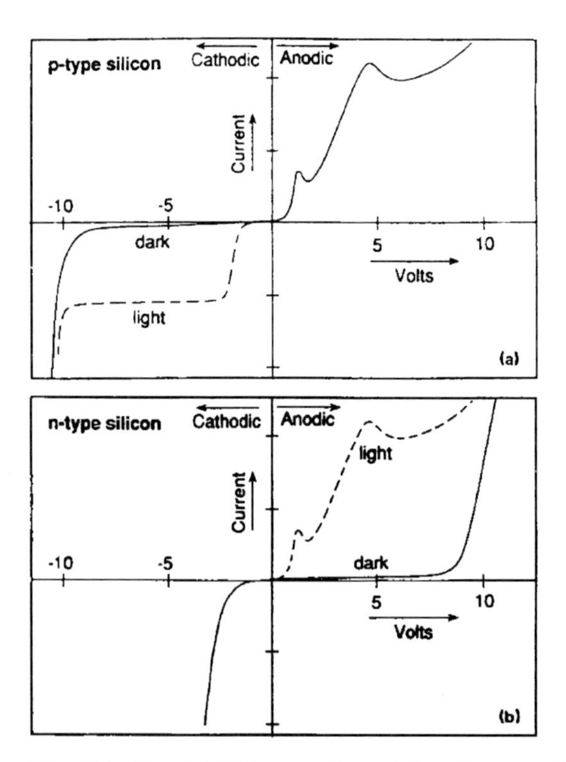

Fig. 3.1. Typical i-V curves for p (**a**) and n-type (**b**) silicon. After [8]

Under cathodic polarization (electrons flow into the electrolyte), for both n- and p-type materials, Si is stable. The only important cathodic reaction is the reduction of water at the Si/HF interface, with the formation of hydrogen gas. This usually occurs only at high cathodic overpotentials, or, using Schottky diode terminology, at reverse breakdown.

Under anodic polarization (holes flow into the electrolyte) Si dissolves. At high anodic overpotentials the Si surface electropolishes and the surface retains a smooth and planar morphology. In contrast, with low anodic overpotentials, the surface morphology is dominated by a vast labyrinth of channels that penetrates deep into the bulk of the Si. PS is formed. Pore formation occurs only in the initial rising part of the i-V curve, for a potential value below that of the small sharp peak (see Fig. 3.1). This current peak is called the electropolishing peak and separates the region of PS formation and of electropolishig. The quantitative values of the i-V curves, as well as the values corresponding to the electropolishing peak, depend on etching parameters and wafer doping. For n-type substrates, this typical i-V behavior is observed only under illumination because hole supply is needed (anodic current).

3.1.2 Dissolution Chemistries

The exact dissolution chemistries of Si are still in question, and different mechanisms have been proposed [8]. However it is generally accepted that holes are required for both electropolishing and pore formation. During pore formation two H atoms evolve for every Si atom dissolved. The H evolution diminishes when approaching the electropolishing regime and disappears during electropolishing. Current efficiencies are about two electrons per dissolved Si atom during pore formation, and about four electrons in the electropolishing regime [9]. The global anodic semireactions can be written during pore formation as:

$$\mathrm{Si} + 6\mathrm{HF} \longrightarrow \mathrm{H_2SiF_6} + \mathrm{H_2} + 2\mathrm{H^+} + 2\mathrm{e^-} \ , \tag{3.1}$$

and during electropolishing as

$$\mathrm{Si} + 6\mathrm{HF} \longrightarrow \mathrm{H_2SiF_6} + 4\mathrm{H^+} + 4\mathrm{e^-} \ . \tag{3.2}$$

The final and stable product for Si in HF is in any case H_2SiF_6 or some of its ionized forms. This means that during pore formation only two of the four available Si electrons participate in the interfacial charge transfer while the remaining two undergo corrosive hydrogen liberation. In contrast, during electropolishing, all four Si electrons are electrochemically active.

Lehmann and Gösele [10] have proposed the most accepted dissolution mechanism (Fig. 3.2). It is based on a surface bound oxidization scheme, with hole capture, and subsequent electron injection, which leads to the divalent Si oxidization state.

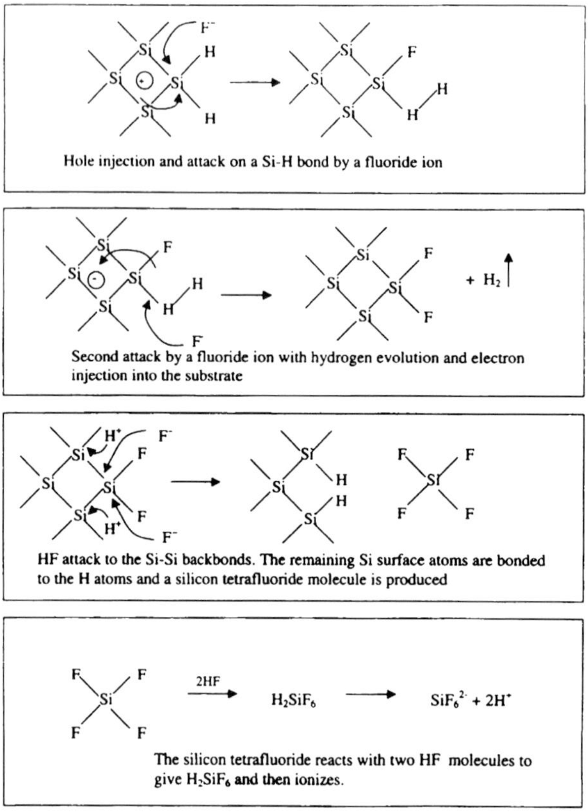

Hole injection and attack on a Si-H bond by a fluoride ion

Second attack by a fluoride ion with hydrogen evolution and electron injection into the substrate

HF attack to the Si-Si backbonds. The remaining Si surface atoms are bonded to the H atoms and a silicon tetrafluoride molecule is produced

The silicon tetrafluoride reacts with two HF molecules to give H_2SiF_6 and then ionizes.

Fig. 3.2. Silicon dissolution scheme proposed by Lehmann and Gösele

3.1.3 Pore Formation

While it is generally accepted that pore initiation occurs at surface defects or irregularities, different models have been proposed to explain pore formation in PS [8]. Some basic requirements have to be fulfilled for electrochemical pore formation to occur [11]:

- Holes must be supplied by bulk Si, and be available at the surface.
- While the pore walls have to be passivated, the pore tips must be active in the dissolution reaction. Consequently, a surface which is depleted of holes is passivated to electrochemical attach, thus: (1) the electrochemical etching is self-limiting and (2) hole depletion occurs only when every hole that reaches the surface reacts immediately. The chemical reaction is not limited by mass transfer in the electrolyte.
- The current density should be lower than the electropolishing critical value.

For current densities above such a value, the reaction is under ionic mass transfer control, which leads to a surface charge of holes and to a smoothing of the Si surface (electropolishing). The behavior at high current densities turns out to be useful to produce PS free-standing layers. Raising the current density above the critical value at the end of the anodization process results in a detachment of the PS film from the Si substrate.

In the low current density regime, where PS forms, some considerations apply:

- A surface region depleted in mobile carriers is formed at the Si/electrolyte interface. This region is highly resistive (comparable to intrinsic Si). The thickness of the depleted region depends on the doping density. It is several μm thick for lightly *n*-type doped Si. It is thin for highly *n*- or *p*-type doped Si, and it does not exist for lightly to moderately *p*-type doped Si.
- The size of the pores is related both to the depletion layer width and to the mechanism of charge transfer.
- In highly doped substrates charge transfer is dominated by tunneling of the carriers, and the pore size reflects the width of the depletion region, being typically around 10 nm.
- In lightly *n*-type doped Si anodized in the dark, generation of carriers occurs at breakdown. The pore dimensions are about 10–100 nm (mesopores), regardless of doping density. Under illumination the pore size is dependent on doping density and anodization conditions, with diameters in the range 0.1–20 μm (macropores).
- A hole depletion is expected in any case if the dimensions of the nanocrystals are about a few nm, independently of the substrate type and doping. In this size region, quantum confinement is effective and the Si band gap is increased. A hole needs to overcome an energy barrier to enter this region. This is highly improbable. The quantum confinement is responsible for pore diameters below 2 nm, denoted as micropores. Micropores can be found on every type of PS sample, but only in moderately and lightly *p*-type doped substrates does pure micro PS exist.
- Both mechanisms coexist during PS formation, resulting in a superposition of micro and meso (or macro) structures, whose average size and distribution depend on substrate and anodization conditions.

3.1.4 Parameters which Affect the PS Formation

Electrolyte. Due to the hydrophobic character of the clean Si surface, absolute ethanol is usually added to the aqueous solution in which HF is sold to increase the wettability of the PS surface and to improve the uniformity of the PS layer in depth. In addition, during the reaction there is H evolution. Bubbles form and stick on the Si surface in pure aqueous solutions, whereas they are promptly removed if ethanol (or some other surfactant) is present. Moreover, it has been found that lateral inhomogeneity and surface roughness can be reduced (increasing electrolyte viscosity) either by diminishing the temperature or introducing glycerol to the composition of the HF solution [12].

Potential. The dissolution is obtained either controlling the anodic current or the potential. Generally, it is preferable to work with constant current, because it allows better control of porosity, thickness and reproducibility of the PS layer.

Cells. The simplest electrochemical cell is a Teflon beaker. The Si wafer acts as the anode and the cathode is generally made of platinum, or other HF-resistant and conductive material. The preferred geometry for the cell is shown in Fig. 3.3. In this cell, the Si wafer is placed on a metal disk and sealed through an O-ring, so that only the front side of the sample is exposed to the electrolyte. When a Si wafer with high resistivity (that is more than few $\mathrm{m}\Omega\,\mathrm{cm}$) is used, an high dose implantation on the back surface of the wafer is required to improve the electrical contact between the wafer and the metal disk. This step is crucial to get lateral homogeneity in the PS layer.

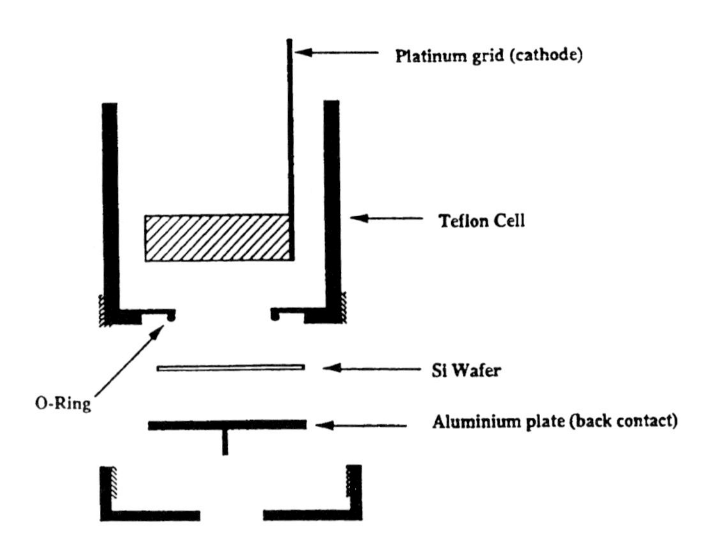

Fig. 3.3. Cross-sectional view of a single tank anodization cell. After [13]

Another type of anodization cell is a double tank geometry with an electrolytic back-side contact. This cell consists of two half-cells in which Pt electrodes are immersed and the Si wafer is used to separate the two half-cells. HF solution, circulated by chemical pumps to remove the gas bubbles and avoid the decrease in the local concentration of HF, is used both to etch the front side and as a back contact. The back-side of the Si wafer acts as a secondary cathode where proton reduction takes place leading to H evolution, while the front side of the wafer acts as a secondary anode, where PS is formed.

3.1.5 Effect of Anodization Conditions

All the properties of PS, such as porosity, thickness, pore diameter and microstructure, depend on anodization conditions. These conditions include HF concentration, current density, wafer type and resistivity, anodization duration, illumination (n-type mainly), temperature, ambient humidity and drying conditions (see Table 3.1).

Table 3.1. Effect of anodization parameters on PS formation

an increase of ... yields a	porosity	etching rate	critical current
HF concentration	decreases	decreases	increases
current density	increases	increases	–
anodization time	increases	almost constant	–
temperature	–	–	increases
wafer doping (p-type)	decreases	increases	increases
wafer doping (n-type)	increases	increases	–

Porosity is defined as the fraction of void within the PS layer and can be easily determined by weight measurements. The wafer is weighted before anodization (m_1), just after anodization (m_2), and after a rapid dissolution of the whole porous layer in a 3% KOH solution (m_3). The porosity is given by the following equation:

$$P(\%) = \frac{(m_1 - m_2)}{(m_1 - m_3)} . \tag{3.3}$$

Guessing the Si density ρ, one can also get the PS layer thickness d:

$$d = \frac{m_1 - m_3}{\rho S} , \tag{3.4}$$

where S is the etched surface.

For p-type doped substrates, and for a given HF concentration the porosity increases with increasing current density, Fig. 3.4. For fixed current density, the porosity decreases with HF concentration (see Fig. 3.5). With fixed HF concentration and current density, the porosity increases with thickness and porosity gradients in depth occur. In n-type doped Si, the layers obtained at low current density have a finer structure and are, therefore, more luminescent.

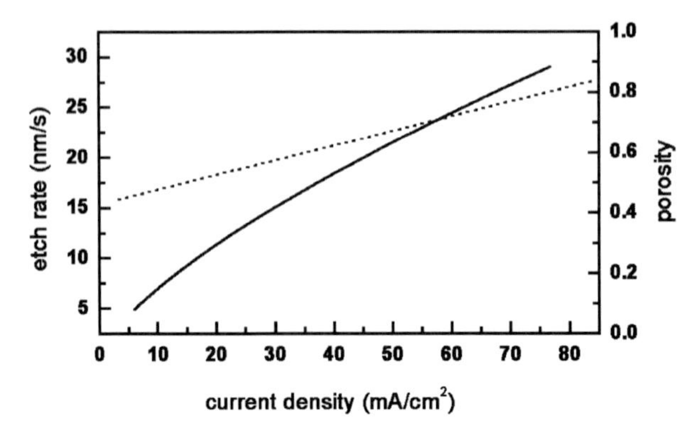

Fig. 3.4. Current density dependence of the etch rate (*left axis* and *full line*) and of the porosity (*right axis* and *dashed line*) for a substrate resistivity of 0.01 Ω cm and an HF concentration of 13%

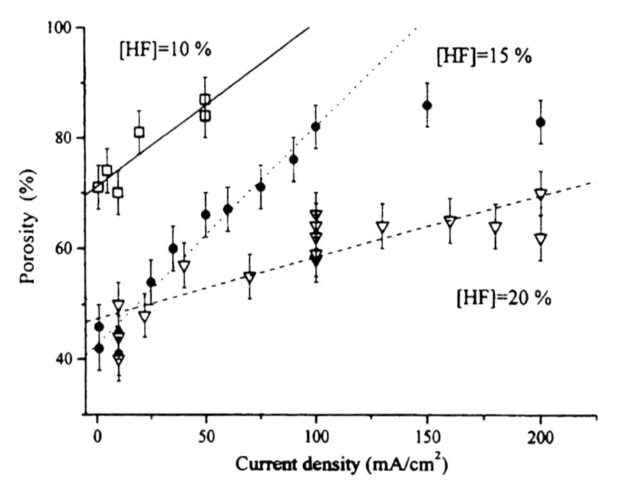

Fig. 3.5. Porosity as a function of current densities for different HF concentrations. Highly doped p-type Si substrate. After [3]

To obtain higher porosities, with samples emitting in the green and blue regions, a two step process has been proposed [14]. The PS layers are prepared by standard anodization which is followed by illumination under open-circuit conditions in the same electrochemical cell for 0–15 min.

The formation of PS is selective with respect to the doping of the substrate. Heavily doped regions are etched faster than low doped regions; in the dark, n-type doped regions embedded into p-type doped regions are not attacked; controlled doping profiles result in controlled PS formation [15].

3.1.6 Morphology of the Resulting Layers

Some general trends on the layer morphology can be derived for different types of starting Si substrates.

Figure 3.6 shows four cross-sectional TEM images of PS samples with different starting substrates [8]. The difference in morphologies is evident. For p-type doped Si both pore size and interpore spacing are very small (Fig. 3.6a), typically between 1 and 5 nm, and the pore network looks very homogeneous and interconnected. As the dopant concentration increases, pore sizes and interpore spacing increase, while the specific surface area decreases. The structure becomes anisotropic, with long voids running perpendicular to the surface, very evident in highly p-type doped Si (p^+), as shown in Fig. 3.6c. For n-type doped Si the situation is more complicated. Generally, pores in n-type doped Si are much larger than in p-type doped Si, and pore size and interpore spacing decrease with increasing dopant concentrations. Lightly doped n-type substrates anodized in the dark have low porosity (1–10%), with pores in the micrometer range. Under illumination higher values of porosity can be achieved, and mesopores are formed together with macropores. The final structure depends strongly on anodization conditions, especially on light intensity and current density. While highly n- and p-type doped Si show similar structures (compare Fig. 3.6c and d), in n-type doped Si the pores form a randomly directed filamentary structure and tend to "pipe" forming large straight channels approaching the electropolishing regime (see Fig. 3.6b).

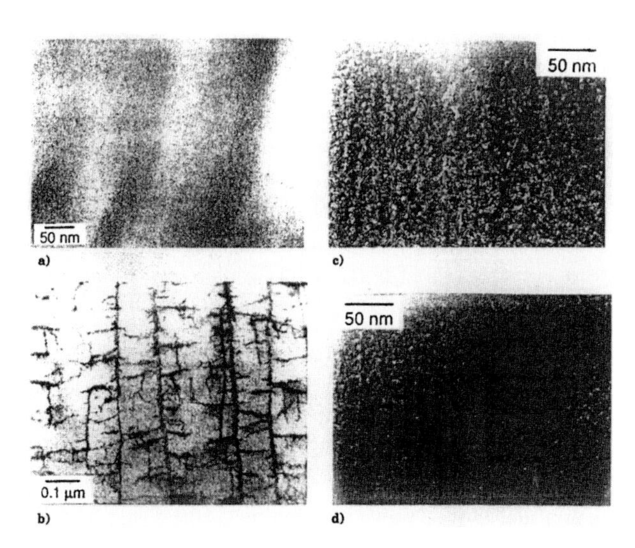

Fig. 3.6. Cross-sectional TEM images showing the basic differences in morphology among different types of samples. (**a**) p-type silicon; (**b**) n-type silicon; (**c**) p^+-type silicon, (**d**) n^+-type silicon. After [8]

3.1.7 Macroporous Silicon Formation

Of interest is the fact that very regular arrays of wide pores with large aspect ratios can be obtained both in n-type and in p-type doped Si [16].

The Si etching in HF is anisotropic due to the different etching rates that develop on the tips and on the walls of the pores. At the tips, the carriers are preferentially collected because the highest curvature radius is responsible for a large electric field. Above a critical carrier density, the tips are no longer able to consume all the holes. Si dissolution then starts also on the pores walls, and electropolishing of Si occurs. As for the details, it is necessary to distinguish between p- and n-type doped silicon (Fig. 3.7). In n-type doped substrates, holes are photogenerated by illuminating the back of the substrate. Due to the applied bias, they diffuse to the surface where they enter the solution at the pore tips. The limiting process responsible of macropore formation is hole diffusion. In the case of p-type silicon, holes are always present, and it is sufficient to induce a current during the etching to produce macropores. These two different mechanisms produce very different structures. With n-type doping, the carrier density among the pores is controlled by the light intensity. High aspect ratios are reachable, because the lateral growth of the pores is perfectly inhibited. With p-type doped substrates, the lateral growth of pores cannot be suppressed but can only be reduced by choosing the opportune substrate resistivity.

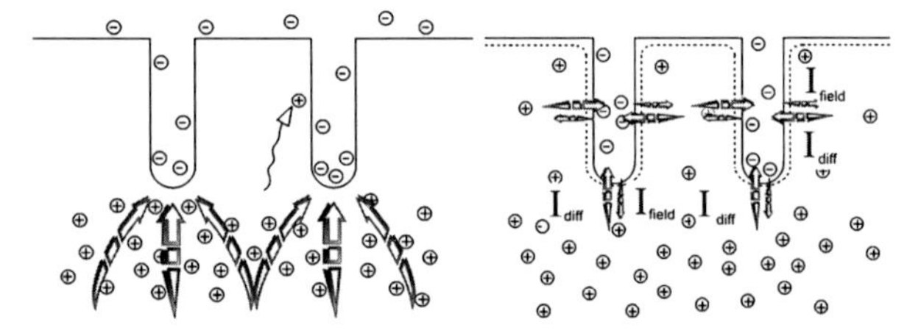

Fig. 3.7. On the *left* is sketched the dissolution mechanism for n-type silicon, in which the semiconductor between two adjacent pores is totally depleted of carriers. On the *right* is shown how the dissolution works for p-type silicon, with the lateral currents responsible for the widening of the pores during etching

The macropores are generated randomly on the surface, while a regular array of macropores is formed when etch-pits are previously prepared on a polished Si surface by photolithography and KOH etch. Regular arrays of pores with pitch ≥ 2 μm, 1 μm wide and more than 100 μm deep can be obtained (Fig. 3.8). The resistivity of the Si should be chosen in agreement with the pitch density, otherwise branching or dying of pores occurs. Also the current density should be adjusted according to the desired porosity. In addition due to the concentration gradients of the electrolyte in the pores

Fig. 3.8. Representative SEM images of two photonic crystals formed by macroporous Si samples obtained on n-type Si (8–12 Ω cm), but with different lattice symmetries

a current density linearly increasing with time has to be used. Limits to macropore fabrication are discussed in [16].

Macropore formation in p-type doped Si is more difficult. First reports claimed that high resistivity wafers and peculiar electrolytes (e.g. water-free acetonitrile) were needed [17]. Later it was discovered that many electrolytes are suitable and that the substrate doping was not critical [18, 19]. The structure of the etched pores is not as good as with n-type doping, but the typical sizes are smaller. The use of lateral pore growth allows one to obtain exotic structures, like pillars or lattices with two elements per unit cell [20].

3.1.8 Drying of the Samples

The drying of PS layers, especially those of high porosities, is a crucial step. After the formation of a highly porous or thick PS layer, when the electrolyte evaporates out of the pores, cracking of the layer is systematically observed. The origin of the cracking is the large capillary stress associated with the evaporation from the pores.

Different methods have been developed to reduce or eliminate the capillary stress [21], including pentane drying, supercritical drying, freeze drying and slow evaporation rates. Pentane drying is the easiest to implement. Pentane has a very low surface tension, and shows no chemical interaction with PS (unlike ethanol). Using pentane as the drying liquid enables one to reduce strongly the capillary tension, but since water and pentane are non-miscible liquids, ethanol or methanol have to be used as intermediate liquids.

Supercritical drying is based on the fact that, when the pressure is raised, the interface between the liquid and the gas phase becomes unstable and, when the pressure is larger than the critical pressure, the gas/liquid interface disappears and a mixture of the two phases appears (supercritical fluid). This is the most efficient drying method [22]. In such a technique, the HF solution is replaced by a suitable "liquid", usually carbon dioxide, under high pressure. The system is then moved above the critical point (31 °C) for CO_2

by raising the pressure and temperature. Then the gas is removed simply by the supercritical liquid. This technique produces layers with very high thickness and porosity values (up to 95%) improving optical flatness and homogeneity as well. However supercritical drying is expensive and complicated to implement, and the other drying methods are normally employed.

3.1.9 Anodical Oxidation

Oxidation is essential to stabilize the properties of PS. Many oxidation methods exist: slow ageing in ambient, anodic oxidation in a nonfluoride electrolyte, chemical oxidation and thermal oxidation.

Following PS production, and keeping the layer wet all the time, electrochemical oxidation is performed in a nonfluoride electrolyte, e.g., a 1:1 solution 1 M of H_2SO_4 with ethanol [23]. Typical current densities are in the range 1–10 mA/cm^2. The monitoring of the potential during the oxidation helps in assessing the time at which the sample is fully oxidized by looking at the step increase in the potential. Rinsing with methanol and/or pentane has to be avoided after oxidation because it destroys the optical quality of the samples.

3.1.10 Porosity Multilayers

Dielectric multilayers are reviewed at length in [3]. The possibility of forming a multilayer structure by using PS of different porosities relies on the basic characteristics of the etching process. The etching is self-limiting and occurs only in correspondence of the pore tips. That is, the already etched structure is not affected by further electrochemical etching of the wafer. Hence porosity multilayers can be obtained by simply changing periodically one of the parameters which affect PS formation. There are basically two types of PS multilayers, classified by the way the porosity is changed from one layer to another. In the first type the current density is changed during anodization [24], whereas in the second the change in porosity is determined by varying the doping level of the substrate [25]. The first method is by far the most convenient. HF concentration and substrate doping have to be carefully chosen to obtain the maximum variations of the refractive index (see Fig. 3.9). Much larger variations of the refractive index are possible for heavily p-type doped substrates, both for the large value of critical current density and the wide range of porosities achievable with a given HF concentration. Moreover, the etch rate is higher and the inner surface is lower for p^+-type doped substrates, and, hence, the additional chemical dissolution due to the permanence in HF is almost negligible. Examples of the results obtained for porosity multilayers are reported in Fig. 3.10.

A different approach is to etch Si wafers under holographic exposure with lateral formation of high and low porosity regions [26]. In this way, lateral periodic patterning of the porosity is obtained.

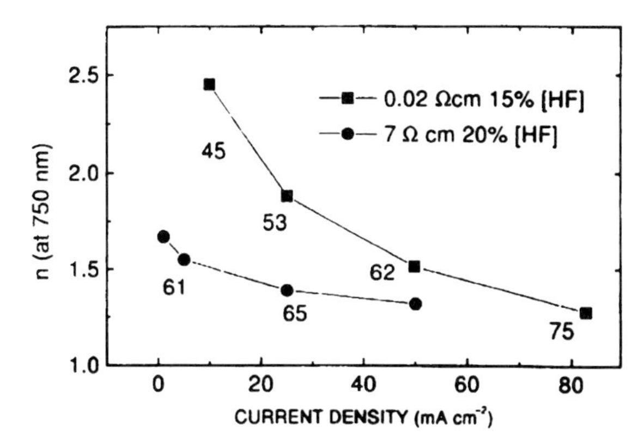

Fig. 3.9. Refractive index as a function of current density and porosity for two different substrate doping levels

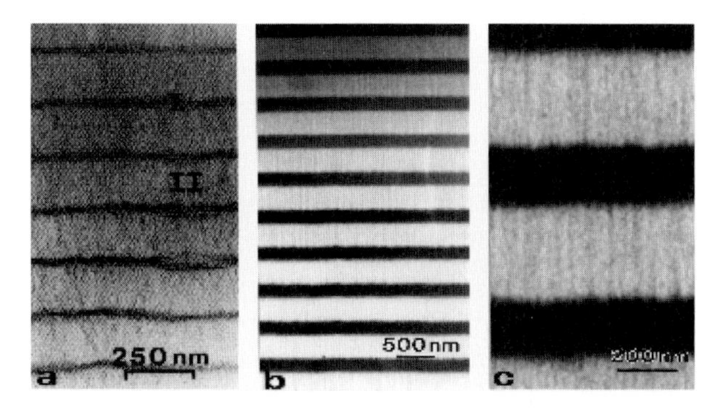

Fig. 3.10. TEM images of part of a PS multilayer produced in (**a**) 1994 (**b**), (**c**) 2000. Dark layers (I) correspond to lower porosity; lighter layers (II) to higher porosity. The multilayer structures shown were both fabricated via current density modulation in p^+-type doped Si wafers. After [27]

3.2 What are the Constituents of PS?

As the PS structure is complicated, most properties are determined both by intrinsic factors (its nanostructure) and by extrinsic factors (the composition of its matrix). In the following we will analyze these factors.

3.2.1 Microstructure of PS

The microstructure of PS has been extensively studied. Focusing the attention on luminescent material, TEM images (Fig. 3.11) show narrow undulating Si columns with diameters less than 3 nm which are crystalline in nature. PS layers, which have been heavily oxidized, show the presence of dispersed Si nanocrystals with dimensions of a few nanometers. A careful study was able to

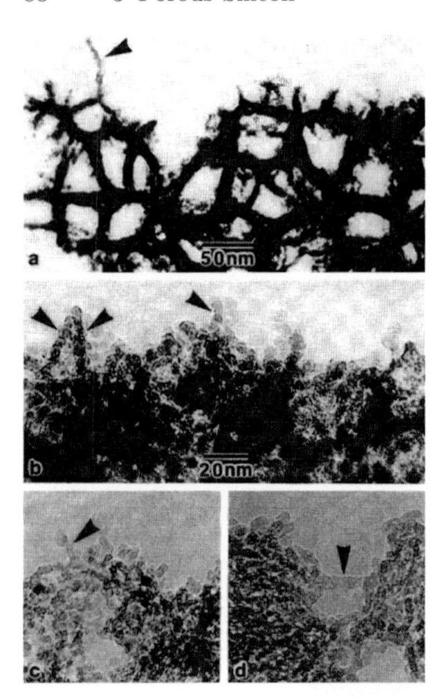

Fig. 3.11. TEM images (bright field, under focus) of thin, high porosity Si layers: (**a**) nonluminescent samples, (**b**)–(**d**) luminescent samples. Nanometer scale, columnar Si structures are arrowed. After [29]

detect nanocrystal sizes as small as 1 nm [28]. Large inhomogeneities in depth are also observed. Associated with the formation of nanostructures (both filaments or nanocrystals), an expansion of the Si lattice in these nanostructures is observed which is, however, difficult to quantify and which depends on the history of the sample under study [30].

A relation between the structure of PS and of the luminescing centers has been put forward through the use of EXAFS (extended X-ray absorption fine structure). Several detection modes have been used: transmission, total electron yield (TEY), and photoluminescence yield (XEOL/PLY). The comparison of the TEY and PLY data has shown that the luminescence originates from the nanocrystalline nature of PS and not by some chemical species adsorbed on its surface [31]. The local structure of the luminescent Si sites has been measured by XEOL [32]. Figure 3.12 shows X-ray absorption near edge structure (XANES) spectra recorded in TEY and XEOL modes at the Si K-edge of PS aged in air, after washing it with HF to remove O passivation and purposely oxidized. TEY and PLY XANES give different information. The TEY data confirm the presence of crystalline structures on a scale of a few nanometers and the presence of O in aged samples. PLY selects only the light-emitting sites. The independence of the lineshape on the oxygen presence demonstrates that the emitting sites are not affect by O.

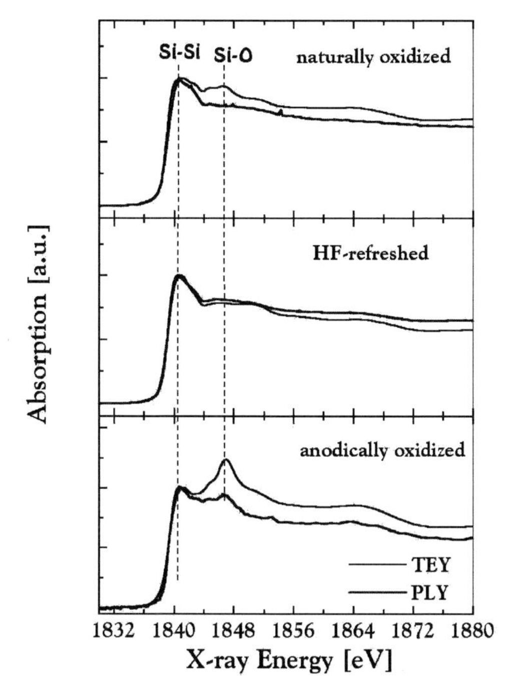

Fig. 3.12. Normalized X-ray absorption spectra near the Si K-edge (XANES) for a porous Si aged sample, for the same sample washed in HF and for anodically oxidized sample, obtained by TEY (*thin line*) and PLY (*thick line*) techniques. Courtesy of N. Daldosso

From similar studies, a model of PS formed in a highly resistive substrate has been elaborated where PS is formed by nanocrystals of particularly small dimensions, in the order of $<0.7-2.5$ nm, with the surface coated by an oxide layer some 1 nm thick [33].

3.2.2 Chemical Composition

The enormous internal surface of PS (more than 500 m^2/cm^3, depending on the dopant level and on the substrate [13]) contains a large quantity of impurities, coming from the electrolyte used for electrochemical etching and from the ambient air.

The original impurity, which is always found in PS layers, is hydrogen. Infrared absorption (IR) experiments have shown the presence of Si-H_x groups ($x = 1, 2, 3$) on the internal PS surface during the etching process [34].

After formation and drying, the Si-H_x groups are still present on the inner surface for weeks and even for months. The atomic ratio H/Si in PS has been determined using secondary ion mass spectrometry (SIMS) and elastic recoil detection analysis (ERDA) [35]. In freshly anodized samples it is as high as 0.1–0.6, depending on the porosity and surface area of the samples. This result means that the surface of freshly etched PS is almost totally covered by SiH_x groups.

Oxygen is the most important nonoriginal impurity, and is normally adsorbed in a few minutes after drying in ambient air. The amount of O can be as high as 1% after 15 minutes of air exposure, as confirmed by electron paramagnetic resonance (EPR) and increases to very high percentage values with ageing.

PS slowly reacts with the ambient air and consequently its chemical composition and its properties evolve continuously with storage time. The oxidization level depends not only on the time elapsed, but also on the ambient conditions. When transferred and stored under UHV and in the dark, PS layers have a very low content of oxygen, undetectable by techniques such as XPS, Auger and Fourier transform infrared spectroscopy (FTIR). On the other hand, when no particular care is taken, oxygen can reach very high levels in PS. A typical Auger spectrum of freshly etched PS is reported in Fig. 3.13 [36]. In the range observed, the Si, Cl and C related transitions are observed. The peak related to O, which should lie around 510 eV, is not detected. This means that the amount of native oxide, if present, is below the detection limit. Moreover, the lineshape of the Si related transition is characteristic of hydrogenated Si, proving that the Si dangling bonds are passivated by H and, eventually, by C. The Auger spectrum of the same sample after a few weeks

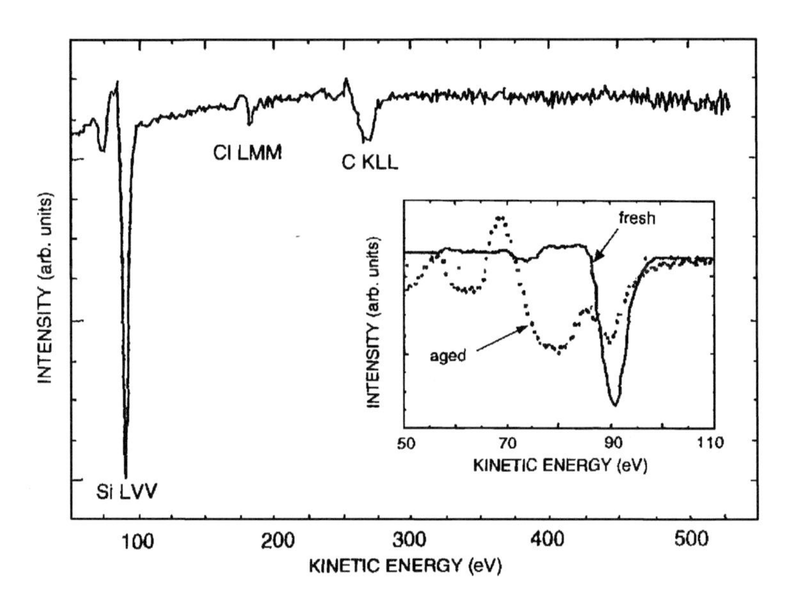

Fig. 3.13. Auger spectrum of a freshly etched PS sample. The *inset* shows the comparison between an aged (*dotted line*) and freshly etched (*solid line*) sample. After [36]

in atmosphere is quite different. The O related transition is present around 510 eV (not shown) and the Si related lineshape is characteristic of Si bonded in a SiO_x complex (inset in Fig. 3.13). A small fraction of hydrogenated Si is still present (the high energy peak in the Si related transition).

The final structure of PS can be summarized in Fig. 3.14: interconnected Si nanocrystals are embedded into an amorphous matrix whose composition changes with time and with the specific medium (atmosphere) in which PS is immersed.

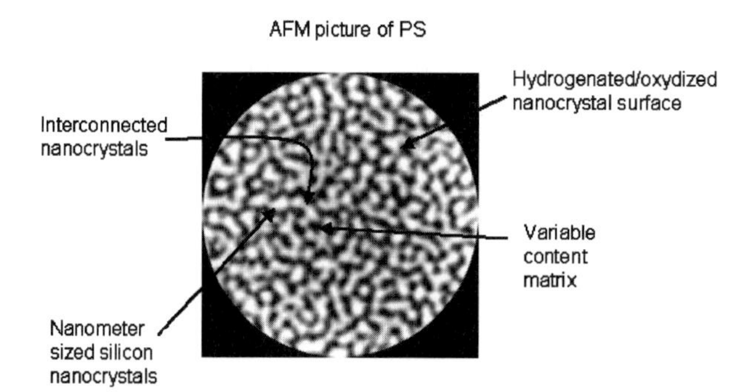

Fig. 3.14. Atomic force microscopy (AFM) image of a PS surface. Courtesy of Hong Ryong Kim

3.3 Electronic and Optical Properties of PS

3.3.1 What is the Nature of the PS Fundamental Band Gap?

The determination of the nature of the electronic structure of PS is critical for the understanding of the luminescence properties of PS. The minimum of the conduction band (E_{CBM}) has been estimated directly through X-ray absorption (XAS) [37]. The estimated increases of the E_{CBM} of PS with respect to bulk Si are indicative of an increasing quantum confinement (QC) shift in the PS samples. A blue shift of E_{CBM} is found when the photon energy of the illumination is raised during sample preparation [37]. This enhanced blue shift of E_{CBM} can hardly be attributed to changes in the chemical environment, but can be simply due to the reduction of the Si nanocrystal dimensions.

The maximum of the valence band E_{VBM} has been evaluated through valence band photoemission [38], Auger LVV spectra [37], soft X-ray emission spectra and ultraviolet photoemission.

All these results go in the same direction: E_{VBM} in PS shifts towards lower energies with respect to the Si bulk value by an amount which is similar or greater than that of E_{CBM}. For example, Fig. 3.15 reports the results obtained by Suda et al. [39].

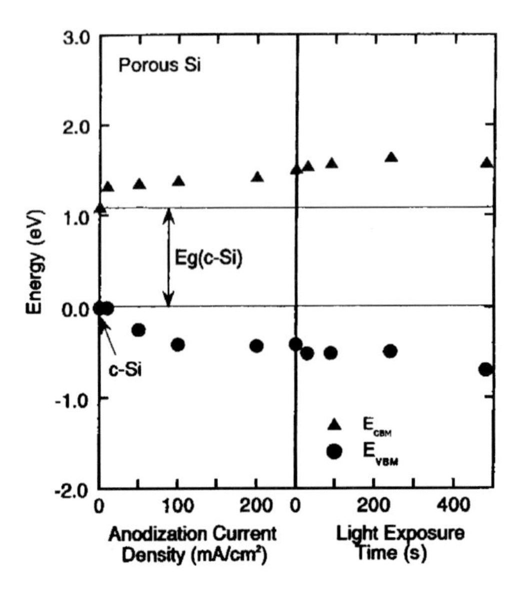

Fig. 3.15. Energy positions of the conduction band minimum (CBM) and valence band maximum (VBM) obtained from porous Si with (*right side of panel*) and without (*left side of panel*) light exposure treatment. After [39]

Few works deal directly with the QC effect on the details of the electronic structure of PS. Ben-Chorin et al. [40] used PL spectroscopy to monitor the behavior of the critical points of the band structure in PS. They have shown that not only the fundamental gap $\Gamma_v - \Delta_c$ is blue shifted with respect to bulk Si, but also the $\Gamma_v - L_c$ and $\Gamma_v - \Gamma_c$ transitions are blue shifted. However, while QC shifts the Γ_v point to a lower energy, the Δ_c point is shifted to a higher energy and the L_c and Γ_c are shifted to lower energies. Thus QC acts differently on the critical points of the PS band structure. Suda et al. [41] have studied the band dispersions of luminescent PS along the Γ–Δ–X symmetry line using angle resolved photoemission (ARPES) and angle integrated photoemission (AIPES). Their results provide another strong indication of the retention of crystalline symmetry in the nanocrystals contained in PS.

The observations that luminescent PS contains nanocrystalline Si structures have promptly produced a large number of theoretical investigations of quantum confinement effects on the electronic properties of Si (see Chap. 2). Since the core of PS is generally described as crystalline Si wires and/or dots the calculations have been performed on idealized models like Si quantum wires [42] and quantum dots [43]. Usually the Si dangling bonds at the wire and dot surfaces are saturated by H; only in a few cases has the presence of O-termination been taken into account [44, 45].

Absorption measurements are often complicated by experimental artifacts and by difficulties of interpretation that arise from the inhomogeneity in PS. Given the distribution of pores and structure sizes always present in PS, the absorption spectrum is the sum of widely different microscopic absorption processes. The absorption coefficient α has been measured in PS by optical transmission [46], photoluminescence excitation (PLE) [47], and photothermal deflection spectroscopy (PDS) [48]. A quantitative evaluation has to take into account the quantity of matter present in the layer. However, what is clear from absorption studies is that the transmission spectra are shifted towards higher energy compared to that of bulk Si and that this shift increases with increasing porosity, as illustrated in Fig. 3.16.

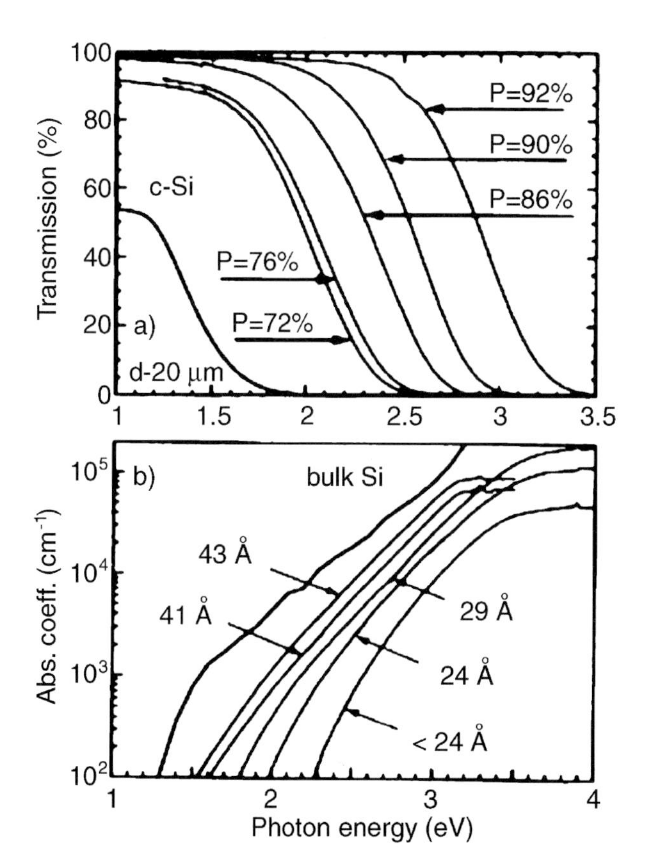

Fig. 3.16. (**a**) Transmission spectra of several 20 μm thick free-standing PS films (5–7 Ωcm *p*-type Si substrate) of various porosities and a 20 μm thick c-Si film. (**b**) Absorption spectra for Si nanocrystals with sizes from 4.3 nm to 2.4 nm deduced from the transmission spectra of (**a**). The absorption of c-Si is shown for comparison. The saturation beyond 3 eV is an artifact of the measurement procedure. After [46]

A more precise lineshape analysis of the absorption coefficient α shows that the energy dependence of α follows a trend like that of an indirect gap semiconductor very similar to that of Si, even though displaced to higher energy (see Fig. 3.17) [49].

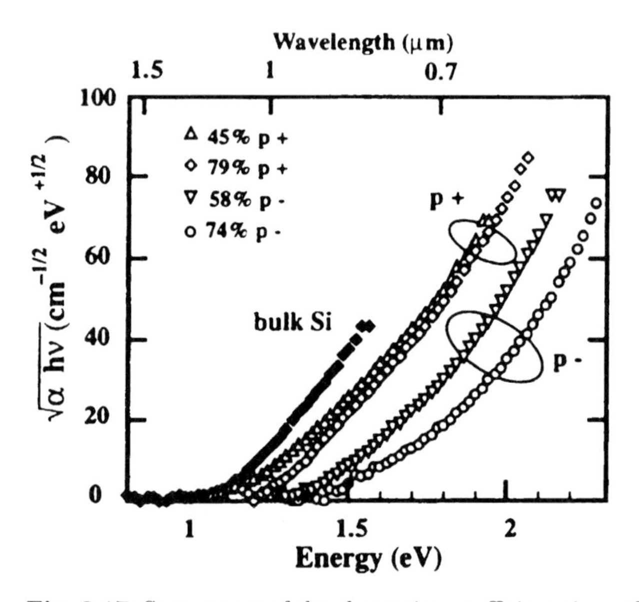

Fig. 3.17. Square root of the absorption coefficient times the photon energy versus photon energy for various porous silicon samples. After [49]

When α for PS is compared to that of Si, of amorphous Si and to that calculated by the Bruggemann approximation for two different porosities [50], it is observed that α of PS is well described by the Bruggemann result for energies higher than the Si direct energy gap. At lower energies, it departs from the Bruggemann result and shows an exponential decrease. This has been described by using the Urbach tail concept typical of Si or of amorphous semiconductors. Values of $\simeq$ 180 meV are usually found for the Urbach energy in PS [47]. The overall behavior of α in PS reflects the crystalline nature of the absorbing center. The Urbach tail is due to the random dispersion of sizes and shapes of these centers.

3.3.2 Dielectric Function and Refractive Index

The experimental determination of the dielectric function of PS is quite a complicated task, because PS is a mixture of Si and air and, in the case of aged samples, of surface oxide. Koshida et al. [51] have used a combination of spectroscopic ellipsometry (SE) in the range 250–800 nm and synchrotron radiation reflectance spectra (SR) in the range 2–27 eV. It is observed that the PS optical constants preserve the structure of bulk Si and are different from that of amorphous Si. The blue shift of the absorption edge of PS is evident in the imaginary part, ε_2, of the dielectric function (Fig. 3.18). In addition, a strong reduction of the absolute value of ε_2 is measured. The larger band gap of PS is reflected in the behavior of the real part, ε_1, of the dielectric function.

Usually reflectance spectra are simulated using a dielectric function model whose parameters, including the layer thickness, are adjusted to fit the measured data [52]. A model with three parameters (the porosity, the percolation strength and a broadening factor) which extends the Bergmann model seems more adequate. Once more, these studies make evident the blue shift for luminescent PS and the strong reduction of ε_2 and $\varepsilon_1$with increasing porosity.

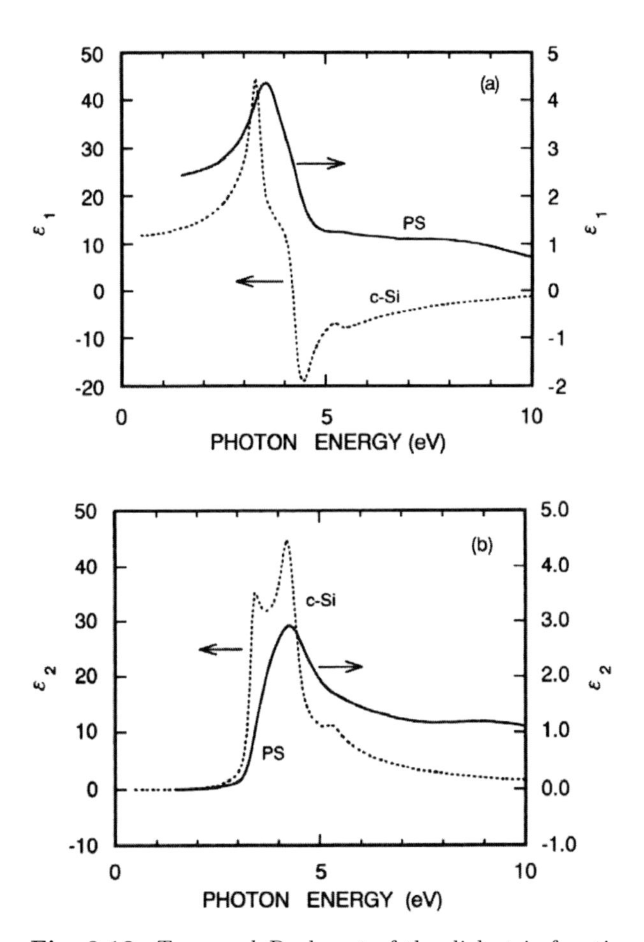

Fig. 3.18. *Top panel*: Real part of the dielectric function of PS (*solid curve*) and c-Si (*dashed curve*). *Bottom panel*: Imaginary part of the dielectric function of PS (*solid curve*) and c-Si (*dashed curve*). After [51]

The experimental refractive index as a function of porosity and of the current density used in the formation of PS for two different substrate-doping levels is reported in Fig. 3.9. It is worth noticing that:

- as the current density is increased, i.e. as the porosity is increased, the refractive index of PS tends to that of air, in agreement with the simple arguments given above;
- for the same porosity the same refractive index is found; this shows that the Bruggemann formula is a good approximation;
- large variations in the refractive index are possible by varying the current density, i.e. the porosity;
- PS formed on heavily doped substrates shows larger refractive index variations than PS formed in lightly doped substrates.
- n is also sensitive to the ageing or treatments the PS samples have suffered. In particular, it decreases during oxidization, because of the lower refractive index of SiO_2 compared to that of Si.

If a crystal has uniaxial symmetry, the optical response of the lattice is characterized by two distinct values of the refractive index. A crystal having such anisotropic dielectric properties is said to be birefringent. In PS optical anisotropy is due to the anisotropic geometry of the pores. Anisotropy of PS depends on the crystalline orientation of the silicon substrate. Birefringence has been reported in (100), (111) and (110) oriented PS [53, 54, 55, 56, 57]. In all these cases, the PS layer can be assumed to be uniaxial, and the direction of the optical axis to be normal to the surface for the (100) and (111) cases, and parallel to the surface for the (110) case. (100) oriented PS birefringence is reported to be positive [53, 54, 55] and values of the order of 10% have been demonstrated [58].

3.4 Photoluminescence of PS

In 1990, PS based structures were reported to luminesce efficiently in the red; soon after also orange, yellow and green PL was observed. The wide spectral tunability that extends from the near infrared through the whole visible range to the near UV (Fig. 3.19) is a characteristic of the PL of PS [59].

Such a broad range of emission energies arises from a number of clearly distinct luminescent bands. The related radiative decay times show a strong wavelength dependence going from ms for the band in the near infrared to ns for the blue band (see Table 3.2) [60].

In addition, PS has been used as an active host for rare earth impurities, e.g. Er, or dye solutions. Direct energy transfer between PS and the impurity or dye is demonstrated.

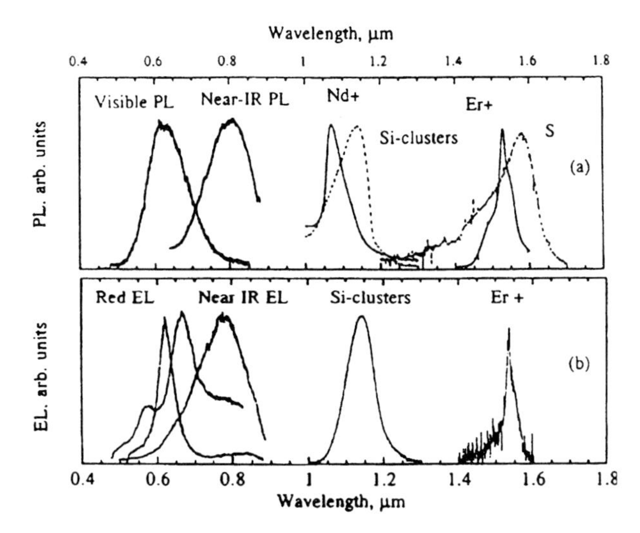

Fig. 3.19. Room temperature photoluminescence and electroluminescence spectra for various PS structures which have been oxidized or implanted with some selected impurities. After [59]

Table 3.2. PS luminescence bands. From [60]

Spectral range	Peak wavelength	Lifetime	Label	PL	EL
UV	~ 350 nm	1 ns	UV band	Yes	No
blue-green	~ 470 nm	1 ns	F band	Yes	No
blue-red	400–800 nm	100 µs	S band	Yes	Yes
near IR	1100–1500 nm	?	IR band	Yes	No

3.4.1 Visible Band

We will focus mainly on the properties of the so-called S-band, where S stands for slow. The S-band has been intensively studied up to now and has the most technological relevance, since it can be electrically excited. The S-band can be tuned from close to the bulk silicon band gap through the whole visible range (Fig. 3.20).

However, while the PL efficiency from red to yellow is high under blue or UV excitation, the blue emission is rather weak [14]. The large spectral width of the S-band comes from inhomogeneous broadening [61], and its spectral position depends on porosity. It is important to note that not only the spectral position, but also the relative intensity of the S-band changes with porosity (Fig. 3.21) [62].

Indeed, the S-band efficiency is not proportional to the inner surface area, but it seems that a "threshold" porosity has to be exceeded to achieve efficient luminescence. Post-anodization chemical etching in HF, corresponding

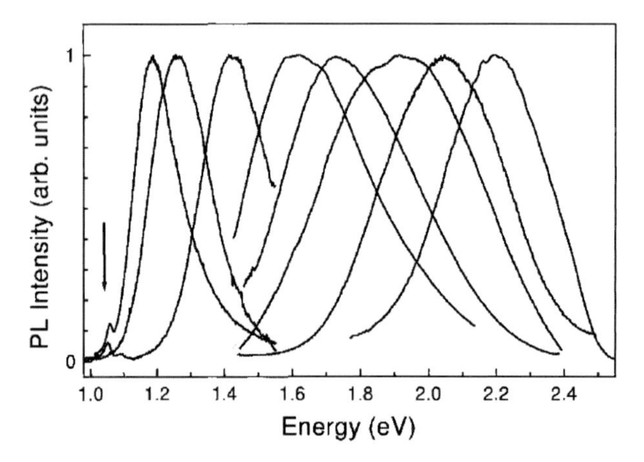

Fig. 3.20. Tunability of the porous silicon PL band. Various etching parameters have been used. The *arrow* shows the spectral position of the lowest energy free exciton transition in bulk silicon. After [63]

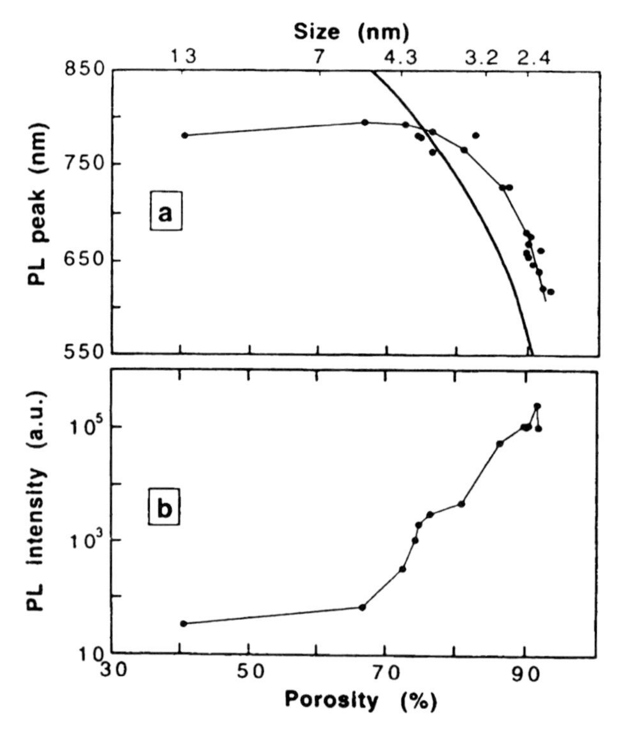

Fig. 3.21. Photoluminescence peak wavelength and intensity versus sample porosity. The *thick line* is obtained from first-principle calculations. After [62]

to a porosity increase, results in a strong rise in PL efficiency and a blue shift of the visible band. External quantum efficiencies higher than 10% are obtainable from high porosity PS layers of all types, but efficiency normally decreases in the order n-type, p-type, n^+-type, and p^+-type doped PS. High porosity is essential for high visible PL efficiency, and it has been proved that the inefficient luminescence observed from inhomogeneous material of low porosity originates from microscopic areas of high porosity. This is further confirmed by studies on isolated PS nanoparticles produced by dispersing a colloidal suspension of PS fragments on a glass coverslip [64]. The isolated nanocrystals show external quantum efficiencies $> 88\%$, while the number of bright to dark nanocrystals in the suspension was only of 2.8%. This means that the average quantum efficiency of a PS layer results from a statistical distribution of high and low quantum efficiency nanocrystals.

The luminescence of PS shows fatigue under measurement. The luminescence fatigue is associated with some photochemical reactions occurring on the inner surface of PS when it is measured. In Fig. 3.22 it is shown that as-formed PS shows a clear degradation of the PL while for oxidized PS the degradation is almost absent. It is a clear indication that the fatigue is associated with the weak H passivation of the nanocrystals surface.

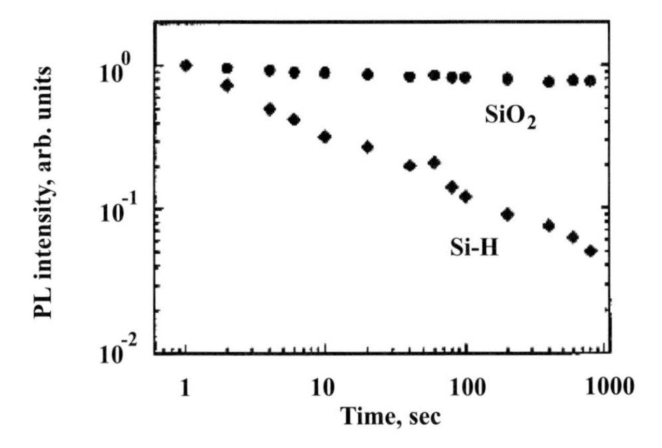

Fig. 3.22. Photoluminescence degradation in as-grown PS (labeled Si-H) and in oxidized PS (labeled SiO_2) versus time. After [59]

To probe the PS inhomogeneous PL band, a standard method is the use of line narrowing experiments where the luminescence is selectively excited with a laser energy within the energy region of the emission. In this way a subset of all the emitting centers is resonantly excited.

When the luminescence is resonantly excited, i.e. the excitation energy is within the luminescence band, and at low temperature, sharp features are observed (Fig. 3.23) [63].

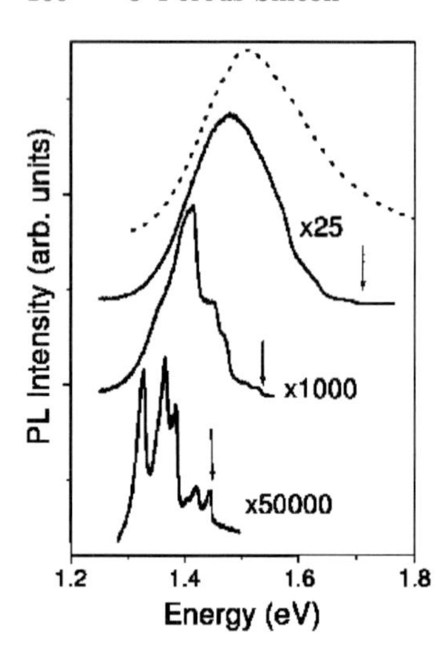

Fig. 3.23. Normalized resonant PL spectra of PS (*solid line*) and normal PS spectrum excited with a laser beam of 2.8 eV energy (*dotted line*). Excitation energies are indicated by arrows. The measurement temperature is 4.2 K. The relative intensities of the spectra are indicated on the right. After [63]

The luminescence onset shows an energy offset with respect to the excitation energy, which reflects the presence of a singlet–triplet exciton state splitting [65]. The offset is not present with two-photon excitation. This is due to the fact that different total angular momentum states are excited in one- and in two-photon experiments. In addition, replicas of this onset are observed at lower energies. These features are a consequence of phonon-assisted luminescence processes (Fig. 3.24).

Indeed the energies of the replicas are multiples of 56 and 19 meV, which are the energies of the TO and TA phonons of bulk Si. Some more comments are possible [60]:

- the number of phonon replicas are explained by an exciton transition in an indirect-gap semiconductor;
- the involved phonons are the same as those participating in optical transitions in crystalline Si;
- the relative strengths of the electron–phonon coupling in PS is very similar to those of Si;
- no coupling is observed to any other phonons.

These experimental results point to the fact that the electronic and the vibronic structures of the luminescence in PS are similar to that in crystalline Si when the QC is taken into account. At large confinement energies (> 0.7 eV), no-phonon quasidirect transitions dominate [63]. It is not clear however the reason for the large intensity drop when scanning the laser en-

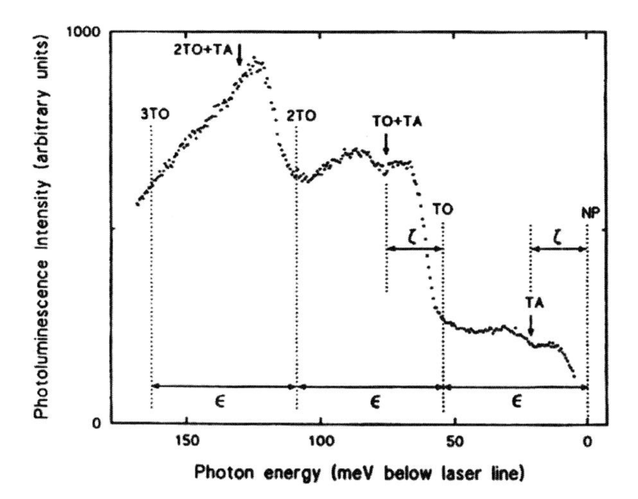

Fig. 3.24. PL spectrum taken with resonant laser excitation at 1.611 eV. *Vertical dotted lines* indicate the position of the no-phonon (NP) onset and its TO and TA phonon replicas. The values of the energies are given. The position of the NP onset is taken at the *laser line*, as the initial onset energy is close to zero at this excitation energy. After [65]

ergy through the emission band and the relative importance of the presence of defect related emission.

Another useful tool is the study of the polarization effects in the luminescence emission. In this way information on the symmetry of the emitting centers can be obtained [63]. When the laser excitation is normal to the PS surface, the luminescence is polarized parallel to the exciting light for all the exciting light polarization. It is interesting to note that the degree of polarization is zero on the IR band while it is different from zero on the S-band (Fig. 3.25). Furthermore, it tends to zero as the energy approaches that of the Si band gap. A maximum is observed for emission polarization parallel to the $\langle 100 \rangle$ directions. These experimental data are interpreted assuming that [60]

- PS is an ensemble of randomly oriented dipoles due to aspherical nanocrystals;
- the nanocrystals are preferentially aligned along the $\langle 100 \rangle$ direction;
- the wavefunction of the recombining carrier is bulk-like, because it is sensitive to the shape of the nanocrystals.

Calculations show that the main peak in the dielectric function is essentially due to Si-bulk like excitations, while the side peak is a truly wire-related feature (see Fig. 2.13) [67]. Evidence of dominating Auger processes when the photoexcited exciton density is high has been reported either by saturation of the luminescence [68] or by hole burning spectroscopy [69].

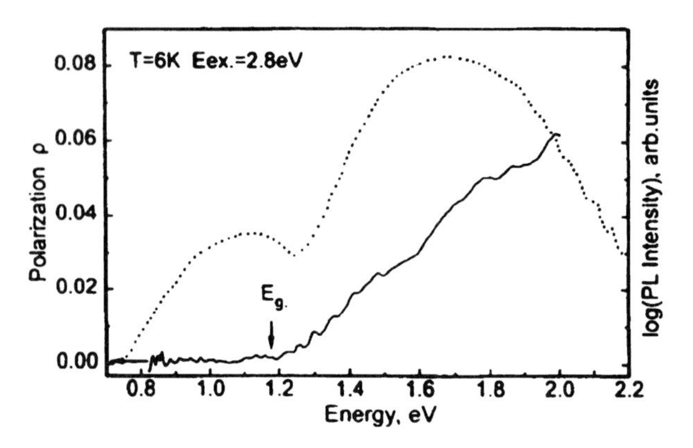

Fig. 3.25. Degree of polarization (*full line*) and luminescence intensity (*dotted line*) at 6K. E_g shows the Si band gap. After [66]

3.4.2 Other Emission Bands

The blue or F-band, due to its fast nanosecond decay time, has been subject of several studies [70]. It is observed only in oxidized PS, and it probably originates from contaminated or defective Si oxide. Annealing in water vapor activated blue emission indicating a possible major role of adsorbed hydroxyl in the emission process [70].

The IR band is weak at room temperature, and becomes much stronger at cryogenic temperatures [71]. In addition its intensity seems to decrease in aged samples. The energy of the IR band follows the energy of the S-band according to the 1/3 rule [72]. Its origin seems to be related to dangling bonds, although no direct correlation has been demonstrated [73].

3.4.3 Luminescence Decay

Since the first time-resolved studies, the PL decay lineshape was found not to be a single exponential, but a multiexponential.

A number of groups have analyzed the form of the PL decay using a stretched-exponential function [74]:

$$I_{\mathrm{PL}}(t) = I_0 \exp\left[-{}^{t}/_{\tau}\right]^{\beta} , \tag{3.5}$$

where $I_{\mathrm{PL}}(\mathrm{t})$ is the luminescence intensity, τ is the PL lifetime and $\beta \leq 1$ is a dispersion exponent. The PL decay at room temperature depends strongly on the detection wavelength, and the extracted decay times increase monotonically with decreasing detection energy, going from a few μs at 2.2 eV to about 100 μs at 1.8 eV (see Fig. 3.26).

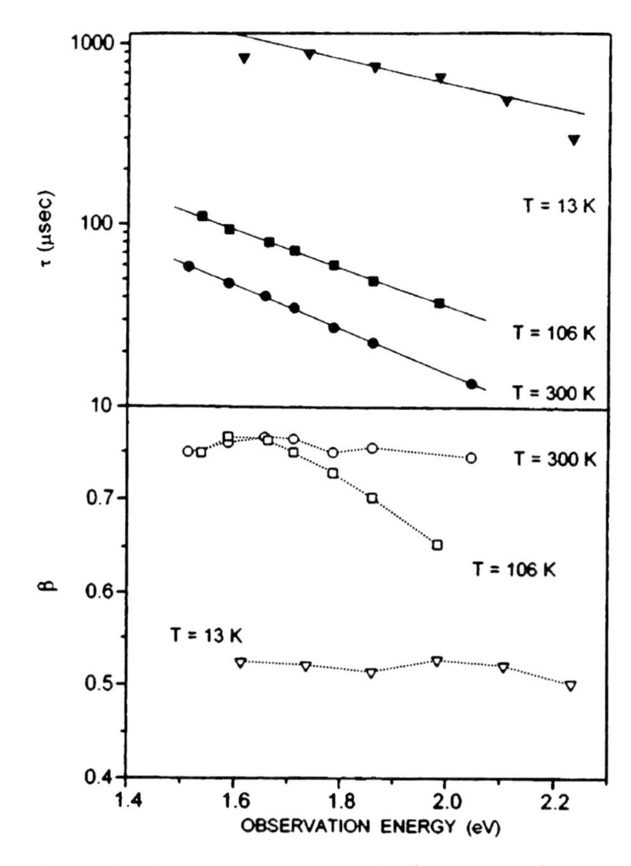

Fig. 3.26. Energy dependence of τ (*upper panel*) and β (*lower panel*) parameters obtained through a least squares fit of luminescence decay data for three different temperatures. After [74]

Such a long decay time compared to the nanosecond radiative lifetimes of direct gap semiconductors suggests that PS is still an indirect gap semiconductor, and the high PL efficiencies in PS compared to bulk Si are due to the strong suppression of nonradiative processes.

The decrease of decay time across the PL band is definitive evidence of the inhomogeneous nature of the PL spectrum. This behavior is in fact explained by two facts:

- The quantum confinement model. In smaller emitting regions, the quantum confinement increases the emission energy and the oscillator strength of radiative transitions, decreasing the radiative lifetime.
- The escape of carriers from Si nanocrystals [75]. The exponential slope of the energy dependence of the decay time is due to the decrease in the effective barrier height as the confinement energy is increased.

Furthermore, the nonexponential decay ($\beta \leq 1$) is evidence of the dispersive motion of the photoexcited carriers [61, 74]. The temperature dependence of both the luminescence decay time and of the luminescence intensity shows two typical regimes (see Fig. 3.27) [75]. For temperature higher than 200 K nonradiative recombinations are dominant. For temperature lower than 200 K, the decay time reflects the excitonic thermalization in the triplet and singlet states [65].

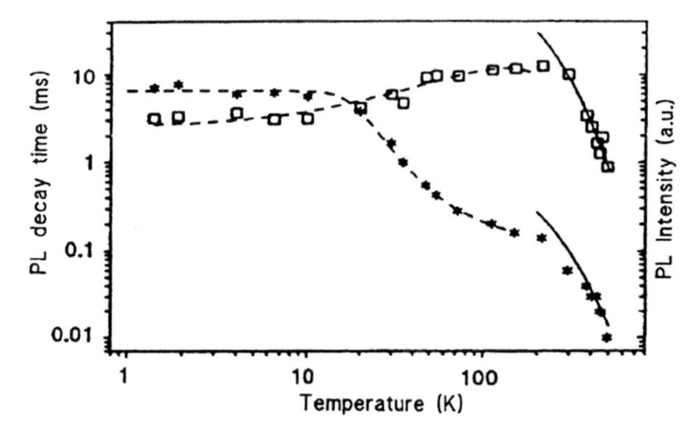

Fig. 3.27. Temperature dependence of the luminescence decay time (*stars*) and of the luminescence intensity (*squares*) recorded simultaneously. The *lines* are only guides for the eyes. After [75]

3.4.4 Models for PS Luminescence

Quantum confinement of carriers in Si nanostructures was the first model proposed to explain PS luminescence [1]. Afterwards, many other alternative explanations have been proposed [60]. The various models can be grouped in six different categories, as illustrated in the scheme of Fig. 3.28. Except for the quantum-confined model, all the others assume an extrinsic origin for the luminescence.

The hydrogenated amorphous silicon model (Fig. 3.28b). It has been proposed that PS luminescence is due to a hydrogenated amorphous phase (a-Si:H) which is formed during anodization. In fact, a-Si:H possesses a PL band in the visible range, and the tunability of the PL from PS can be in principle explained with alloying effects and with the variation of H and O percentages. In addition, the time resolved measurements indicate that the disorder plays a key role in the recombination dynamics [76].

Against this model, recent TEM studies where sample damage has been minimized showed that there is little amorphous Si in PS. Moreover, the

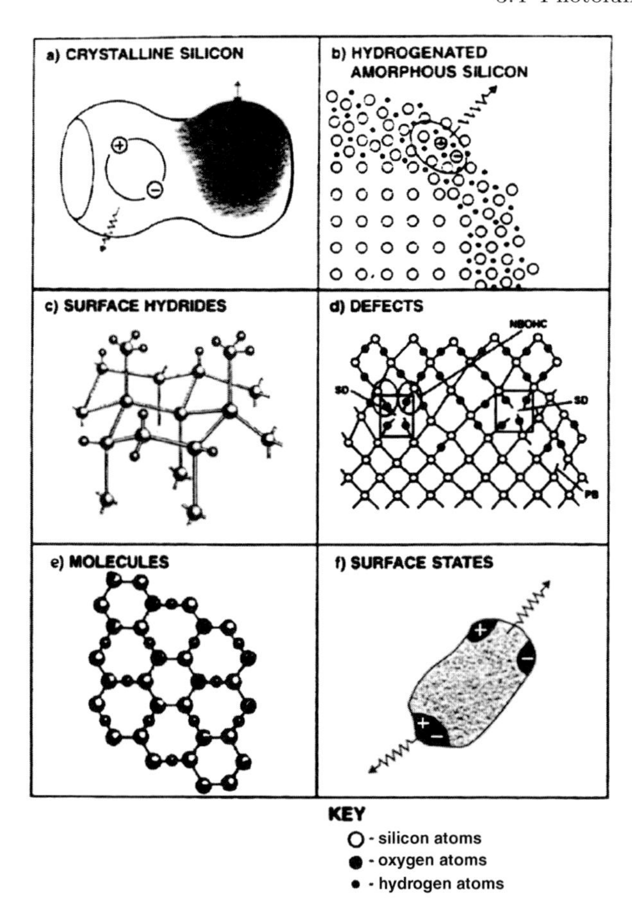

Fig. 3.28. The six groups of models proposed to explain PS luminescence. (**a**) Section of an undulating crystalline quantum wire. A surface defect gives an undulation nonradiative, while an exciton localized in the neighboring undulation recombines radiatively. (**b**) Crystalline Si covered by a layer of hydrogenated amorphous Si, where radiative recombination occurs. (**c**) Si surface passivated by SiH_x terminations. Radiative recombinations occur at the Si-H bonds. (**d**) Partially oxidized Si containing defects proposed as radiative centers. (**e**) Siloxene molecule proposed to exist on the large inner PS surface and act as luminescence center. (**f**) Si dot with surface states that localize carriers and holes separately (*upper part*) or together (*lower part*). After [60]

a-Si:H PL band is strongly quenched going from cryogenic to room temperature, while the PS visible band is enhanced as the temperature is raised. Finally strong spectroscopic evidence exists that the luminescence has both an electronic and a vibrational nature similar to that of crystalline Si.

The surface hydrides model (Fig. 3.28c). Since the PS luminescence decreases dramatically if H on the surface is thermally desorbed, and since the PL intensity can be recovered with immersion in HF, which restores the H coverage, SiH_x surface species were suggested to be responsible for luminescence in PS.

There is much evidences against this model. FTIR studies have demonstrated that luminescence is totally quenched when the majority of H is still on the PS surface, and the luminescence loss is probably related to the formation of dangling bonds, an efficient nonradiative decay channel. Other evidence comes from the simple fact that when a good quality oxide layer replaces the hydride coverage, the PL process is still efficient. Finally the spectroscopic evidence supports the crystalline nature of the emitting centers.

Defect models (Fig. 3.28d). In defect models the luminescence originates from carriers localized at extrinsic centers, that is defects in the silicon or silicon oxide that covers the surface [77, 78].

However, luminescent nanocrystalline Si can be created in many different ways, and passivated either with H or with O and it is then very unlikely that the same impurity or defect is always present. In any case the defects in the silicon oxide are ruled out, because SiO_2 is not present in fresh PS. Also the tunability of the PL band is difficult to justify, because the emission from defects is expected to be almost insensitive to the size structure. This has been, for example, demonstrated by first principle calculations of the role of defect states on the optical properties of Si quantum wells embedded in SiO_2 [79]. Also the spectroscopic evidence is against this model.

Siloxene model (Fig. 3.28e). Siloxene, a Si:H:O based polymer, supposely created during PS anodization, was proposed as the origin of PS luminescence [6]. This model is supported by the fact that the optical properties of siloxene are somewhat similar to those of PC. Siloxene possesses a visible-red PL band and the IR spectrum closely relates to that of aged PS.

There is much evidences against this model. It is now well demonstrated that freshly etched PS has no detectable content in oxygen. PS can still be luminescent above 800°C, while siloxene or other related molecules are totally decomposed at such a high temperature. In addition to the spectroscopic evidence, which points to the crystalline nature of the emission, synchrotron radiation measurements show that SiO groups play no role in the emission process [31].

Surface states models (Fig. 3.28f). The enormous inner surface of PS (about 10% of the Si atoms in PS are surface atoms) has led to the proposal that it is involved in the luminescence process [80, 81]. In this model, absorption occurs in quantum-confined structures, but radiative recombination involves localized surface states. Either the electron, or the hole, or both or none can be localized. Hence, a hierarchy of transitions is possible which explains the various emission bands of PS. The energy difference between absorption and emission peaks is well explained in this model, because photoexcited carriers relax into surface states. The dependence of the luminescence from external factors or from the variation of the PS chemistry is naturally accounted for by surface state changes.

Evidence against the attribution of the PL process to surface states is the resonantly excited PL results [60]. PL arises from exciton coupling with momentum-conserving phonons. This means that the exciton wavefunction is extended over many Si atoms and not strongly localized, as it should be in the case of deep surface states. Furthermore, the values of the exchange splitting energy extracted from temperature dependent lifetime measurements also suggest that carriers are not localized on the atomic scale, but in the whole volume of the Si nanocrystals, and that luminescence does not originate from localized states in the gap, but from extended states. Finally, the polarization measurements point to the extended nature of the luminescence states.

However a major question mark is still present: why is the intensity of the resonantly excited PL low?

Quantum confinement model (Fig. 3.28a). Quantum confinement in crystalline Si was the first model proposed to explain the efficient photoluminescence of PS [1]. Quantum confinement effects result in an enlargement of the band gap, in the relaxation of the momentum-conserving rule, and in the size dependence of the PL energy that naturally explains the efficient luminescence, the up-shift and the tunability of the PL band in PS. Many other experimental results support the quantum confinement model. Structural characterization has proven that PS is crystalline in nature. Observations of nanocrystal of nanometric dimensions have been reported. The band gap up-shift is clearly visible in absorption spectra, and the luminescence blue shift after further chemical dissolution in HF is easily explained by further reduction of nanocrystal dimensions. Fresh structures can be theoretically modeled as quantum wires, while aged structures are usually more dot-like [82, 83].

Qualitative agreement with experiments has been obtained for the observed splitting of the lowest lying exciton states and for calculated radiative lifetimes. Moreover the agreement between the observed and calculated decay times becomes better when phonon-assisted recombination channels are included [84].

Figure 3.29 shows the dependence on the confinement energy of the most important quantities which describe the recombination processes of a confined exciton in silicon nanocrystals. Each quantity follows the expected strongly nonlinear enhancement over a large dynamic range with increasing confinement energy and decreasing crystallite size. All these quantities can be traced continuously and smoothly from bulk silicon PL up to the visible spectrum, where the confinement energy of more than 1 eV is as large as the funda-

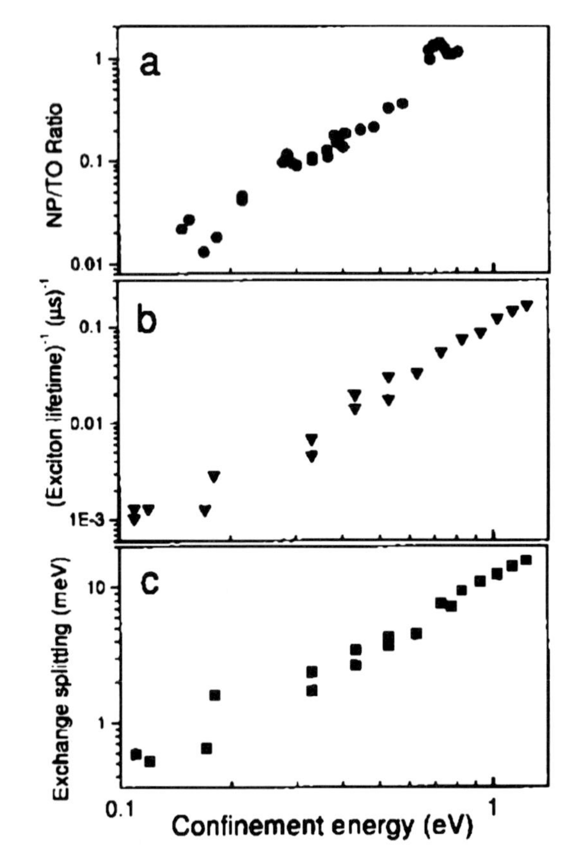

Fig. 3.29. Influence of quantum confinement on the basic properties of excitons localized in Si nanocrystals. Confinement energy dependence of (**a**) relative strength of no-phonon and TO-phonon assisted exciton recombination channels; (**b**) inverse lifetime of the optically active exciton state; (**c**) exciton exchange splitting. After [63]

mental silicon band gap itself. The spatial confinement of excitons in small nanocrystals breaks the k-conservation rule for indirect transitions. Thus no-phonon assisted transitions, which are forbidden in bulk silicon, have the same probability as TO phonon-assisted processes at a confinement energy of > 0.8 eV (Fig. 3.29a). However, small Si nanocrystals do not become a direct semiconductor. A consequence of the more direct nature of the recombination is an increase of the oscillator strength, i.e. a shortening of the lifetime by two orders of magnitude over the whole spectral range (Fig. 3.29b). The enforced enhancement of the overlap between the confined electron and hole wavefunctions leads to an enhancement of the exchange splitting between absorbing and emitting exciton states from 658 μeV in bulk Si to more than 15 meV for small nanocrystals (Fig. 3.29c).

However, it is becoming increasingly clear that, even in the quantum confinement framework, the emission peak wavelength is not only related to size effects. Freshly etched and very high porosity samples which have not been exposed to the air have luminescence peak energies in the 3 eV range while as soon as they get into contact with air their luminescence peak moves to the usual 2 eV range [59]. Absorption studies reveal a fundamental absorption edge which is systematically larger than the emission peak position measured in luminescence (see Fig. 3.30) [85].

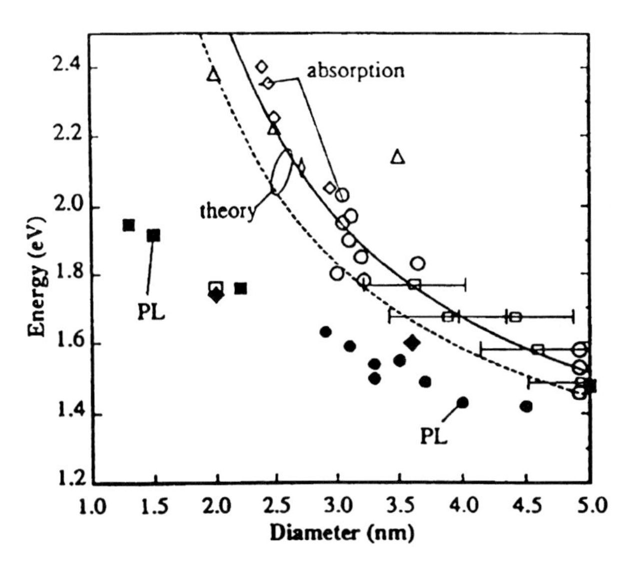

Fig. 3.30. Compilation of optical band gaps of silicon crystallites and PS samples obtained from optical absorption (*unfilled symbols*) and luminescence (*filled symbols*). The lines represent calculated values with (*dashed line*) or without (*full line*) excitonic correction. After [85]

A change in the surface passivation, as well as dielectric effects, can produce wavelength shifts. In addition, all the spectroscopic studies reported in this section are not able to discriminate between true extended states and shallow localized states, in which the carrier wavefunction extends over several lattice parameters. For these reasons, even though it is certain that QC plays a fundamental role in determining the peculiar properties of PS, some complements to the pure quantum confinement model are needed in order to take into account surface effects, especially when the nanocrystals have sizes smaller than 3 nm. Shallow surface trap states due to the formation of Si = O bonds have been shown to be able to fit the PL peak energy vs sizes data (see Fig. 3.31 [45]). Very similar radiative surface states have also been suggested to explain optical gain measurements in oxide passivated silicon nanocrystals [4].

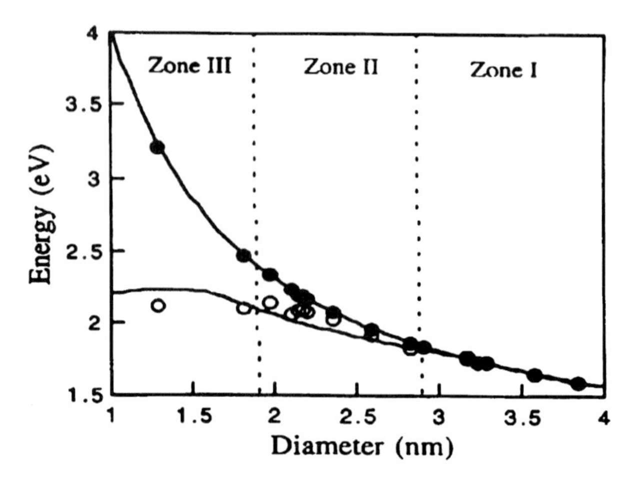

Fig. 3.31. Comparison between experimental and theoretical PL energies as a function of nanocrystal sizes. The *upper line* is the free exciton band gap and the *lower line* the lowest transition energy in the presence of a Si = O bond. The *filled* and *empty dot symbols* refer to the peak PL energy obtained from as-grown and air exposed samples, respectively. After [45]

3.5 Electrical Properties

To yield optimized light emitting devices, the injection of free carriers into the active centers needs to be understood. The transport paths and mechanisms, the role of the quantum confinement, and the nature of the moving particles are still unclear. Several models have been proposed which differ on the transport paths and mechanisms. These range from transport in the Si nanocrystals [86] and diffusion [87] or tunneling [88] between the Si nanocrystals or on their surface, to transport in the amorphous and disordered matrix surrounding the nanocrystals [89], or through both [90]. As the mechanisms suggested there are band transport [91], activated hopping in the band tail [89], trap controlled hopping through nanocrystals [87], activated deep states hopping, Poole–Frenkel processes and activated hopping in fractal networks [92].

3.5.1 PS Diode

Current flow through metal/PS devices is not explained by the modified Schottky equation

$$I = I_V \left[\mathrm{e}^{\left(\frac{qV - IR_S}{mKT} \right)} - 1 \right] \tag{3.6}$$

where R_S is a series resistance and m an ideality factor. In fact, extremely high values of R_S and of m were measured [93]. A more sophisticated model is needed where two junctions exist: one at the PS–metal interface and the

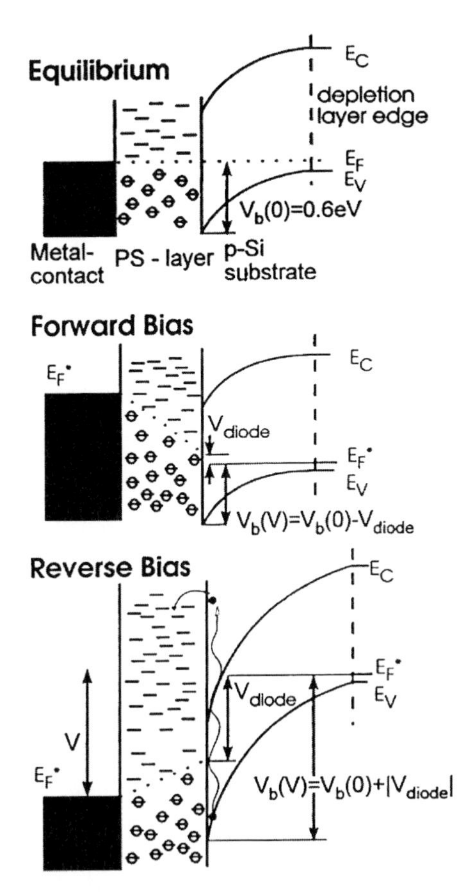

Fig. 3.32. A schematic band diagram of a PS device in equilibrium, forward and revere bias. The *shaded area* marks the metal contact. The *short lines* are localized electron states, some of which are occupied (*circles*). The *heavily dotted lines* mark the position of the Fermi level (quasi-Fermi level) in equilibrium (under an applied voltage). The *wavy line* at the *lower part* marks the possible mechanism of photocurrent enhancement at high photon energies. After [96]

other at the silicon–PS interface. Diodes with a thick PS layer show an ohmic I–V characteristic, indicating that the current is limited by transport through PS, which acts as a voltage dependent resistor [94]. Diodes with a thin PS layer shows a rectifying characteristic. This is independent of the metal used to form the top electrode [95]. In addition, a large density of states associated to mid-gap defects present in PS avoids band bending at the metal–PS interface. Significant photoresponse signal is observed only when the silicon–PS junction is illuminated [95]. As the Fermi level in PS is pinned at the mid-gap significant band bending and depletion occur only inside Si. Photocurrent experiments show that the associated energy barrier is about 0.6 eV

[96]. Considering all these facts the schematic band diagram of a diode under reverse, forward and neutral conditions shown in Fig. 3.32 is obtained. From the point of view of electroluminescence applications Fig. 3.32 shows that whilst hole injection can be efficient, electron injection is complicated by the low excess energy of electrons, which are at the Fermi level. Hole should travel all the way up to the metal contact to recombine with an electron.

3.5.2 Conduction in PS

Bare PS is insulating with a resistivity which is five orders of magnitude higher than that of intrinsic Si. Free carrier depletion is a consequence *(1)* of the quantum confinement induced widening of the band gap, which reduces the thermal generation of free carriers, and *(2)* the trapping of free carriers. Trapping can occur during the preparation of PS either because the binding energy of dopant impurities is increased or because surface states form.

The DC conductivity σ of PS shows an activation behavior, i.e. $\sigma(T) = \sigma_0 \exp(-J_{\mathrm{A}}/KT)$, where $E_{\mathrm{A}} \approx 0.5$ eV, i.e. half of the energy of the band gap deduced by luminescence. 0.5 eV is comparable to the activation energies found in intrinsic Si. The disordered nature of the PS skeleton and its local crystalline structure are expected to have strong geometrical effects on the conductivity. In addition, disorder induced localization of free carriers affects the free carrier motion in a similar way as in amorphous silicon. Hence the measured activation energies can be related to activation of carrier over mobility edges where the carriers can move in extended states or to typical energy barriers for carrier hopping. In the first case the activation energy should reflect the energy difference between the Fermi energy and the mobility edge. In the second case the activation energy is related to a typical barrier height separating neighboring localized states.

All the various experimental data about the temperature dependence of the DC conductivity have been represented and analyzed according to the Meyer–Neldel rule [97]:

$$\ln(\sigma_0) = B_{\mathrm{MNR}} + \frac{E_A}{E_{\mathrm{MNR}}}, \tag{3.7}$$

where B_{MNR} and E_{MNR} are constants and are the signature of a specific transport mechanism. It has been found that two transport mechanisms can coexist in the same PS network. One, which is very similar to what is found in amorphous silicon (extended state transport), and another which is due to activated band-tail hopping. A tunneling contribution is added to conduction in the amorphous tissue, which coats the Si nanocrystals. As far as the tunneling contribution is concerned, evidence is presented that this occurs via a Coulomb blockade mechanism. Additional evidence of this Coulomb blockade tunneling has been reported by looking at the exhaustion of the electroluminescence [98].

3.6 Light Emitting Diodes

Even though the interest in PS renewed after the observation of its emission properties, the potential application areas of PS are much wider than simple light emission. A list is reported in Table 3.3. In this section we will concentrate on light emitting diodes.

Table 3.3. Potential application areas of PS. Adapted from [22].

Application area	Role of PS	Key property
optoelectronics	LED	efficient electroluminescence
	waveguide	tunability of refractive index
	field emitter	hot carrier emission
	optical memory and switching	nonlinear properties
micro-optics	Fabry–Perot filters	refractive index modulation
	photonic band gap structures	regular macropore array
	all optical switching	highly nonlinear properties
energy conversion	antireflection coatings	low refractive index
	photoelectrochemical cells	photocorrosion
environment	gas sensing	ambient sensitive properties
microelectronics	microcapacitor	high specific surface area
	insulator layer	high resistance
	low-k material	electrical properties
wafer technology	buffer layer in heteroepitaxy	variable lattice parameter
	SOI wafers	high etch selectivity
micromachining	thick sacrificial layer	controllable etching parameters
biotechnology	tissue bonding	tunable chemical reactivity
	biosensing	enzyme immobilization

It is well know that a silicon based pn junction emits light both in forward and in reverse bias conditions. A pn junction shows an output power efficiency of about 10^{-4} at 1.1 μm in forward bias, and of about 10^{-8} in the visible and in reverse bias while in the avalanche breakdown regime [99]. The spectral features are various, the origin of the luminescence is not yet clear and several models have been presented [100]. Very recently a record efficiency of 1% has been achieved in a forward biased textured solar cell [101] (see Sect. 1.2.2).

Electroluminescence in PS based devices is observed both in the IR and in the visible with various efficiencies depending on the diode structure and on the PS contact method. A steady increase in the external quantum efficiency (see also 1.2) is observed, see Fig. 3.33. Device efficiency is now $> 1\%$, about 6 orders of magnitude higher than that of bulk Si diodes. Modulation speed is in the range of 10 MHz, peak emission is centered at 0.4–0.8 μm, the spectral width is of the order of 200 meV and the stability is of the order of weeks. Complete integration of PS based LED with CMOS driving circuits has been demonstrated [2].

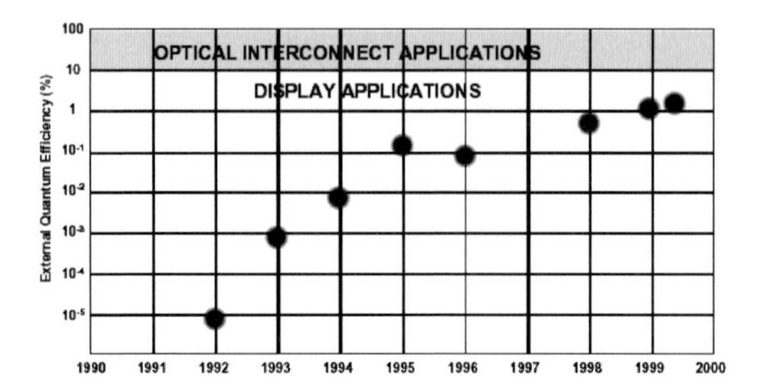

Fig. 3.33. Chronological rise in PS LED efficiency. After [27]

3.6.1 Electroluminescent Devices Based on Simple Solid Contact

The most straightforward way to form a LED with PS is to deposit on top of it a thin layer of metal or of a transparent semiconductor such as ITO. Metal deposition is indeed the way most electroluminescent devices have been fabricated. Various attempts have been reported, and for all of them the turn-on voltages are usually high (10 V or higher) while the efficiencies are low (10^{-5}). The emission is in the red, while when anodization is performed under light illumination blue emission can also be observed. The stability of the device is generally poor and the response time long. The low efficiency of metal/PS devices prompted several groups to look for strategies to improve the device performance. The main objective was to increase the contact area for enhancing the charge injection:

- PS doping: improvements in the EL characteristics are observed when metals are diffused into PS [102].
- Spin-on doping: In order to avoid thermal budgets too heavy for the PS layer during the doping, alternative approaches have been attempted. Spin-on doping is one of these [103].
- Organic covering/filling: One way of increasing the contact area is to use a conducting polymer as a PS contact, since it may be possible to penetrate the PS microstructure with such a material. Various polymers have been used [104, 105, 106].
- Alumina contact: One group has reported on a device structure which shows, with external efficiency of the order of 10^{-5}, high stability (> 100 hr) [107] and fast modulation (200 MHz) [108]. The device is based on PS formation in the transition regime where the produced PS layer is oxidized

during the formation. Al is then evaporated or sputtered on the sample and transparent windows are formed via anodic oxidation of Al into alumina. The high stability of the device is thought to be due to the protective action of both the oxidation during the etch and of the alumina covering.

- Oxidation studies: The best performances for this type of LED have been obtained with an electrochemical oxidation of an electrochemical etched *p*-type layer [109]. This improvement is based on the electrical isolation of those conducting paths which are not luminescent, such as those formed by sequences of large nanocrystals. Indeed anodic oxidation is obtained by using an electrochemical process: the oxidation develops firstly on those PS regions where the current flows easily, i.e. through large nanocrystals. External quantum efficiencies larger than 0.8% are found, even though the LEDs used are based on PS/metal contact. Figure 3.34 shows the comparison between the I-V and EL-I characteristics of an oxidized and an as-grown sample [110]. Oxidation develops a thin layer of SiO_x on the surface of the Si nanostructures causing their electrical isolation. As this occurs for the largest structures, which are also less efficient in emission, for a fixed voltage, the overall current flow through the LED is reduced, while the EL is almost constant.

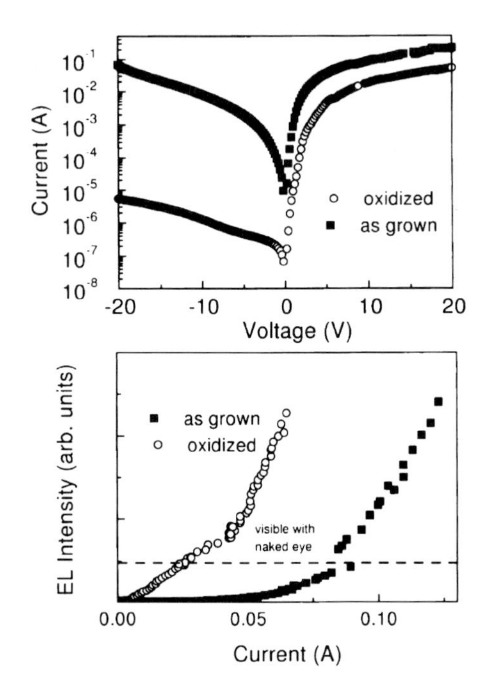

Fig. 3.34. I-V (*top panel*) and EL-I (*bottom panel*) characteristics of as-grown and anodically oxidized LED. After [110]

3.6.2 Homojunction Devices

The recognition that the injection efficiency of the metal/PS junction and that the transport in the PS layer is low, suggested the use of different device structures. One of them is based on the electrochemical etch of a Si pn junction with light assistance [111]. Metal deposition or ITO provides the external contact. External quantum efficiencies higher than 0.1% have been reported both for CW [111] and for pulsed operation [112]. The threshold voltage is very low (as low as 0.7 V) [111]. The stability of the structure is usually only some hours when operated in air. Comparison among various device structures shows that the best results are obtained when a p^+n junction is realized instead of a n^+p junction, i.e. it is easier to inject electrons from the substrate instead from the top-metal contact. The largest external efficiency (> 1%) reported to date is produced by a PS homojunction LED, which was electrochemically oxidized [109].

3.6.3 Heterojunction Devices

A different approach was considered where the junction is formed between a crystalline semiconductor and PS. Mostly devices with a silicon/PS junction were realized [113, 114]. Two different ways have been approached: one uses a grown-in np junction and exploits the doping selectivity of the PS formation [114]; the other uses the deposition of a polysilicon film on top

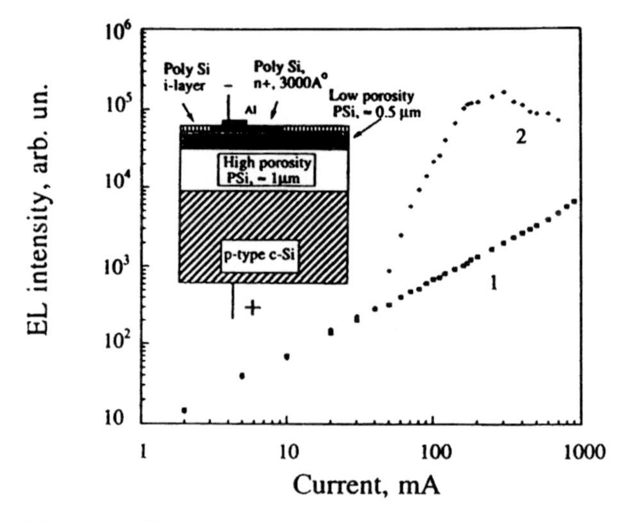

Fig. 3.35. Electroluminescence intensity as a function of current for samples annealed in pure nitrogen (1) and in diluted oxygen (2). The *curves* are normalized to yield the same EL at low current. The *inset* shows a schematic of the PS-based bipolar LED. After [113]

of the PS layer [113]. The first device structure aims to maximize the injection efficiency into the luminescent nanocrystals. The complete structure of the second device is shown in Fig. 3.35. In this device, a crucial step is the thermal annealing of the PS layer in diluted oxygen, which causes the formation of a so-called silicon rich silicon oxide (SRSO) layer. Based on this device structure, high power efficiencies (0.1%), long stability (weeks) and high modulation frequencies (MHz) are obtained [59].

3.7 Conclusions

The field of light emission from silicon for optoelectronic applications is very wealthy and many hopes still exist that silicon photonics can be finally obtained. Several different approaches are now actively pursued [115]. Porous silicon acts as a prototype and benchmark of many ideas, which are now actively looked for in other more sophisticated nanocrystals.

References

1. L.T. Canham: Appl. Phys. Lett. **57**, 1046 (1990)
2. K.D. Hirschman, L. Tsybeskov, S.P. Duttagupta, P.M. Fauchet: Nature **384**, 338 (1996)
3. L. Pavesi: La Rivista del Nuovo Cimento **20**, 1 (1997)
4. L. Pavesi, L. Dal Negro, C. Mazzoleni, G. Franzo, F. Priolo: Nature **408**, 440 (2000)
5. MRS Symposia Proceedings No. 256, 283, 358, 452, 486 (Materials Research Society, Pittsburgh, 1992-1993-1995-1997-1998); *Optical properties of low dimensional silicon structures*, ed. by D.C. Benshael, L.T. Canham and S. Ossicini, NATO ASI Series vol. 244 (Kluwer Academic Publisher, Dordrecht 1993); special issues of Journal of Luminescence **57** (1993) and **80** (1998); special issues of Thin Solid Films **255** (1995) and **297** (1997); special issue of phys. stat. sol. (a) **182** (2000); special issue of Optical Materials **17** (2001); special issue of phys. stat. sol. (a) **182** (2003)
6. M. Stutzmann, J. Weber, M.S. Brandt, H.D. Fuchs, M. Rosenbauer, P. Deak, A. Hopner, A. Breitschwerdt: In: *Festkorperprobleme/Advances in Solid State Physics*, vol. 32 ed. by U. Rossler (Vieweg, Braunschweig, Wiesbaden 1992) pag. 179; F. Buda, J. Kohanoff: Prog. Quant. Electr. **18**, 201 (1994); R.L. Smith, S.D. Collins: J. Appl. Phys. **71**, R1 (1992); K.H. Jung, S. Shih, D.L. Kwong: J. Electrochem Soc. **140**, 3046 (1993); L. Pavesi: In: *Amorphous Hydrogenated Silicon, Solid State Phenomena part B* **44-46**, 261 (1995); Y. Kanemitsu: Physics Reports **263**, 1 (1995); G.C. John, V.A. Singh: Physics Reports **263**, 93 (1995); L. Pavesi: Microelectronics Journal **27**, 437 (1996); P.M. Fauchet: J. Luminescence **70**, 294 (1996); R.T. Collins, P.M. Fauchet, M.A. Tischler: Physics Today **50**, 24 (1997); W. Theiss: Surface Sci. Rep. **29**, 91 (1997); A.G. Cullis, L.T. Canham, P.D.J. Calcott: J. Appl. Phys. **82**, 909 (1997); L. Pavesi: La Rivista del Nuovo Cimento **20**, 1 (1997); L. Pavesi,

V. Mulloni: 'Porous silicon'. In: *Silicon-based Microphotonics: from basics to applications*, ed. by O. Bisi, S.U. Campisano, L. Pavesi and F. Priolo (IOS press, Amsterdam 1999) pp. 87–161; O. Bisi, S. Ossicini, L. Pavesi: Surface Sci. Rep. **38**, 5 (2000); D. Kovalev, H. Heckler, G. Polisski, F. Koch: phys. stat. sol. (b) **215**, 871 (1999)

7. (a) *Porous Silicon*, ed. by Zhe Chuan Feng and R. Tsu (World Scientific Publishing Co., New York 1995); (b) *Porous Silicon Science and Technology*, ed. by J.-C. Vial, J. Derrien (Les Editions de Physique, Paris 1995); (c) *Structural and Optical Properties of Porous Silicon Nanostructures*, ed. by G. Amato, C. Delerue and H.-J. von Bardeleben (Gordon and Breach Science Publishers, Amsterdam 1997); (d) *Properties of Porous Silicon*, ed. by L.T. Canham (IEE INSPEC, The Institution of Electrical Engineers, London 1997); (e) *Light Emission in Silicon: from Physics to Devices*, ed. D.J. Lockwood, Semiconductors and Semimetals vol. 49 (Academic Press, London 1998); (f) *Silicon-Based Microphotonics: from Basics to Applications*, ed. by O. Bisi, S.U. Campisano, L. Pavesi, F. Priolo (IOS press, Amsterdam 1999)
8. R.L. Smith, S.D. Collins: J. Appl. Phys. **71**, R1 (1992)
9. V. Lehmann, H. Foll: J. Electrochem. Soc. **137**, 653 (1990)
10. V. Lehmann, U. Gösele: Appl. Phys. Lett. **58**, 856 (1991)
11. U. Gösele, V. Lehmann: 'Porous silicon quantum sponge structures: formation mechanism, preparation methods and some properties'. In: *Porous Silicon*, ed. by Zhe Chuan Feng, R. Tsu (World Scientific Publishing Co., New York 1995) pp. 17–132
12. S. Setzu, S. Letant, P. Solsona, R. Romestain, J.C. Vial: J. Luminescence **80**, 129 (1999)
13. A. Halimaoui 'Porous silicon formation by anodisation'. In: *Properties of Porous Silicon*, ed. by L.T. Canham (IEE INSPEC, The Institution of Electrical Engineers, London 1997) pp. 12–23
14. H. Mizuno, H. Koyama, N. Koshida: Appl. Phys. Lett. **69**, 3779 (1996)
15. H. Higa, T. Asano: Jpn. J. Appl. Phys. **35**, 6648 (1996)
16. V. Lehmann, U. Grunin: Thin Solid Films **297**, 13 (1997)
17. E.K. Propst, P.A. Khol: J. Electrochem. Soc. **141**, 1006 (1994)
18. E.A. Ponomarev, C. Levy-Clement: J. Electrochem. Soc. Lett. **1**, 1002 (1998)
19. B. Wehrspohn, J.N. Chazalviel, F. Ozanam: J. Electrochem. Soc. **145**, 2958 (1998)
20. P. Bettotti, L. Dal Negro, Z. Gaburro, L. Pavesi, A. Lui, M. Galli, M. Patrini, F. Marabelli: J. Appl. Phys. **92**, 6966 (2002)
21. D. Bellet: 'Drying of porous silicon'. In: *Properties of Porous Silicon*, ed. by L.T. Canham (IEE INSPEC, The Institution of Electrical Engineers, London 1997) pp. 38–43
22. L.T. Canham, A.G. Cullis, C. Pickering, O.D. Dosserm, T.I. Cox, T.P. Lynch: Nature **368**, 133 (1994)
23. A. Bsiesy, A. Gaspard, R. Herino, M. Lingeon, F. Muller, J.C. Oberlin: J. Electrochem. Soc. **138**, 3450 (1991)
24. L. Pavesi, V. Mulloni: J. Luminescence **80**, 43 (1999)
25. M.G. Berger, R. Arens-Fisher, S. Frohnhoff, C. Dieker, K. Winz, H. Luth, H. Munder, M. Artzen, W. Theiss: Mater. Res. Soc. Symp. Proc. **358**, 327 (1995); M.G. Berger, M. Thonissen, R. Arens-Fisher, H. Munder, H. Luth, M. Arntzen, W. Theiss: Thin Solid Films **255**, 313 (1995)

26. G. Lerondel, R. Romestain, J.C. Vial, M. Thonissen: Appl. Phys Lett. **71**, 196 (1997)
27. L.T. Canham: 'Nanostructured silicon as an active optoelectronic material'. In: *Frontiers of Nano-Optoelectronic Systems* ed. by L. Pavesi, E. Buzaneva, NATO Science Series II. Mathematics, Physics and Chemistry vol. 6 (Kluwer Academic Publishers, Dordrecht 2000) pp. 85–98
28. Z. Yamani, O. Gurdal, A. Alqal, M.H. Nayfeh: J. Appl. Phys. **85**, 8050 (1999)
29. A.G. Cullis, L.T. Canham: Nature **353**, 335 (1991)
30. D. Bellet, G. Dolino: Thin Solid Films **276**, 1 (1996)
31. T.K. Sham, D.T. Jang, I. Coulthard, J.W. Lorimer, X.H. Feng, K.H. Tang, S.P. Frigo, R.A. Rosenberg, D.C. Houghton, B. Bryskiewicz: Nature **363**, 331 (1993)
32. N. Daldosso, F. Rocca, G. Dalba, P. Fornasini, R. Grisenti: J. Porous Materials **7**, 169 (2000)
33. S. Schuppler, S.L. Friedman, M.A. Marcus, D.L. Adler, Y.H. Xie, F.M. Ross, T.D. Harris, W.L. Brown, Y.J. Chabal, L.E. Brus, P.H. Citrin: Phys. Rev. Lett. **72**, 2648 (1994)
34. R.L. Smith, S.-F. Chuang, S.D. Collins: J. Electron. Mater. **138**, 533 (1988)
35. L.T. Canham, M.R. Houlton, W.Y. Leong, C. Pickering, J. M Keen: J. Appl. Phys. **70**, 422 (1991)
36. L. Calliari, M. Anderle, M. Ceschini, L. Pavesi, G. Mariotto, O. Bisi: J. Luminescence **57**, 83 (1993)
37. S. Eisebitt, J.E. Rubensson, J. Lüning, W. Eberhardt, T. van Buuren, T. Tjedje: 'Soft X-ray emission and absorption spectroscopy on porous silicon'. In: *The Physics of Semiconductors*, ed. by D.J. Lockwood (World Scientific Singapore 1995) pp. 2133–2136
38. T. van Buuren, T. Tiedje, S.N. Patitsas, W. Weydanz: Phys. Rev. B **50**, 2719 (1994)
39. Y. Suda, T. Ban, T. Koizumi, H. Koyama, Y. Tezuka, S. Shin, N. Koshida: Jpn. J. Appl. Phys. **33**, 581 (1994)
40. M. Ben-Chorin, B. Averboukh, D. Kovalev, G. Polisski, F. Koch: Phys. Rev. Lett. **77**, 763 (1996)
41. Y. Suda, K. Obata, N. Koshida: Phys. Rev. Lett. **80**, 3559 (1998)
42. S. Ossicini: 'Porous silicon modelled as idealised quantum wires'. In: *Properties of Porous Silicon*, ed. by L.T. Canham (IEE INSPEC, The Institution of Electrical Engineers, London 1997) pp. 207–211
43. C. Delerue, M. Lannoo, G. Allan: 'Porous silicon modelled as idealised quantum dots'. In: *Properties of Porous Silicon*, ed. by L.T. Canham (IEE INSPEC, The Institution of Electrical Engineers, London 1997) pp. 212–215
44. S. Ossicini, O. Bisi: Solid State Phenom. **54**, 127 (1997)
45. M. Wolkin, J. Jorne, P.M. Fauchet, G. Allan, C. Delerue: Phys. Rev. Lett. **82**, 197 (1999)
46. J. von Behren, T. van Buuren, M. Zacharias, E.H. Chimowitz, P.M. Fauchet: Solid State Commun. **105**, 317 (1998)
47. A. Kux, M. Ben-Chorin: Phys. Rev. B **51**, 17535 (1995)
48. G. Vincent, F. Leblanc, I. Sagnez, P.A. Badoz, A. Halimaoui: J. Luminescence **57**, 217 (1993)
49. I. Sagnez, A. Halimaoui, G. Vincent, P.A. Badoz: Appl. Phys. Lett. **62**, 1155 (1993)

50. D. Kovalev, G. Polisski, M. Ben-Chorin, J. Diener, F. Koch: J. Appl. Phys. **80**, 5978 (1996)
51. N. Koshida, H. Koyama, Y. Suda, Y. Yamamoto, M. Araki, T. Saito, K. Sato, N. Sata, S. Shin: Appl. Phys. Lett. **63**, 2774 (1993)
52. W. Theiss: Surface Sci. Rep. **29**, 91 (1997)
53. F. Ferrieu, A. Halimaoui, D. Bensahel: Solid State Commun. **84**, 293 (1992)
54. P. Basmaji, G. Surdutovich, R. Vitlina, J. Kolenda, V.S. Bagnato, H. Mohajerimoghaddam, N. Peyghambarian: Solid State Commun. **91**, 91 (1994)
55. I. Mihalcescu, G. Lerondel, R. Romestain: Thin Solid Films **297**, 245 (1997)
56. H. Kryzanowska, M. Kulik, J. Zuk: J. Lumin. **80**, 183 (1999)
57. D. Kovalev, G. Polisski, J. Diener, H. Heckler, N. Künzner, V. Yu. Timoshenko, F. Koch: Appl. Phys, Lett. **78**, 916 (2001)
58. C.J. Oton, Z. Gaburro, M. Ghulinyan, L. Pancheri, P. Bettotti, L. Dal Negro, L. Pavesi: Appl. Phys. Lett. **81**, 4919 (2002)
59. P.M. Fauchet: J. Luminescence **80**, 53 (1999)
60. A.G. Cullis, L.T. Canham, P.D.J. Calcott: J. Appl. Phys. **82**, 909 (1997)
61. L. Pavesi: J. Appl. Phys. **80**, 216 (1996)
62. P.M. Fauchet, J. von Behren: phys. stat. solidi (b) **204**, R7 (1997)
63. D. Kovalev, H. Heckler, G. Polisski, F. Koch: phys. stat. sol. (b) **215**, 871 (1999)
64. G.M. Credo, M.D. Mason, S.K. Buratto: Appl. Phys. Lett. **74**, 1978 (1999)
65. P.D.J. Calcott, K.J. Nash, L.T. Canham, M.J. Kane, T. Brumhead: J. Phys. Cond. Matter **5**, L91 (1993)
66. D. Kovalev, M. Ben-Chorin, J. Diener, F. Koch, A.L. Efros, M. Rosen, N.A. Gipius, A. Tikhodev: Thin Solid Films **276**, 120 (1996)
67. F. Buda, J. Kohanoff, M. Parrinello: Phys. Rev. Lett. **69**, 1272 (1992)
68. I. Mihalcescu et al.: Phys. Rev. B **51**, 17605 (1995)
69. D. Kovalev, H. Heckler, B. Averboukh, M. Ben-Chorin, M. Schwartzkopff, F. Koch: Phys. Rev. B **57**, 3741 (1998)
70. L. Tsybeskov, Y.V. Vandyshev, P.M. Fauchet: Phys. Rev. B **49**, 7821 (1994)
71. P.M. Fauchet, E. Ettedgui, A. Raisanen, L.J. Brillson, F. Seiferth, S.K. Kurinec, Y. Gao, C. Peng, L. Tsybeskov: Mater. Res. Soc. Symp. Proc. **298**, 271 (1993)
72. F. Koch: Mater. Res. Soc. Symp. Proc. **298**, 319 (1993)
73. S. Ossicini: phys. stat. sol. (a) **170**, 377 (1998)
74. L. Pavesi, M. Ceschini: Phys. Rev. B. **48**, 17625 (1993)
75. J.C. Vial, A. Bsiesy, F. Gaspard, R. Herino, M. Ligeon, F. Muller, R. Romestain, R.M. Macfarlane: Phys. Rev. B **45**, 14171 (1992)
76. L. Pavesi: Solid State Phenomena **44-46**, 261 (1995)
77. G.G. Qin, Y.Q. Jia: Solid State Commun. **86**, 559 (1993); M. Sacilotti et al.: Electron. Lett. **29**, 790 (1993)
78. S.M. Prokes: Appl. Phys. Lett. **62**, 3244 (1993)
79. E. Degoli, S. Ossicini: Surface Sci. **470**, 32 (2000)
80. F. Koch, V. Petrova-Koch, T. Muschik, A. Nikolov, V. Gavrilenko: Mater. Res. Soc. Symp. Proc. **283**, 78 (1993)
81. F. Koch, D. Kovalev, B. Averbouck, G. Polisski, M. Ben-Chorin, A.L. Efros, M. Rosen: J. Luminescence **70**, 320 (1996)
82. S. Ossicini, C.M. Bertoni, M. Biagini, A. Lugli, G. Roma, O. Bisi: Thin Solid Films **297**, 154 (1997)

83. S. Ossicini, E. Degoli, M. Luppi: phys. stat. sol. (a) **182**, 301 (2000)
84. M.S. Hybertsen: Phys. Rev. Lett. **72**, 1514 (1994)
85. M. Lannoo, G. Allan, C. Delerue: 'Theory of optical properties and recombination processes in porous silicon'. In: *Structural and Optical Properties of Porous Silicon Nanostructures*, ed. by G. Amato, C. Delerue, H.-J. von Bardeleben (Gordon and Breach Science Publishers, Amsterdam 1997) p. 187
86. R. Tsu, D. Babic: Appl. Phys. Lett. **64**, 1806 (1994)
87. H. Eduardo Roman, L. Pavesi: J. Phys: Cond. Matter **8**, 5161 (1996)
88. A. Diligenti, A. Nannini, G. Pennelli, F. Pieri: Appl. Phys. Lett. **68**, 687 (1996)
89. J. Kocka, J. Oswald, A. Fejfar, R. Sedlacik, V. Zelezny, Ho The-Ha, K. Luterova, I. Pelant: Thin Solid Films **276**, 187 (1996)
90. I. Balberg: Phil. Mag. B **80**, 691 (1999)
91. H. Koyama, N. Koshida: J. Luminescence **57**, 293 (1993)
92. M. Ben-Chorin, F. Möller, F. Koch, W. Schirmacher, M. Eberhard: Phys. Rev. B **51**, 2199 (1995)
93. N. Koshida, H. Koyama: Appl. Phys. Lett. **60**, 347 (1992)
94. M. Ben-Chorin, F. Moller, F. Koch: Phys. Rev. B **49**, 2981 (1994)
95. N.J. Pulsford, G.L.J.A. Rikken, Y.A.R.R. Kessener, E.J. Lous, A.H.J. Venhuizen: J. Luminescence **57**, 181 (1993)
96. M. Ben-Chorin, F. Möller, F. Koch: J. Appl. Phys. **77**, 4482 (1995)
97. Y. Lubianiker, I. Balberg: Phys. Rev. Lett. **78**, 2433 (1997)
98. N.D. Lalic, J. Linnros: J. Appl. Phys. **80**, 5971 (1996)
99. J. Kramer, P. Seitz, E.F. Steigmeier, H. Auderset, B. Delley, H. Baltes: Sensors and Actuators A **37-38**, 527 (1993)
100. N. Akil, S.E. Kerns, D.V. Kerns Jr., A. Hoffmann, J.-P. Charle: Appl. Phys. Lett. **73**, 871 (1998)
101. M.A. Green, J. Zhao, A. Wang, P.J. Reece, M. Gal: Nature **412**, 805 (2001)
102. K. Nishimura, Y. Nagao, N. Ikeda: Jpn. J. Appl. Phys. **36**, L643 (1997)
103. S. Sen, J. Siejka, A. Savtchouk, J. Lagowski: Appl. Phys. Lett. **70**, 2253 (1997)
104. N. Koshida, H. Koyama, Y. Yamamoto, G.J. Collins: Appl. Phys. Lett. **63**, 2655 (1993)
105. K.-H. Li, C. Tsai, J.C. Campbell, M. Kovar, J.M. White: J. Electron. Mater. **23**, 409 (1994)
106. D.P. Halliday, E.R. Holland, J.M. Eggleston, P.N. Adams, S.E. Cox, A.P. Monkman: Thin Solid Films **276**, 299 (1996)
107. S. Lazarouk, P. Jaguiro, S. Katsouba, G. Masini, S. La Monica, G. Maiello, A. Ferrari: Appl. Phys. Lett. **68**, 2108 (1996)
108. M. Balucani, S. La Monica, A. Ferrari: Appl. Phys. Lett. **72**, 639 (1998)
109. B. Gelloz, T. Nagakawa, N. Koshida: Appl. Phys. Lett. **73**, 2021 (1998); and J. Appl. Phys. **88**, 4319 (2000)
110. L. Pavesi et al.: J. Appl. Phys. **86**, 6474 (1999)
111. A. Loni, A.J. Simmons, T.I. Cox, P.D. J. Calcott, L.T. Canham: Electronics Lett. **31**, 1288 (1995)
112. J. Linnros, N. Lalic: Appl. Phys. Lett. **66**, 3048 (1995)
113. L. Tsybeskov, S.P. Duttagupta, K.D. Hirschman, P.M. Fauchet: Appl. Phys. Lett. **68**, 2058 (1996)
114. L. Pavesi, R. Guardini, P. Bellutti: Thin Solid Films **297**, 272 (1997)
115. P. Ball: Nature **409**, 974 (2001)

4 Silicon Nanostructures: Wells, Wires, and Dots

The discovery of room temperature photoluminescence (PL) from porous Si (see Chap. 3) prompted a great deal of attention to quantum confinement effects on low-dimensional Si structures. In fact, despite the interesting properties of porous silicon its use in devices is hampered by its large internal surface which is highly reactive and causes, therefore, ambient dependent properties. For this reason, other systems have been developed which are based on low-dimensional Si. Low-dimensional silicon exists in three flavors: two-dimensional (2D) quantum wells or quantum slabs, one-dimensional (1D) quantum wires and zero-dimensional (0D) quantum dots. Silicon quantum dots are more often referred to as silicon nanocrystals due to their crystalline character. All these materials systems have been widely investigated in the last ten years.

In this chapter we will discuss the work done on silicon–insulator multi-quantum wells and superlattice structures, on silicon nanopillars and wires, and on silicon nanoparticles and nanocrystals. In all case details of the fabrication methods and structures, electronic and optical properties, photo- and electroluminescence will be presented and discussed.

4.1 Silicon–Insulator Two-Dimensional Systems

Silicon-based heterostructures form one of the most promising classes of Si low-dimensional systems because of their compatibility with conventional silicon-based integrated circuit technology. In particular silicon–insulator multiquantum well and superlattices, where calcium fluoride (CaF_2) or silicon dioxide (SiO_2) are used as insulating materials, have been studied from both the experimental and theoretical points of view. In these systems the thickness of the silicon layer lies in the nanometer range and the goal is to have the possibility to control the size and dimensionality of the Si nanostructures in order to control their absorption and luminescence properties. In addition to crystalline systems, amorphous silicon multilayers have also been investigated for optoelectronic applications. These are based on alloys either of a-Si:H or a-$Si_{1-x}C_x$:H or a-$Si_{1-x}N_x$:H and of their multiple combinations [1]. Luminescence at room temperature has been measured and

LED have been developed [2, 3]. However we will not detail these approaches because they are not based on crystalline silicon.

The first system we discuss is formed by silicon sandwiched between adjacent layers of adsorbed oxygen. This system shows robustness, stability and interesting transport and photoluminescence characteristics [4]. The Si/O superlattices were grown by ultrahigh-vacuum molecular beam epitaxy (MBE) [4, 5]. Oxygen adsorption, usually between 20 and 50 Langmuir, on a clean Si surface near room temperatures, was followed by silicon growth at a relatively low substrate temperature, with reflection high-energy electron diffraction (RHEED) as a guide to monitor the growth. Normally 1–2 nm-thick silicon was deposited at a low growth rate and the process was cyclically repeated in order to grow superlattices. A sample was usually made by nine Si/O periods.

The left part of Fig. 4.1 shows high-resolution cross-sectional transmission electron microscopy (TEM) of a single Si quantum well formed by the deposition of a single 1.1 nm thick Si layer embedded between two 10 Langmuir thick O layers. The deposition was made on a Si buffer layer at 550 °C. TEM and *in situ* reflection high-energy electron diffraction (RHEED) and

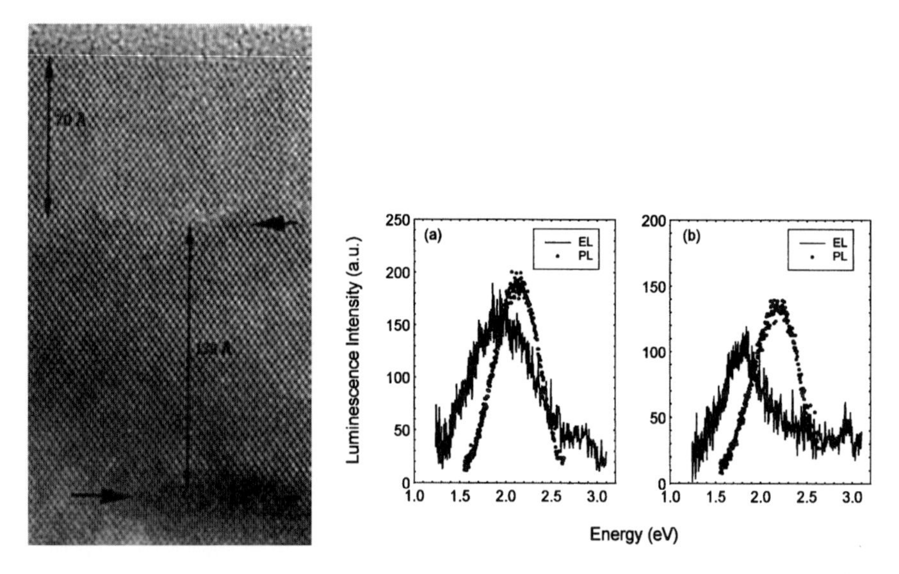

Fig. 4.1. *Left*: High-resolution TEM of 12–13 nm Si buffer deposited at 550 °C. Two 10 L oxygen exposures, separated by 1.1 nm Si deposition, separate the buffer from the top 8 nm of Si. After [5]. *Right*: EL (*solid curve*) and PL (*dotted curve*) spectra measured at room temperature of (**a**) a superlattice grown at 550 °C with a Si layer thickness of 1.2 nm, and then annealed in $H_2 + N_2$, at reverse bias $V = 14$ V; and (**b**) a superlattice grown at 300 °C with the other conditions as in the previous sample. After [6]

Raman measurements indicate that the deposited Si layers are epitaxial and crystalline.

The samples were, successively, annealed under various conditions, mainly for removing dangling bond defects [6]. For these samples both photoluminescence (PL) and electroluminescence (EL) have been observed [6, 7, 8]. The PL and EL spectra are shown in the right part of Fig. 4.1. They are as broad as in the case of porous silicon. The broadness is attributed to variation in the Si layer size. Some red-shift is observed between the EL and PL main peaks. Variations in the thickness of the Si nanolayers cause changes in the energy positions of the main peaks. These changes are small and suggest that quantum confinement is not the only reason for the visible emission; silicon–oxygen bonds also play an important role.

The EL efficiency is similar to that of freshly made porous silicon and has been tested for over seven months without degradation [6], as witnessed by Fig. 4.2. Indeed it was shown that the Si/O superlattice can serve as a barrier as well as isolations for Si devices.

For applications in silicon photonics, these results are interesting and encouraging, even though the growth is quite complicated.

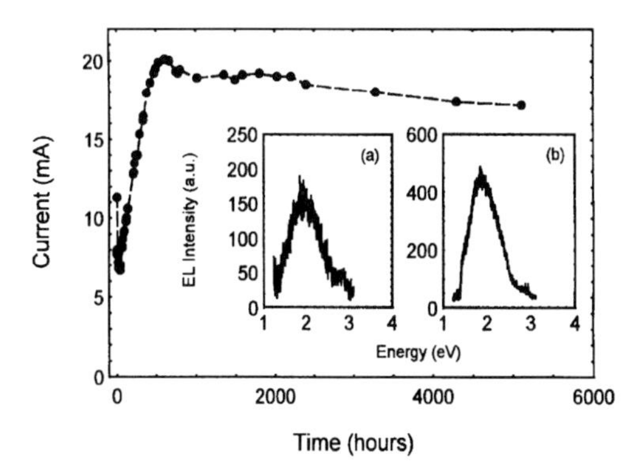

Fig. 4.2. The dependence of operating current with time under constant reverse bias of 10.4 V. The *insets* (**a**) and (**b**) are the EL from the same device after it has been tested for a few days and for seven months, respectively. After [6]

4.1.1 Si-CaF_2

Calcium fluoride (111) layers can be grown epitaxially on silicon (111) layers using molecular beam epitaxy (MBE) due to the similar face centered cubic structure and a very small lattice mismatch (0.6 %). CaF_2 is an insulator with a large band gap (about 12 eV). Thus Si thin layers embedded between CaF_2 layers have been proposed, from a theoretical point of view, as a prototype Si-based system with known microscopic structure and promising optoelectronic properties [9, 10, 11, 12]. In Sect. 2.2 the details of the calculations have been presented. Here we discuss the experimental determination of the structural and optical properties of nanocrystalline Si/CaF_2 multilayers.

Fabrication Techniques and Structure. Nanocrystalline Si/CaF_2 multi-quantum wells (MQWs) or superlattices (SLs) can be engineered with atomic layer precision at the required thickness. They are synthesized by MBE in an ultrahigh vacuum system. CaF_2 and Si are evaporated by means of a standard effusion cell and an electron gun evaporator, respectively. The growth rates of CaF_2 and Si were in the range of a few tenths Å/s [13, 14]. Typical samples consisted of a stack of bilayers containing a CaF_2 and a Si layer with nominal thickness in the nanometer range. Depending on the thickness of the CaF_2 layer, a sequence of uncoupled Si wells (MQW, with CaF_2 thickness larger than 1.2 nm) or a sequence of electronically coupled Si wells (superlattice, with CaF_2 thickness smaller than 1.2 nm) can be grown. This influences the electroluminescence properties. The Si layer thickness is varied in order to study the quantum confinement effect.

The main difficulty in fabricating layered Si/CaF_2 heterostructures is to find a substrate temperature at which both materials grow in a two-dimensional way on each other. The growth of CaF_2 on Si is 2D for the usual range of temperatures (300–750 °C), conversely, at these relatively high temperatures, where the system approaches thermodynamic equilibrium, Si grows as islands on CaF_2. This is due to the different free energies of the Si(111) and CaF_2(111) faces. However if both Si and CaF_2 are grown at room temperature, Si/CaF_2 multilayers are grown with good crystalline order as evidenced by RHEED patterns. Moreover the Si growth proceeds in a 2D way. In this way, Si/CaF_2 multilayers with up to 100 layers were grown.

In order to check the multilayered periodicity of the deposited films XRD measurements under grazing incident were performed. The method probes Bragg reflections due to the large unit cell. Figure 4.3, on the left, shows the XRD intensity under grazing incidence versus the scattering angle obtained on a multilayer consisting of 100 periods deposited on a CaF_2 substrate. The presence of the two satellite peaks indicates unambiguously that the structure is periodic. Moreover, since only reflections near the (0, 0, 0) primary beam can be observed, the nanocrystalline nature of the multilayers is confirmed.

To have a more quantitative idea of the local structure, i.e. the presence of short-range order, number, type and distances of nearest neighbors in the Si layers, EXAFS (extended X-ray absorption fine structure) measurements have also been performed. In the right part of Fig. 4.3 the magnitude of the Fourier transforms of the EXAFS spectra above the Si K edge for samples of different Si thickness are shown. The comparison between amorphous Si (a-Si), crystalline Si (c-Si) and the Si/CaF_2 multilayer demonstrates the presence of structures that correspond to the presence of shells up to the third coordination shell. This is an indication that the Si grain size is about 1.5 nm. The peak at 1.3 Å indicates the presence of Si-O bonds showing that

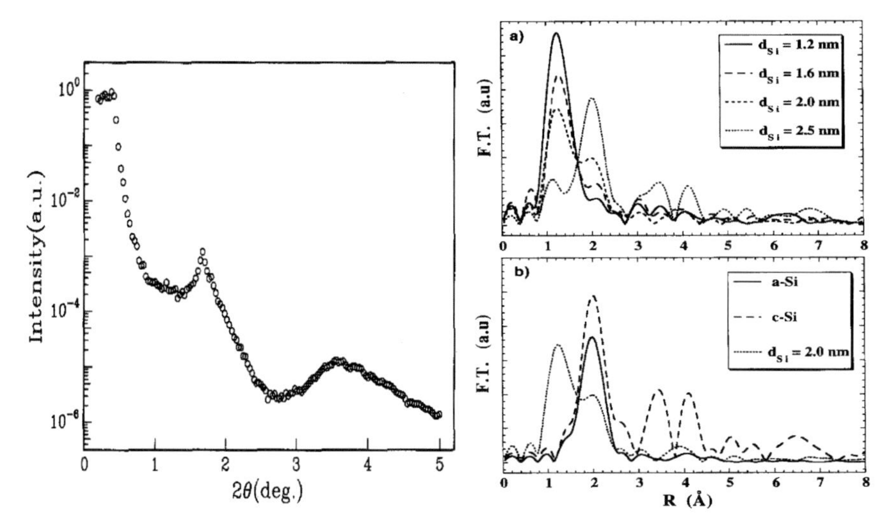

Fig. 4.3. *Left*: XRD intensity under grazing incidence versus scattering angle of a 100-period Si/CaF_2 multilayer. After [13]. *Right*: (**a**) Fourier transforms (FT) of the EXAFS spectra for four Si/CaF_2 multilayers of 50 periods. The Si layer thickness depends on the sample. (**b**) FT of the EXAFS spectra of amorphous silicon (a-Si), crystalline silicon (c-Si) and a multilayer of 50 periods. After [14]

the multilayers contain oxygen, due to migration along the grain boundaries. The structural picture that emerges from these studies is the growth of Si crystallites of nanometric dimensions whose diameter is of the order of the Si layer thickness.

Electronic and Optical Properties, Photo- and Electroluminescence. The optical absorption spectra for Si/CaF_2 multilayers have been obtained from transmission measurements. For this purpose the multilayers were deposited on CaF_2 substrates; due to the high energy band gap, CaF_2 substrates are indeed transparent in the investigated wavelength range.

Figure 4.4 (left part) shows the optical absorption spectra for multilayers having different Si layers thickness and a constant CaF_2 layer thickness. The spectra have been obtained from fresh multilayers, in order to avoid oxidation of the samples. The optical absorption decreases with decreasing Si layer thickness. From these spectra it is possible to obtain the band gap values (E_g) using Tauc's linear relationship. As witnessed by Fig. 4.4 (right part), E_g increases as the Si layer thickness decreases, going from 1.76 eV for the 5 nm thick Si layer to 2.80 eV for the 1 nm thick Si layer. This is a clear indication of quantum confinement effects. The observed linear dependence of $(\alpha E)^{1/2}$ versus E is indicative of an indirect band gap semiconductor. The optical band gaps, obtained by zero-extrapolation of Tauc's relation, can be fitted by a power dependence on the Si layer thickness.

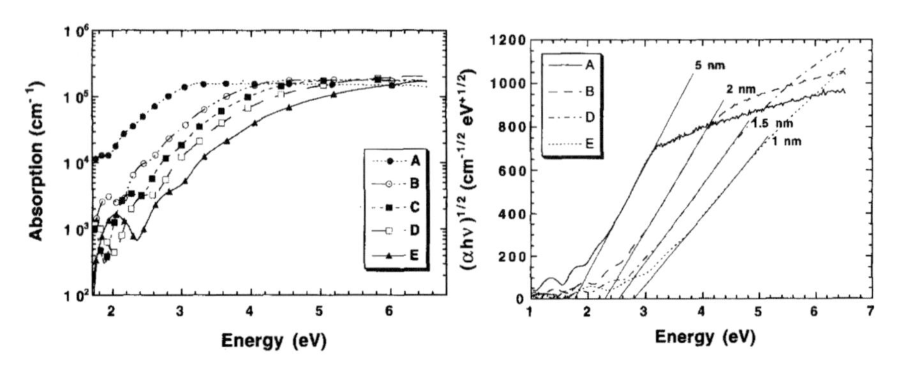

Fig. 4.4. *Left*: Optical absorption coefficient versus photon energy at 300 K for a series of Si/CaF_2 multilayers having different Si layer thickness. Oscillations on the lower energy side are due to the interference fringes in the multilayered structures. *Right*: Tauc's plot for the same Si/CaF_2 multilayers. Extrapolation to $\alpha = 0$ yields the optical pseudogap. Labels refer to the Si layer thickness. After [15]

These Si/CaF_2 multilayers show an intense luminescence [13, 14, 15, 16], whereas 500 nm thick CaF_2 or Si layers do not present any detectable PL. Normalized PL spectra for three Si/CaF_2 multilayers with different Si thicknesses are presented in the left part of Fig. 4.5; the PL spectra show a blue-shift for decreasing Si layer thickness; moreover when the Si thickness exceeds 25 Å the PL disappears.

The PL intensity increases with decreasing temperature, without PL peak shifts. The lifetime at room temperature is of the order of 5 μs. It is thought that the excited carrier recombinations are governed by the tunneling escape from bright crystallites towards dark crystallites where carriers can recombine

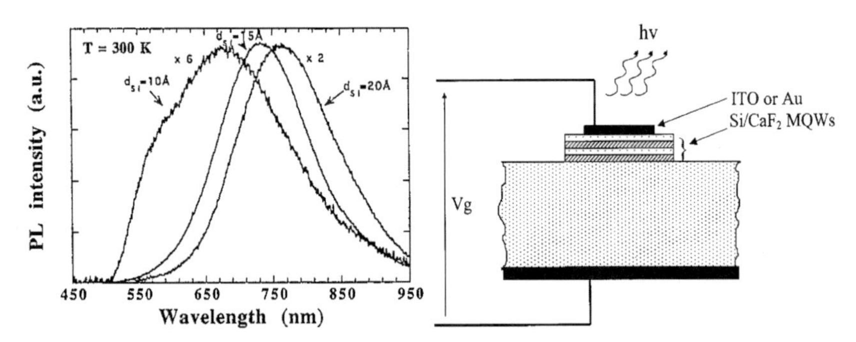

Fig. 4.5. *Left*: RT normalized PL spectra for three Si/CaF_2 multilayers consisting of Si layers thickness of 10, 15 and 20 Å, respectively. Excitation of the PL was provided by the 457.9 nm line of an Ar^+ ion laser with a mean power intensity of 100 mW. After [16]. *Right*: Schematic representation of the device structure. After [20]

nonradiatively [17]. The PL quantum efficiency can be improved from 0.01% to 1% if the structures are annealed [18, 19]. It is worth noting that between absorption and PL energies for the same sample there is a huge Stokes shift of the order of about 1 eV. This leads to the conclusion that the PL is due to recombinations at localized states. This is due to similar mechanisms which, as we have already discussed, are typical of Si nanocrystals, see Sect. 2.4.

To investigate the feasibility of realizing efficient light emitting systems, devices were fabricated by depositing a semitransparent Au or indium–tin oxide as the gate metal, using the lift-off technique in order to define the device area, and by forming a back ohmic contact by conventional techniques as shown in the right part of Fig. 4.5 [20]. The electrical properties of the devices were very sensitive to the CaF_2 barrier thicknesses [21, 22, 23]. When the thickness is above 1 nm, the whole stack of multilayers behaves as an insulator. The current transport is very low (below a few nanoamperes) and the *C-V* curves are similar to those from a metal–insulator structure. Charge exchange between the Si substrate and the Si thin layers is possible, giving rise to a frequency dependent impedance; the leakage current density is quite important, reaching 10^{-8}–10^{-6} A cm^{-2}. The electrical response is dominated by the presence of a continuous distribution of deep traps that have energies in the range of 0.3–0.8 eV and allow the formation of a space charge region near the injecting interface. This kind of sample does not give any electroluminescence.

The charge carrier transport across single tunneling devices (tunneling structures: CaF_2 barrier and one or two Si wells) has also been investigated. The *I-V* characteristics show a shift of the voltage at zero-current and a negative differential resistance (NDR) behavior for reverse bias conditions (see the left part of Fig. 4.6) [21].

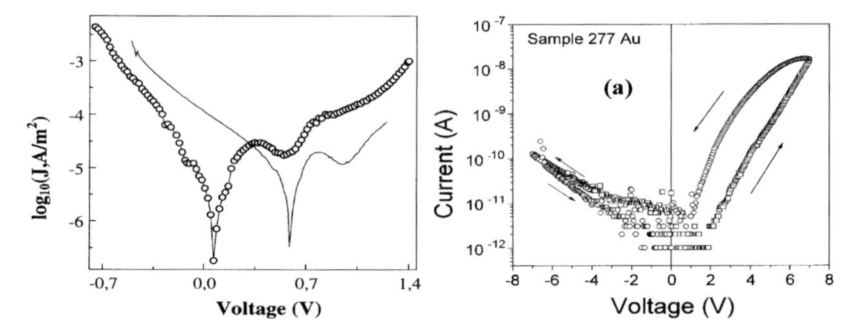

Fig. 4.6. *Left*: *I-V* characteristics of one Si well (*line*) and two Si well (*open circles*) devices. After [21]. *Right*: Typical *I-V* curve from a superlattice consisting of 100 layers of Si(1.6 nm)/CaF_2(0.56 nm). A strong hysteresis effect is observed in both polarities, being more pronounced at positive gate bias. *I-V* curves were obtained by scanning the voltage from zero to the maximum positive value and back to zero and from zero to negative and again back to zero. After [23]

This is an attractive peculiarity of these quantum wells, because there is the possibility to form a device with a NDR based on resonant tunneling of the charge carriers. Simple models based on equivalent circuits and considering capture and release of electrons in and from trap states have been developed to explain these results [24, 25, 26].

In multilayers, where the CaF_2 thickness is less than 1 nm, an important current through the structure is observed and electroluminescence can be measured. All samples present instabilities in current flow and an important hysteresis effect between the up and down sweep of a positive voltage on the gate [23]. This effect is less pronounced in reverse bias. A typical result is shown in the right part of Fig. 4.6. The hysteresis effect disappears at temperatures below 280 K.

The observed variation of the conductance at different temperatures is reported in the left part of Fig. 4.7. It has been interpreted in terms of a Poole–Frenkel type mechanism of carrier transport through the multilayer, due to the linear and symmetric behavior at high voltages [22]. Electroluminescence starts to be visible with the naked eye when the current is of the order of 0.1 A cm^{-2}.

Figure 4.7, right, shows typical EL spectra. The corresponding PL spectrum from the same area of the device prior to final processing is given for comparison. It should be noted that some variations in the structure of the active region of the diode occur during LED processing. In particular, Si nanocrystals form in the multilayers. Voltage and laser power tunability of both EL and PL has been observed and has been attributed to a combination of Auger quenching of radiative recombinations and size dependent carrier injection or ionization of nanocrystals [23]. EL stabilization and de-

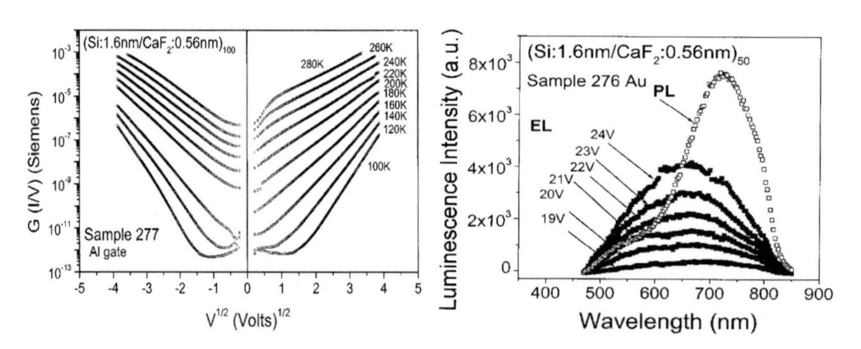

Fig. 4.7. *Left*: Conductance ($G = I/V$) versus $V^{1/2}$ at selected temperatures. At low electric fields the characteristics are quite different at opposite polarities indicating that the electrodes control the injection properties and therefore the transport. The bias step was 50 mV below and 100 mV above 4 V. After [22]. *Right*: EL spectra from a sample composed of 50 bilayers of Si (1.6 nm)/CaF_2 (0.56 nm) at different gate voltages. The PL spectrum from the same sample obtained at low laser excitation power (10 mW) is shown for comparison. After [23]

vice efficiency remain the main issues in order to obtain useful devices for optoelectronic applications, based on the above structures.

4.1.2 Si-SiO_2

The interface between Si and SiO_2 is one of the most studied interfaces in nature and is the one present in almost all electronic devices worldwide. Indeed SiO_2 is the dielectric which forms the gate oxide in CMOS logic. Thus, quite naturally, two-dimensional Si structures sandwiched between SiO_2 have been studied to see whether this system has superior emission properties. By changing the size of the two-dimensional Si layer, one can control the delocalization volume of excitons and thus the absorption and luminescence properties of the Si nanostructure.

Fabrication Techniques and Structure. Several techniques have been used for the preparation of single and multiple quantum wells of Si separated by SiO_2; their structural characterization has been found by a large variety of experimental techniques.

Single crystalline Si quantum wells (c-Si/SiO_2 QW) were prepared using thermal oxidation of a silicon-on-insulator (SOI) wafer [27]. SOI wafers were prepared by separation by implantation of oxygen (SIMOX). Very thin and flat 2D Si layers are clearly shown by TEM images.

The same technique was used [28, 29, 30] to fabricate c-Si/SiO_2 QWs where c-Si layers (0.6–1.7 nm in thickness) were formed between a thin SiO_2 (nearly 30 nm) layer at the surface and a thick buried SiO_2 (nearly 400 nm) layer. A typical TEM image of a 2.7 nm c-Si/SiO_2 single quantum well sample is shown in Fig. 4.8. The fluctuation of the Si layers is about 0.5 nm; the c-Si well region is a single crystalline sheet.

Successive Si/SiO_2 superlattices were grown by using MBE at room temperature for silicon deposition and subsequent ozone exposure for *ex situ* oxidation of each layer [31, 32]. Excellent control of both Si and SiO_2

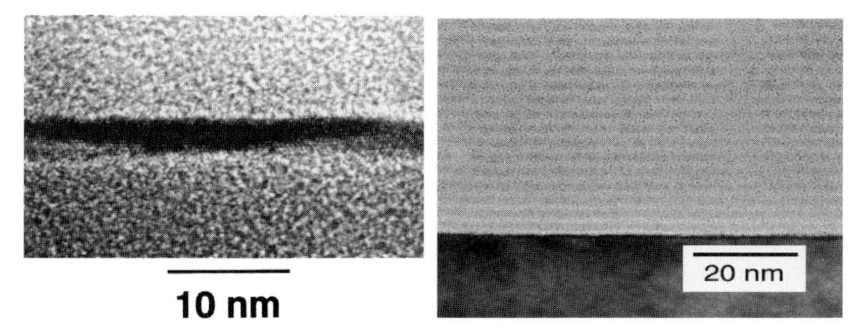

Fig. 4.8. *Left*: Cross-section TEM image of a c-Si/SiO_2 single quantum well. After [28]. *Right*: Cross-section TEM image of a a-Si/SiO_2 multilayer sample. After [29]

thicknesses was achievable. In this case both Si and oxide layers in the superlattices were amorphous (a-Si/SiO_2). Similar amorphous superlattices were fabricated by using *in situ* oxidation by an rf-plasma source [33, 34]. a-Si/SiO_2 QW and multilayer structures were also obtained by an e-beam deposition technique [30, 35, 36, 37]. The SiO_2 layer thickness was 3.0 nm and the a-Si layer thickness was varied from 0.8 to 3.0 nm (see right part of Fig. 4.8). Indeed a-Si/SiO_2 superlattices with 100–525 periods and a 2–3 nm periodicity were deposited by radio frequency magnetron sputtering onto a silicon and quartz substrate [38, 39]. A similar technique was used to grow a Si/SiO_2 superlattice with four periods of 1.8 nm Si and 1.5 nm SiO_2 layers [40].

Nanocrystalline Si/SiO_2 superlattices were fabricated by controlled recrystallization of amorphous superlattices. The silicon layers in this case were relatively thick, from 2.5 to 25 nm [41, 42]. The multilayer structure is not damaged by annealing, recrystallization occurs within the Si layer, and the shape of the Si nanocrystals is almost spherical (see Fig. 4.9, left part). The presence of a nanocrystalline phase is confirmed by electron diffraction patterns. Moreover X-ray diffraction shows that the Si nanocrystal size equals the initial a-Si layer thickness.

Rapid thermal annealing (RTA) of superlattices deposited by MBE on fused quartz and crystalline Si substrates has also been employed [43, 44].

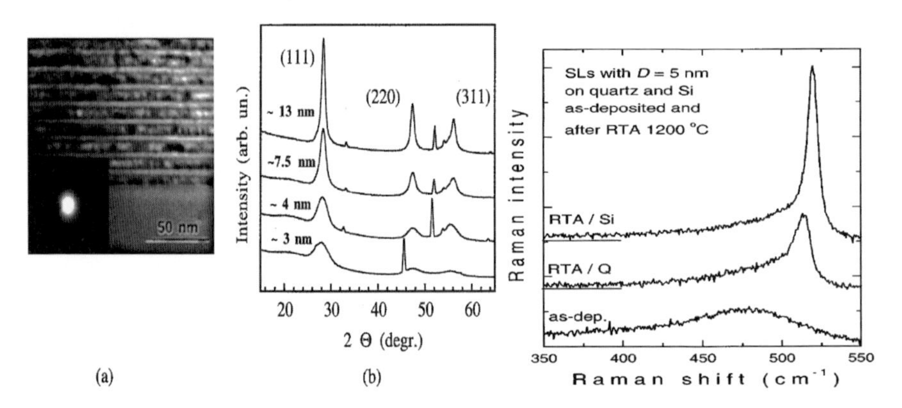

Fig. 4.9. *Left*: TEM image (**a**) and selected-area electron diffraction pattern [(**a**), *inset*] of a recrystallized 77 Å, c-Si/35 Å, SiO_2 superlattice (after thermal annealing at 900 °C and 15 min of 1050 °C furnace annealing). The nanocrystal sizes, estimated by X-ray diffraction (**b**) are equal to the initial a-Si layer thickness. The radiation source is Cu Kα, and the size of the nanocrystals is obtained using the Scherrer multiple peak method. After [41, 42]. *Right*: Raman spectra of Si/SiO_2 superlattices on fused quartz and crystalline Si substrate as deposited and after RTA at 1200 °C. Substrate-induced background scattering is subtracted. The as-deposited samples on quartz and Si show similar Raman spectra. Note the different Raman spectra of the annealed samples on Si and quartz substrates. After [43, 44]

As-deposited Si layers are amorphous as evidenced by Raman scattering (see Fig. 4.9, right part). After RTA at 1200 °C, all samples exhibit some crystallization; the crystallites coexist with a more disordered network. The disorder is larger for samples deposited on quartz, when compared with those deposited on Si substrate.

c-Si/SiO_2 QWs were grown using a combination of room temperature UV-ozone oxidation and wet-chemical etching [45]. Si layer thicknesses between 0.9 and 2.2 nm were obtained. Auger electron spectroscopy and Ar^+ sputtering show (see Fig. 4.10) that (i) a well-defined ultrathin Si film can be readily made, (ii) the transition from oxide to Si occurs rapidly and (iii) the Si/SiO_2 interface is very sharp. Moreover the single-crystal structure of the ultrathin silicon films is confirmed by X-ray photoelectron diffraction (XPD).

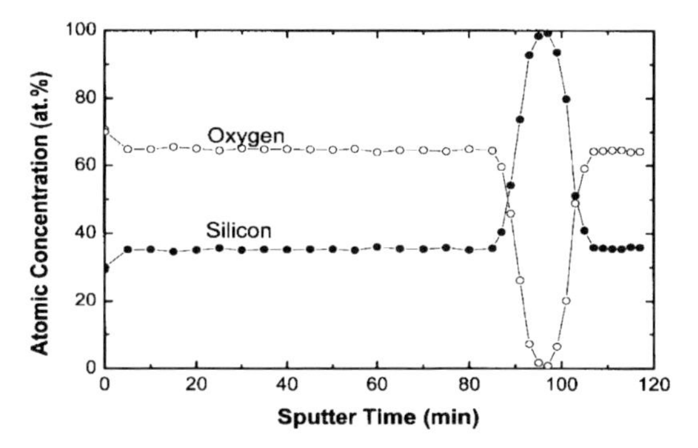

Fig. 4.10. Auger electron spectroscopy depth profile of SiO_2/Si/SiO_2 structures using 1 KeV Ar^+-ion sputtering. The sputtering rate is 1 nm/min, calibrated based on thermal SiO_2 with known thickness. After [45]

Si/SiO_2 superlattices have also been grown by low pressure chemical vapor deposition (LPCVD). A process was developed, based on a cycle of Si-deposition, thermal oxidation and wet etching processes, which allows one to prepare Si/SiO_2 multilayers with Si-layers thinner than 2 nm using an industrial reactor. All processing was performed in a complementary metal-oxide-semiconductor (CMOS) line, in a clean room environment, and it was fully compatible with integrated circuit fabrication [46, 47, 48, 49]. The approach, depicted in the left part of Fig. 4.11, is based on the following steps: (i) use poly-Si deposition conditions which give thin reproducible and uniform layers and (ii) reduce these Si layers to the thickness needed by means of an oxidation process. By sequentially repeating the poly-Si deposition and oxidation, multilayered structures can be obtained. The outlined process has been employed to prepare both Si quantum wells [46] and Si/SiO_2 multilayers [47]. TEM images show some evidence of isolated Si nanocrystals.

A similar process was used in [50, 51, 52]. Structural characterization in this case shows a sharp interface between Si and SiO_2 and that the size

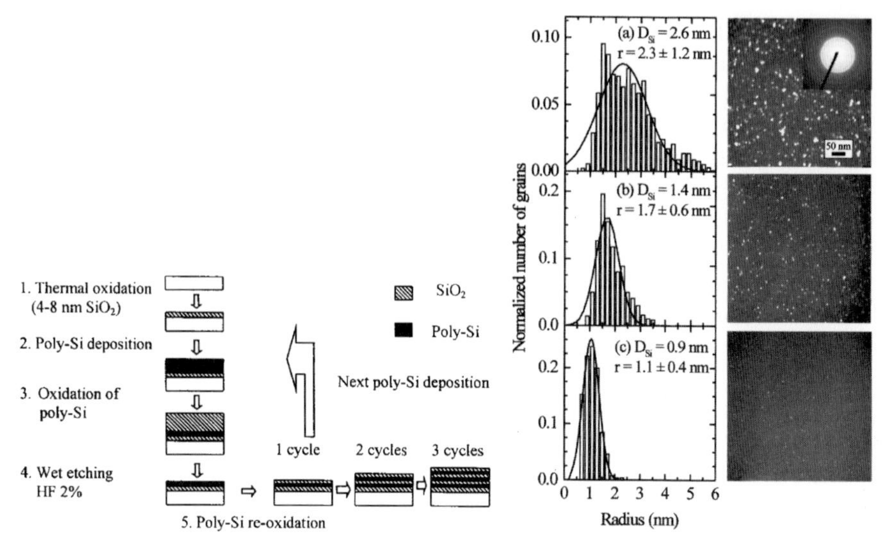

Fig. 4.11. *Left*: Schematic representation of the process used to obtain sub-3 nm poly-Si layers. Steps 2–5 are repeated 1, 2, or 3 times to obtain 1, 2, or 3 periods of the Si/SiO_2 layers. After [46]. *Right*: TEM plan-view micrograph (*right-hand side*) and relative nanocrystal size distributions (*left-hand side*) for Si/SiO_2 superlattices annealed at 1200 °C for 1 h, with (**a**) silicon layer thickness of 2.6 nm, (**b**) 1.4 nm, and (**c**) 0.9 nm. The corresponding diffraction pattern is also shown as an inset for one of the samples. After [53]

of the Si nanocrystallites in the vertical direction, that of the deposition, is approximately that of the Si film thickness, while in the plane their shape is in general slightly deviated from the spherical case.

Alternatively nanocrystalline silicon/silicon dioxide superlattices were fabricated by plasma enhanced CVD (PECVD) and subsequent annealing at temperature ranging between 1000 and 1250 °C [53]. The thickness of the SiO_2 layers was maintained constant (8.5 nm), while the thickness of the Si layers was varied between 0.9 and 2.6 nm. Figure 4.11 (right part) shows the TEM plan-view micrographs and relative nanocrystal size distributions. The TEM plan-view analysis allow one to determine the effect of the temperature on the process of nanocrystal formation. The obtained Si nanocrystals are concentrated in an ordered, planar arrangement, each plane being spaced with respect to the adjacent ones by at least 5 nm of SiO_2 along the depth axis. Moreover, within a single plane, the nanocrystals have an average distance of about 3 nm, suggesting that in this case they are more isolated than those produced with other techniques.

In summary we can conclude that it is possible to grow a periodic Si/SiO_2 multilayer, where the thickness of the Si nanocrystals is controlled by the thickness of the Si layers and where thermal treatments can be used to obtain good crystalline structures of the desired dimensions and interconnectivity.

Electronic and Optical Properties, Photo- and Electroluminescence. The most evident effect of quantum confinement is the modification of the electronic energy band gap of the two-dimensional Si confined layers. The changes in the band gap can be related to the shift of the conduction band minimum (CBM), of the valence band maximum (VBM) or of both. The shift in the CBM can be determined using soft X-ray Si $L_{2,3}$ edge absorption spectroscopy (XANES).

In $L_{2,3}$ absorption spectroscopy one excites the occupied Si core shell p electron to the empty conduction band. As a consequence of the dipole selection rule, the absorption onset is related to the Si CBM s-like electrons. Even if the absolute CBM position cannot be measured very accurately, the relative CBM shift with respect to a thick film can be precisely determined. For the determination of the VBM shift one can measure the occupied valence band density of states by X-ray photoelectron spectroscopy (XPS). The results for the CBM and VBM shifts relative to a-Si/SiO_2 superlattices is shown on the left in Fig. 4.12. The figure shows that the VBM shifts to lower energy, whereas the CBM moves to higher energy. There is a clear dependence of both VBM and CBM shifts on the Si thickness. The CBM shift can be described by the usual effective mass theory. In the same samples, visible to near-infrared PL was observed only for superlattices whose Si thickness was smaller than 3 nm. The variation in the PL peak energy with Si thickness is

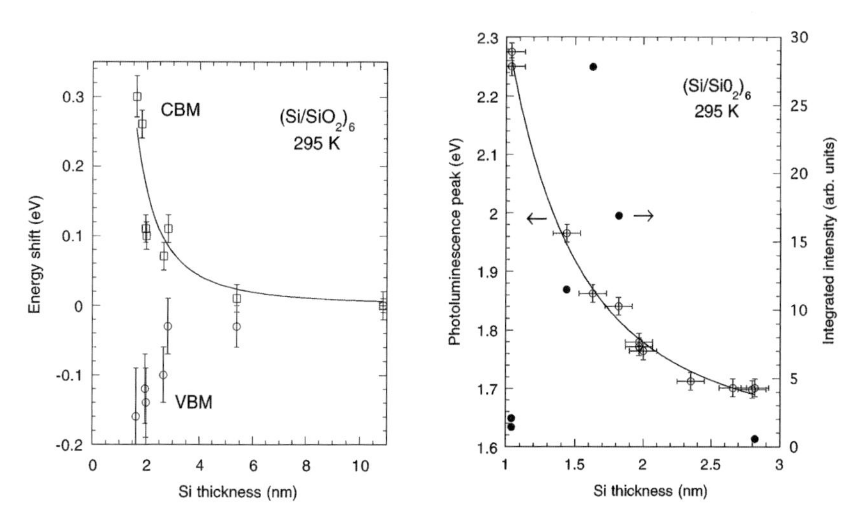

Fig. 4.12. *Left*: The shifts in conduction (CBM) and valence (VBM) bands at room temperature for Si/SiO_2 superlattices as a function of Si layer thickness. The *solid line* is the fit by effective mass theory. After [31]. *Right*: The photoluminescence peak energy (*open circles*) and integrated intensity (*full circles*) at room temperature in Si/SiO_2 superlattices as a function of Si layer thickness. The *solid line* is a fit by effective mass theory. After [32]

depicted in the right part of Fig. 4.12. This blue-shift can also be reasonably fitted using the formula

$$E(\mathrm{eV}) = E_\mathrm{g} + 0.72\frac{1}{d^2}. \tag{4.1}$$

The best fit is obtained with $E_\mathrm{g} = 1.60$ eV, which is typical of a-Si. The similarity between the values obtained for band gap widening and the PL shift points toward the direct nature of the gap.

The changes in the energy gap have also been measured for c-Si/SiO$_2$ multilayers using XANES and XPS [45]. The results for CBM and VBM shifts are shown in Fig. 4.13 (left part), where the insets depict two CB and VB spectra respectively. Dashed lines are related to bulk Si, solid lines to a Si quantum well of 0.9 nm thickness. Similar to what is observed for a-Si superlattices the CBM and VBM show shifts for thicknesses below 3 nm. The dependence of these shifts on the Si thickness seems to be less strong than for the a-Si case. This fact is confirmed by the thickness dependence of the PL energy at low temperatures in c-Si/SiO$_2$ and a-Si/SiO$_2$ quantum wells [30], see the left part of Fig. 4.13. The PL peak energies in the amorphous structures are higher than those in the crystalline ones, because the PL energy in bulk a-Si is higher than in bulk c-Si. The PL peak energy is blue-shifted in both structures by decreasing the Si well thickness.

Another method for the determination of the VBM shift is the use of transmission spectra. These show a blue-shift of the transitions when the Si

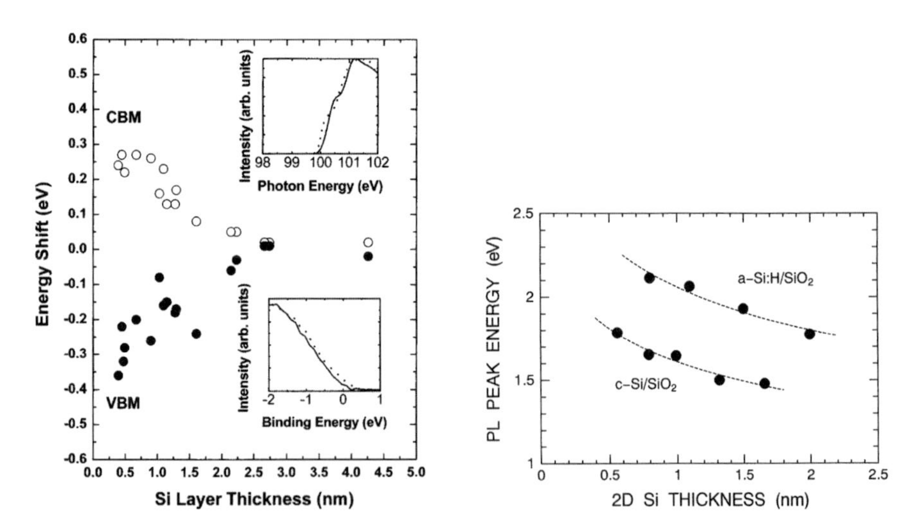

Fig. 4.13. *Left*: CMB (*open circles*) and VBM (*solid circles*) energy shifts as a function of Si layer thickness for c-Si/SiO$_2$ SLs. The *top inset* shows the CB as probed by XANES and the *bottom inset* shows the VB as measured by XPS. After [45]. *Right*: Thickness dependence of the PL energy in c-Si/SiO$_2$ and a-Si/SiO$_2$ quantum wells at low temperatures. The *broken lines* are a guide for the eyes. After [35]

layer thickness decreases from 5 nm to 1.5 nm as witnessed by Fig. 4.14 (left part) [46]. Similar results have been obtained by other groups [41, 42].

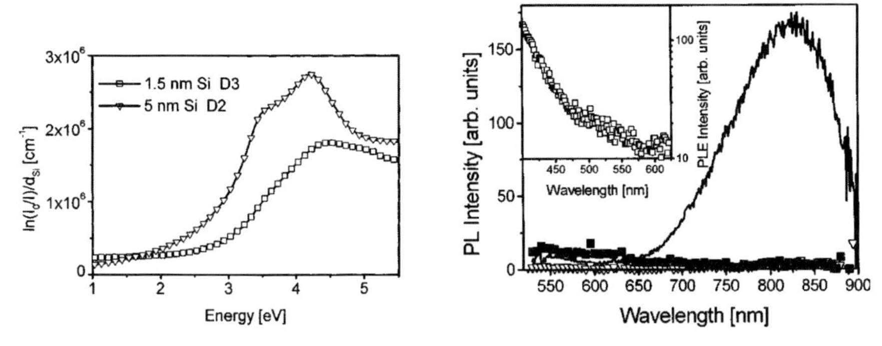

Fig. 4.14. *Left*: Spectral dependence of the transmission (at 300 K) of two different samples with Si layer thickness 1.5 nm and 5 nm respectively. *Right*: PL spectra at 300 K of $SiO_2/Si/SiO_2$ quantum wells grown on fused silica substrates (laser excitation 514 nm, 50 mW): sample with Si layer thickness 1.5 nm (*solid line*); Si layer 5 nm (*closed squares*), fused silica substrate (*inverted triangle*). *Inset*: PLE spectrum (300 K) of the 1.5 nm sample (observation wavelength 825 nm). The spectrum is corrected for the spectral shape of the excitation power of the Xe-arc lamp used. After [46]

However, an exact estimation of the band gap for indirect semiconductors from absorption measurements is very difficult. Using PL excitation spectroscopy (PLE) the indirect band gap for a 1.5 nm thick Si layer has been estimated to be about 1.6 eV, which means a band gap widening of about 0.5 eV (see the right part of Fig. 4.14). Larger (3–20 nm) Si layers show no shift of the transitions to higher energies and hence the quantum confinement effects are negligible in this thickness range. Comparing the PLE energy value for the band gap to the PL peak energy one obtains a small Stokes shift of about 0.1 eV between absorption and emission [46]. Larger shifts have been estimated using transmission measurements versus PL energies by other groups [41, 42].

Among the large number of published results [27, 28, 29, 30, 31, 32, 33, 34, 40, 43, 44, 45, 46, 47, 50, 51, 53, 54], PL spectra can be classified according to two categories:

- spectra dominated by a peak at 1.65 eV which increases in intensity with decreasing Si layer thickness without showing any shift with the Si layer dimension;
- spectra where a thickness dependent shift of the PL is accompanied by a concurrent intensity increase.

This confusing behavior can be well understood using site-selective excitation spectroscopy [28, 30]. Figure 4.15 shows (left part) the thickness

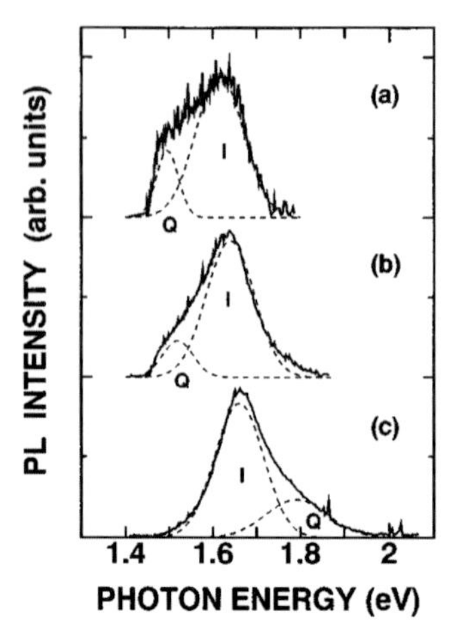

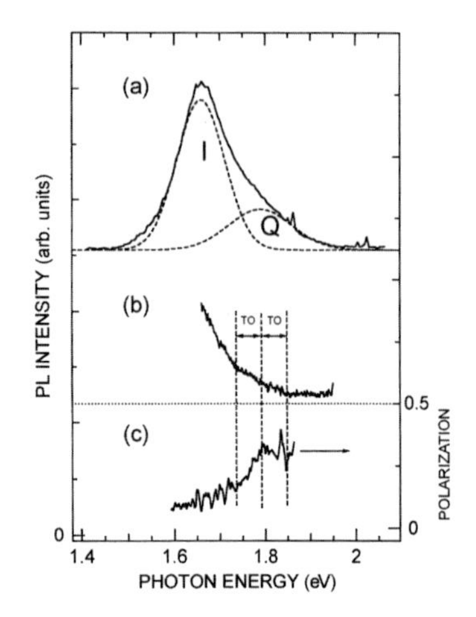

Fig. 4.15. *Left*: Photoluminescence spectra of c-Si/SiO_2 single quantum wells under 488 nm laser excitation at 2 K: (**a**) 1.7, (**b**) 1.3, and (**c**) 0.6 nm thickness. The asymmetric PL spectra can be fitted by two Gaussian bands: the Q-band and the I-band. After [28]. *Right*: PL spectra of c-Si/SiO_2 single quantum well with 0.6 nm Si thickness at 2 K. The excitation laser energies of PL (**a**) and resonantly excited PL (**b**) and (**c**) are 2.540 and 1.959 eV respectively. After [35]

dependence of PL spectra in 2D c-Si/SiO_2 systems under 488 nm laser excitation at 2 K [28].

The asymmetric PL spectra in the red and infrared spectral region can be fitted by two Gaussian bands, the weak PL band (denoted as Q) and the strong PL band (denoted as I). The PL peak energy of the I band is almost independent of the Si well thickness and appears at about 1.65 eV, whereas the peak energy of the Q band depends strongly on the thickness of the Si well. The Q band blue-shifts with decreasing Si thickness, moving from lower to higher energies with respect to band I. Site-selectively excited PL spectra taken at 2 K for the 0.6 nm Si well clearly show that no c-Si phonon-related structures are observed when the excitation energies are within the I band. On the contrary, when the excitation is within the Q band a steplike structure is observed (see the right part of Fig. 4.15). This structure corresponds to TO-phonon related fine structures [28, 30]. From these results one can conclude that the I band is caused by radiative recombination at localized interface states, whereas the Q band is due to quantum confinement effects. These conclusions have been confirmed by theoretical results [55, 56, 57, 58, 59], see also Sect. 2.2.

Of particular interest is the time decay behavior of the PL for the Si/SiO_2 SLs produced by PECVD (see Fig. 4.16) [53]. In fact, single exponential decay lineshapes are observed, in contrast with literature data on Si nanocrystals where nonexponential decays are always reported. Since the nonexponential behavior has been attributed to a strong nanocrystal–nanocrystal interaction

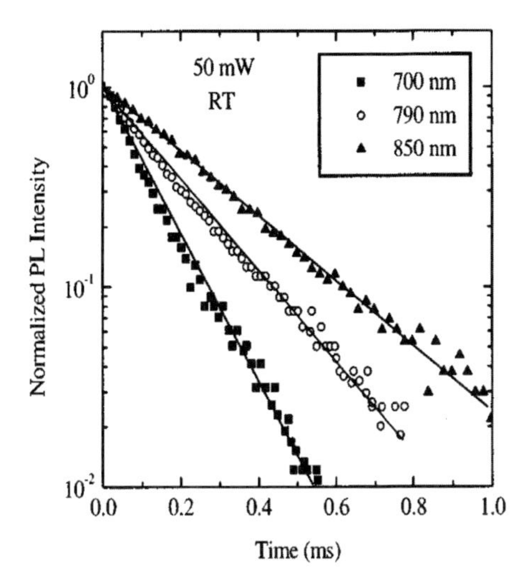

Fig. 4.16. Room temperature measurements of the time decay of the PL signal at 700 (*squares*), 790 (*circles*), and 850 (*triangles*) nm for Si/SiO_2 SLs with Si thickness of 0.9 nm after annealing at 1200 °C for 1 h. The pump power of the excitation laser was 50 mW. The *continuous lines* are single exponential fits to the experimental data with lifetimes of 0.1 ms at 700 nm, 0.2 ms at 790 nm and 0.3 ms at 850 nm. After [53]

with the energy traveling along the sample, one can deduce that in this case the nanocrystals are almost isolated. Indeed lifetimes are also particularly long, 0.8 ms at 17 K and 0.25 ms at 300 K, demonstrating that nonradiative processes are quite weak.

Silicon light emitting devices with (SiO_2/Si) periodic multilayer structures have been fabricated [46, 51, 60, 61, 62]. The insertion of the Si/SiO_2 multilayers in MOS capacitor structures leads to strong variations of the electrical characteristics of the MOS device. Electrical injection improvements and voltage breakdown reductions have been observed, see Fig. 4.17 [46, 60]. In addition, these devices in the forward bias condition showed stable electroluminescence, with modulation frequencies larger than 1 MHz and external efficiency still too low for practical applications. The typical measured external quantum efficiency (EQE) was of the order of 5×10^{-7} at 0.65 µm. The main problem in the low efficiency is carrier injection. Based on the comparison between PL and EL results, the mechanism of EL is mainly ascribed to electron–hole recombination of carriers generated by hot electrons in the anodic region, which overwhelms the carrier recombination process in the thin poly-Si layers.

Room and low temperature EL in the visible range has also been observed from a single layer of Si nanocrystals sandwiched by two thin SiO_2 layers [51]. The EL peak energy exhibits tunability from the red to the yellow depending

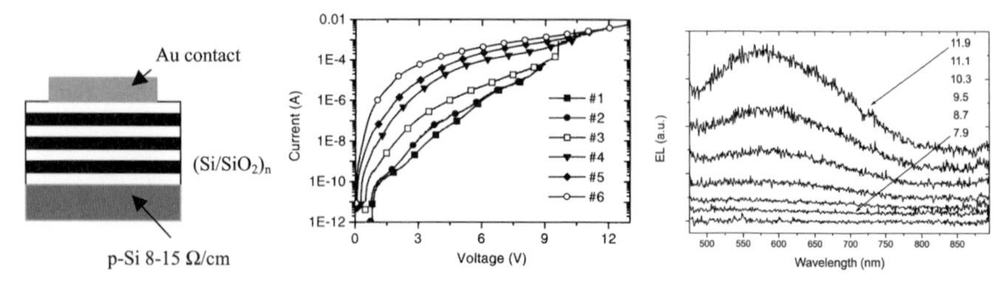

Fig. 4.17. *Left*: Sketch of the device. *Middle*: *I-V* characteristics of a LED made by a SL with various Si(1.1 nm)/SiO_2(6 nm) periods. The voltage is applied to the back contact with respect to the front contact, and swept from 0 to the final value. The different curves correspond to successive measurements, in the order shown in the plot legend. The first two measurements were performed with a final voltage lower than the breakdown point (9.7 V); the third measurement (*hollow squares*) as well the others had a final voltage higher than the breakdown. *Right*: EL spectra of a (1nm Si/4.5 nm SiO_2)$_2$ double layer on a p-Si substrate with a 8 nm SiO_2 gate oxide and an Au top contact taken in forward bias. The numbers beside the spectra give the applied voltage. The *lowest line* is the reference signal in the dark. After [46, 60]

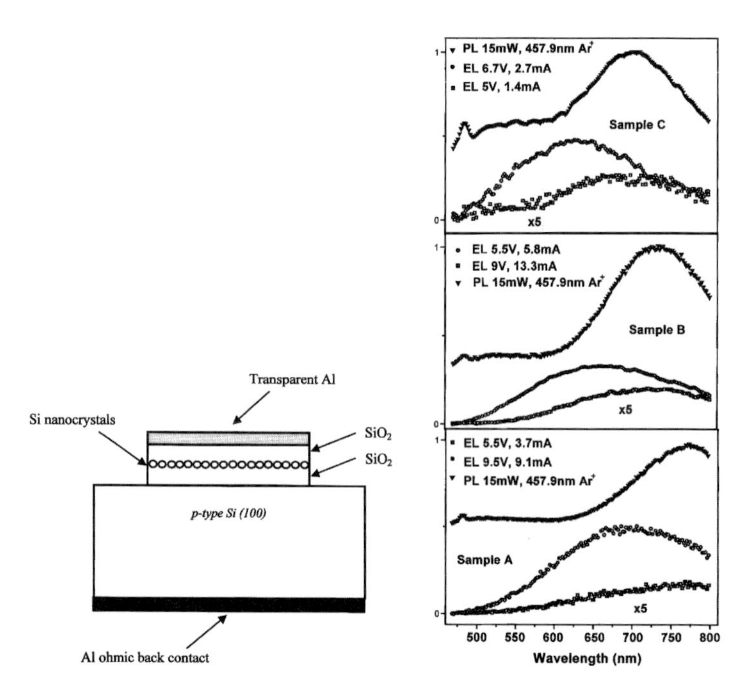

Fig. 4.18. *Left*: Light emitting diode structure. *Right*: PL and EL spectra from samples A, B, and C with Si nanocrystals of 2 nm, 1.5 nm and 1 nm dimensions respectively. PL measurements were done with a 15 mW, 458 nm Ar^+ laser line. For low voltages, EL spectra show peaks at the same wavelength as PL spectra. For higher voltages EL peaks are blue-shifted. After [51]

on the excitation voltage. By decreasing the temperature at constant excitation voltage, an increase in the EL intensity together with a blue-shift in the EL energy peak position is detected, as witnessed by Fig. 4.18. At high voltages a saturation in the intensity and a peak broadening is observed. Carrier injection is the most important limiting factor for the effective light emission. Current flow through a leaky SiO_2 offers the possibility to produce by impact excitation both electrons and holes into the nanocrystalline layers, consequently EL shows the same characteristics as PL. However optical pumping is much more effective than electrical pumping. Nonradiative Auger recombination, Coulomb charging effects and quantum confined Stark effects could be at the origin of this behavior.

Resonant carrier tunneling (RCT) and negative differential conductivity (NDC) has been observed in Si/SiO_2 multilayers at low temperatures [63]. RTC and NDC were both consequences of the interface roughness that creates additional lateral components in the vertical carrier transport. Differential AC conductivity as a function of the applied bias and its frequency dependence with applied longitudinal magnetic field have been measured (see Fig. 4.19). The presence of RCT and NDC has been interpreted in terms of hole tunneling via quantized valence band states of Si nanocrystals of different sizes and shapes [63].

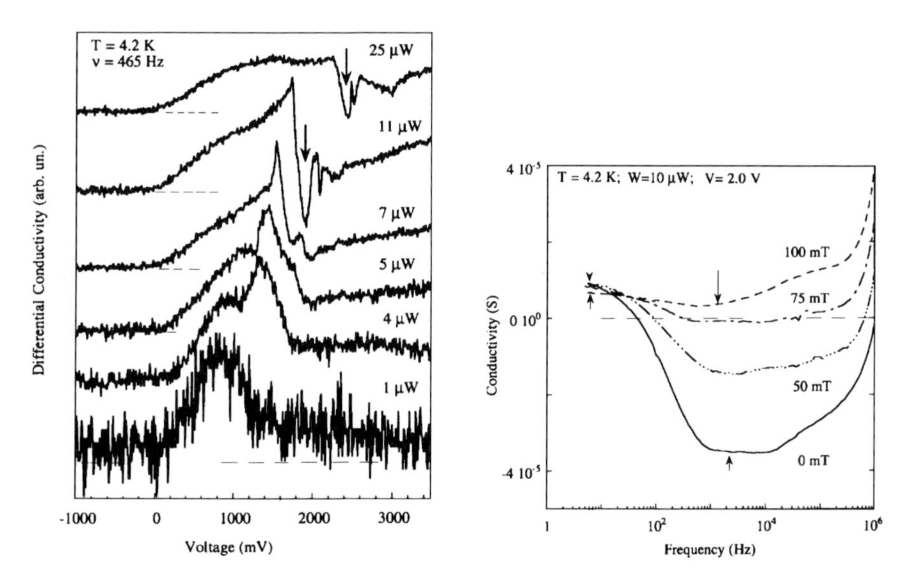

Fig. 4.19. *Left*: AC differential conductivity versus applied bias measured at 465 Hz and 4.2 K under different levels of photoexcitation power. The NDC regime is denoted by *arrows*. *Right*: Frequency dependence of the AC differential conductivity measured a 4.2 K with the applied bias of 2 V and light power of 10 mW for the specified longitudinal magnetic field. *Arrows* show the magnetoresistance effect at low frequency. The zero level is shown by a *dashed line*. After [63]

Current–voltage characteristics for Si/SiO_2 SLs have been computed through model calculations [64]. From the numerical results, a number of simple conclusions useful to optimize physical parameters in order to achieve the maximum electroluminescence efficiency has been derived. Recently Monte Carlo simulation of the electron transport in these superlattices has been performed. The simulation shows that, for oblique fields, the transport properties along the vertical direction are strongly enhanced by the in-plane component of the electric field, thus higher vertical drift velocities can be easily obtained [65]. Lateral carrier injection is, thus, a possibility to improve EQE.

The use of Si/SiO_2 multilayers for Fabry–Perot type planar optical microcavities will be discussed in Chap. 6.

4.2 Silicon Nanopillars and Wires

The synthesis and characterization of silicon nanopillars and nanowires (Si NWs) have been motivated by the efforts to produce new optoelectronic devices. Very recently, Si-NW based simple field effect transistors have been demonstrated: a suitably doped Si NW can be assembled in order to have pn junctions, bipolar transitors and complementary inverters [66, 67, 68]. Moreover direct evidence of quantum confinement effects has been found by investigating the role of the Si quantum wires size [69]. In this section we will show how Si nanopillars and NWs can be fabricated, their structural characteristics and their optoelectronic properties.

4.2.1 Fabrication Techniques and Structure

Si NWs have been successfully synthesized by a variety of methods: photolithography and electron-beam (e-beam) lithography combined with etching, scanning tunneling microscopy (STM), laser ablation techniques and vapor–liquid–solid (VLS) methods. The structure of the Si NW has been investigated using TEM, SEM, XRD, NEXAFS and Raman spectroscopy.

Photolithography followed by etching was the first technique employed for the production of Si NWs [70, 71, 72, 73]. Nassiopoulos and coworkers [72] used conventional optical lithographic and reactive ion etching techniques (RIE) in order to produce silicon pillars. The first step is the pattern definition done by optical lithography. In order to push down the resolution limit of the process over- and underexposure were used. The second step is highly anisotropic reactive ion etching that allows one to obtain Si pillars with sizes below 0.1 μm, The final step is high temperature thermal oxidation and oxide removal of the produced SiO_2 skin. After oxide removal by an HF dip, Si pillars of diameter below 10 nm and with an aspect ratio as high as 50:1 were present in some areas of the produced pattern [74]. Figure 4.20 shows SEM images of these Si pillars obtained directly after etching and after oxide removal.

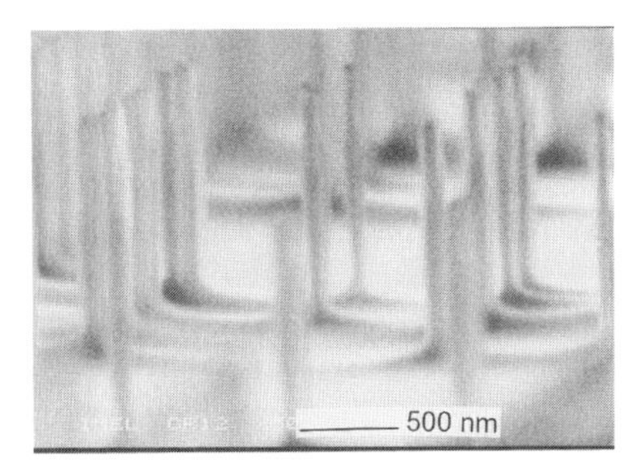

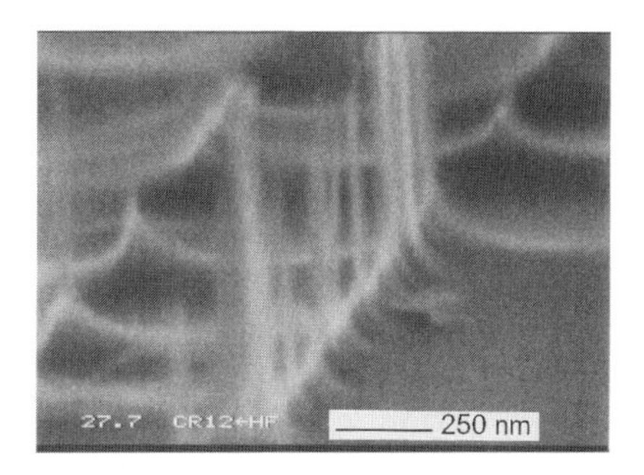

Fig. 4.20. SEM images of Si pillars. *Left*: Pillars obtained directly after etching. *Right*: after an oxidation cycle at 900 °C and removal of the oxide skin. After [74]

The limitation of optical lithography can be overcome by using e-beam lithography. Several groups have thus obtained Si pillars with diameters around and below 10 nm [75, 76, 77]. Ordered arrays of Si nanopillars have been fabricated via reactive ion etching [78]. Self-assembled polymer spheres are used as masks to deposit hexagonal arrays of Ag islands on a Si substrate. Etching with SF_6 and CF_4 after polymer removal results in hexagonal arrays of Si NWs, whose aspect ratios depend on etching time and applied rf power. Similar results can be obtained using rough Ag films deposited on the Si substrate [79, 80]. A scheme of the fabrication process is sketched in Fig. 4.21. In the same figure the SEM image makes it evident that pillars with diameter as little as 5 nm can be produced and that silver islands remain on the top of the Si columns. Raman spectra of these NWs have been interpreted

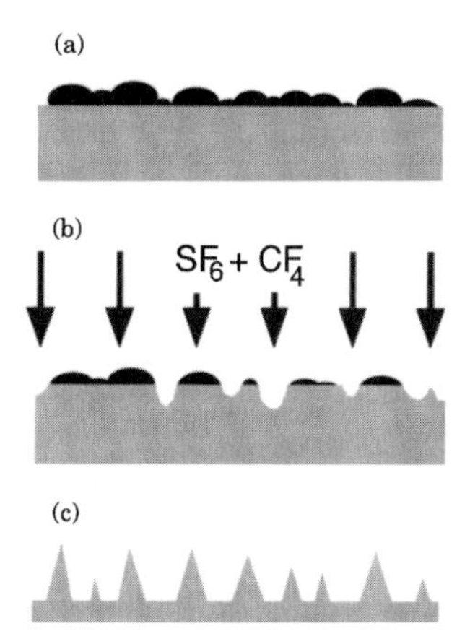

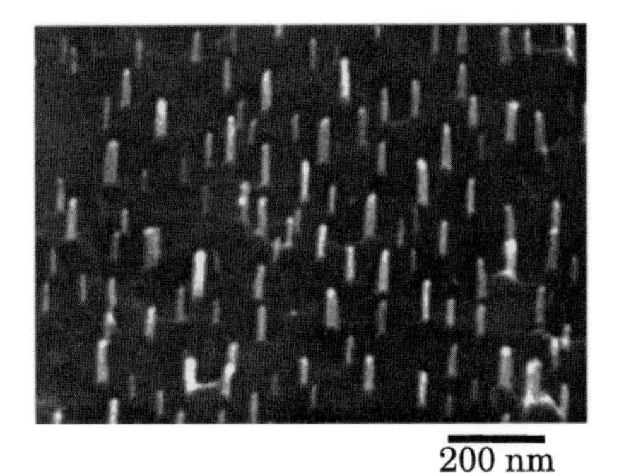

Fig. 4.21. *Left*: Fabrication process. (**a**) A rough Ag film is deposited on the substrate. (**b**) The plasma reduces the size of the Ag islands by sputtering, exposing regions of the Si surface to the etching gas. (**c**) after the etching process Si pillars are produced. After [79]. *Right*: SEM of a Si sample etched for 3 min. After [78]

through a phonon confinement model showing that the confinement effect is important for wire diameters less than 22 nm [81].

Few groups have used STM to grow Si NWs. In fact using STM, Si NWs were grown either by applying a voltage at a constant current between a Si substrate and a gold STM tip [82] or during indentation of the STM tip onto the Si(111) surface [83].

The laser ablation method has been, on the contrary, employed and implemented by several groups. In [84] a combination of laser ablation cluster formation and VLS growth were used. In the process, depicted in Fig. 4.22, the laser ablation is used to prepare a nm-diameter catalyst cluster that imposes limits to the size of the Si NW produced by VLS growth. Uniform Si NWs (Fig. 4.22, middle) with diameters ranging from 6 to 20 nm and lengths of 1 to 30 μm were prepared. The proposed growth model (Fig. 4.22, right) for the Si NW is made of four steps [84]:

- (A) Laser ablation of the $Si_{1-x}Fe_x$ target creates a dense, hot vapor of Si and Fe species.
- (B) The hot vapor condenses into small clusters as the Si and Fe species cool through collisions with the buffer gas. The furnace temperature is controlled to maintain the Si-Fe nanoclusters in a liquid state.
- (C) NW growth begins after the liquid becomes supersaturated in Si and continues as long as the Si–Fe nanoclusters remain in a liquid state and Si reactant is available.
- (D) Growth terminates when the NW passes out of the hot reaction zone (in the carrier gas flow) onto the cold finger and the Si–Fe nanoclusters solidify.

Using gold as catalysts Si NWs, whose crystalline cores remain highly crystalline for diameter as small as 2 nm, were obtained [85].

Large-scale synthesis of Si NWs was achieved by using laser ablation at high temperature using Si powder with Fe impurities together with nanosized Ni and Co powder [86, 87]. Figure 4.23 is a typical TEM image of the morphol-

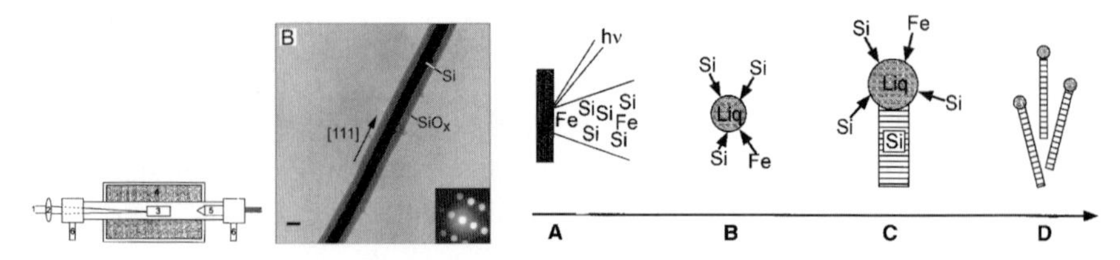

Fig. 4.22. *Left*: Growth apparatus. The output from a pulsed laser (1) is focused (2) onto a target (3) located within a quartz tube; the reaction temperature is controlled by a furnace (4). A cold finger (5) collects the product as it is carried in the gas flow that is introduced (6, *left*) through a flow controller and exits (6, *right*) into a pumping system. *Middle*: Diffraction contrast TEM of a Si NW. Scale bar 10 nm. *Right*: The proposed model for NW growth. After [84]

ogy of extremely pure Si NWs with uniform distribution of sizes. The average diameter is about 15 nm and the length varies from a few tens to hundreds of μm. Their core is crystalline Si as witnessed by the selected-area electron diffraction pattern (SAED) results. A variation of this method is based on physical evaporation [88]. Si ultrapure powders are mixed with a small percentage of Fe powder and hot-pressed at about 150 °C to form a plate that is placed inside a quartz tube in vacuum. The system is heated at 1200 °C for a long time under an argon flow at high pressure (about 100 torr). TEM images of the obtained sponge-like product clarify that the nanowires are covered by a silicon oxide layer (about 2 nm) that can be etched away in HF. The diameter of the remaining wires is of the order of 10 nm. Figure 4.23, middle, shows the XRD spectrum with up to nine peaks, clearly related to c-Si. The derived average lattice parameter for the wires is only 0.1% larger that that of bulk Si. The peaks observed in Raman spectra (Fig. 4.23, right) are due to the first-order and second-order optical phonon mode of c-Si and to the second-order transverse acoustic phonon mode slightly downshifted as a consequence of the confinement (Fig. 4.23). Varying the ambient pressure between 150 and 600 torr it is possible to control the diameter of the wires [89].

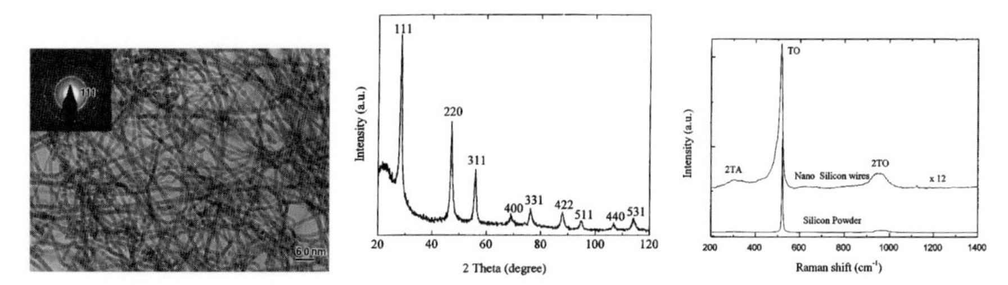

Fig. 4.23. *Left*: TEM image of the Si nanowires. The inset shows a ring pattern of SAED similar to that of bulk Si. After [87]. *Middle*: XRD spectrum. After [88]. *Right*: Raman spectrum of the Si nanowires compared with that of bulk Si. After [88]

The laser ablation technique was used in a variety of configurations to obtain undoped and doped Si NWs [90, 91, 92, 93, 94, 95, 96, 97, 98, 99].

Using laser ablation at high temperature of a sintered Si powder mixed with 0.5% Fe, Si NWs have been produced with diameters ranging from 3 to 43 nm and lengths up to a few hundreds μm [90]. Besides the most abundant smooth-surface nanowires, four other forms, named spring-shaped, fishbone-shaped, frog-egg shaped, and pearl-shaped NWs were observed, see Fig. 4.24 [91].

In a refinement of this technique, it was demonstrated that a metal catalyst is not necessary [93]. SiO_2 was discovered to be a special and effective catalyst that largely enhances Si NW growth [94]. The morphology and structure of Si NWs have been systematically investigated by HRTEM and, contrary to

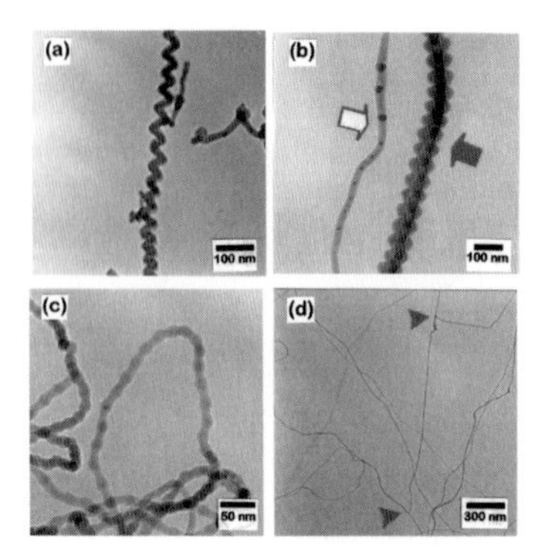

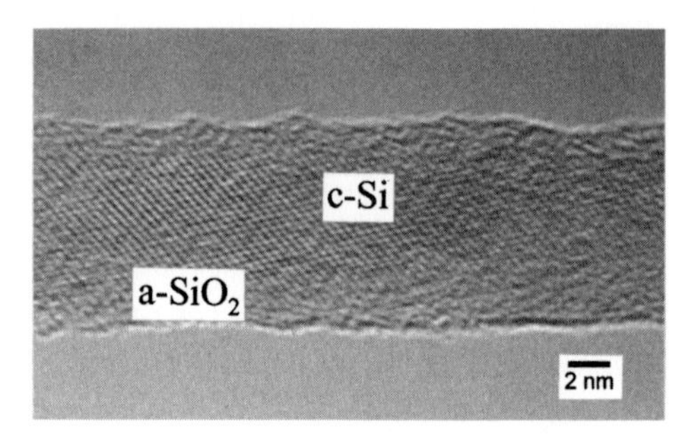

Fig. 4.24. *Left*: TEM image of a Si NW. (**a**) Spring-shaped; (**b**) fishbone-shaped (*solid arrow*) and frog-egg shaped (*hollow arrow*); (**c**) pearl-shaped; (**d**) the polysites of nucleation (*arrows*) extend the NW. After [91]. *Right*: HRTEM morphology of a single Si NW. After [92]

the situation where metals are used as a catalyst, now no evidence of metal is found at the tips of the NW. The diameter of the Si NW can be modified and controlled using laser ablation in different ambient gases [95]. The diameter distributions peak at around 13.2, 9.5 and 6 nm for He, Ar and N_2 as ambient gas respectively. These NWs show PL, which depends on their dimensions. By ablating a silicon monoxide powder target with a pulsed KrF excimer laser at 1200 °C in an Ar atmosphere bulk-quantity Si NWs with negligible contamination of nanoparticles have been obtained [92]. In HRTEM images (see Fig. 4.24) a crystalline core of few nm diameter is observed and the interplanar spacing corresponds to that of the 111 plane of bulk Si. Doping of the NWs is particularly important for their use in nanoscale electronics. Boron doped and phosphorus doped Si NWs have been produced and investigated by several techniques [96, 97, 98]. In particular P-doped Si NWs have been studied using STM and NEXAFS in both total electron yield (TEY) and X-ray fluorescence yield (FLY) mode, see Fig. 4.25 [98]. The first probes the surface of the wire, the second the core. Using HF to remove the surface oxide one observes that both TEY and FLY show the disappearing of the Si oxide related peak and the appearance of the crystalline Si edge; moreover it is also observed that the NW core has a slightly larger interatomic distance on average with respect to bulk Si. This fact demonstrates that the Si NWs are encapsulated within silicon oxide and that the core is c-Si doped with P. Disperse silicon NWs in liquid were also produced by a mild etching treatment [98]. Finally, in the last few years, Si NWs have been successfully synthesized by various groups using other methods [100, 101, 102, 103].

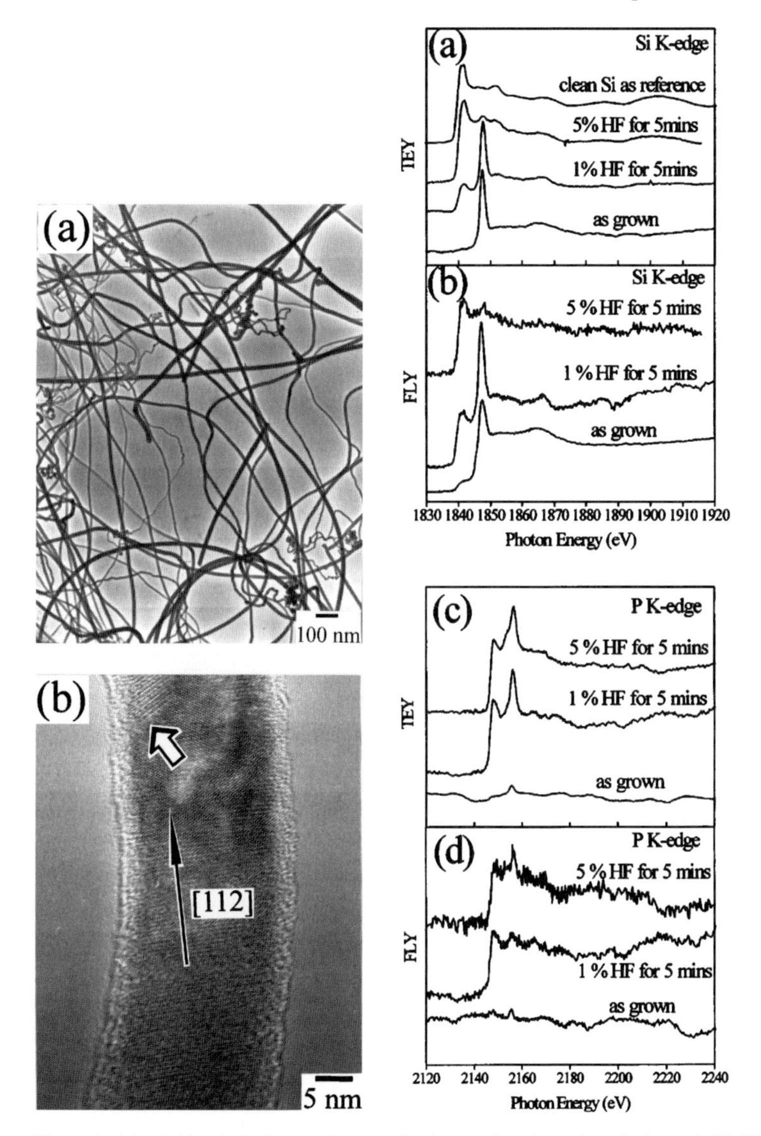

Fig. 4.25. *Left*: (**a**) General morphology of a heavily P-doped Si NW shown by TEM; (**b**) HRTEM image: the typical growth direction lies along the [112] direction. *Right*: TEY and FLY spectra of K-edge NEXAFS after sequential HF etching: (**a**) TEY of Si K-edge; (**b**) FLY of Si K-edge; (**c**) TEY of P edge; (**d**) FLY of P edge. After [98]

4.2.2 Electronic and Optical Properties, Photo and Electroluminescence

Since the diameters of synthesized Si NWs are larger than the dimensions expected for quantum confinement to be effective (less than 5–10 nm), direct measurements of band-gap opening are scarce in the literature.

The first PL study of Si quantum wires was performed in [74, 104]. The NWs had diameters ranging between 5 and 10 nm and lengths of about 0.5 μm. PL measurements show a broad PL band centered at about 500–600 nm [74]. The PL emission was found to depend both on the excitation wavelength and on the polarization of the laser beam, as shown in Fig. 4.26. Energy and efficiency of the PL increased with decreasing wavelength. This has been interpreted as an effect due to the size distribution of the Si NW: an increase in Si NW diameter results in a decrease of the band gap and thicker structures interact with longer wavelength, while thinner structures become off-resonant.

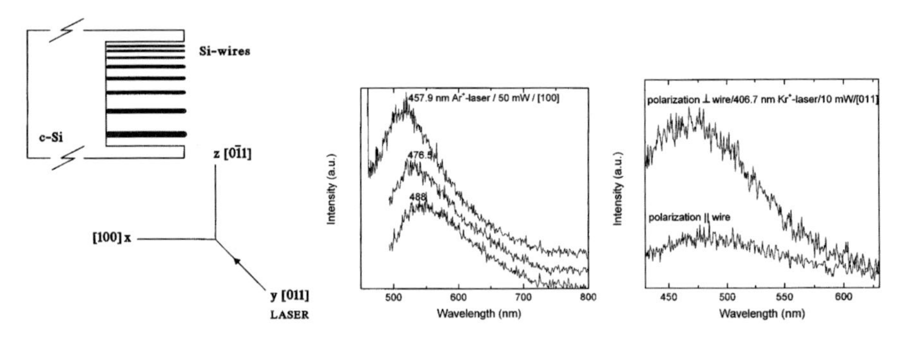

Fig. 4.26. *Left*: Si NW grown on c-Si(100) and measured in a backscattering geometry with the incident light along [011]. *Middle*: Dependence of PL emission on the excitation wavelength. *Right*: Dependence of PL emission on the polarization state of the initial laser beam. After [104]

Concerning the polarization dependence, excitation with a laser beam polarized parallel to the wire axis results in a PL efficiency three times smaller than in the case of excitation with the laser beam perpendicular to the wire axis (in the direction of the confinement). This is in agreement with theoretical calculations for the polarization dependence of PL in Si quantum wires, see Sect. 2.3.

PL characteristics under ultraviolet photoexcitation have been investigated for Si NWs of extreme uniformity, whose diameters were around 10–15 nm [69, 88]. The room-temperature PL spectrum of the as-grown wires shows (see Fig. 4.27) three broad emission bands corresponding to dark red (1.52 eV, 816 nm), green (2.40 eV, 517 nm) and blue (2.95 eV, 420 nm) [69]. To study the origin of these bands the dependence of the PL from the dimensions of the wires has been studied by oxidizing the NW in different conditions, to achieve a progressive shrinking of the NW. At moderate oxidation the intensity ratios of the various peaks are varied, and upon further oxidation a blue-shift of the red band is observed. For wires oxidized for 5 min at 700 °C the three bands are now peaked at 804, 515 and 420 nm. After ten minutes the read peak increases in intensity and is far shifted to 795 nm.

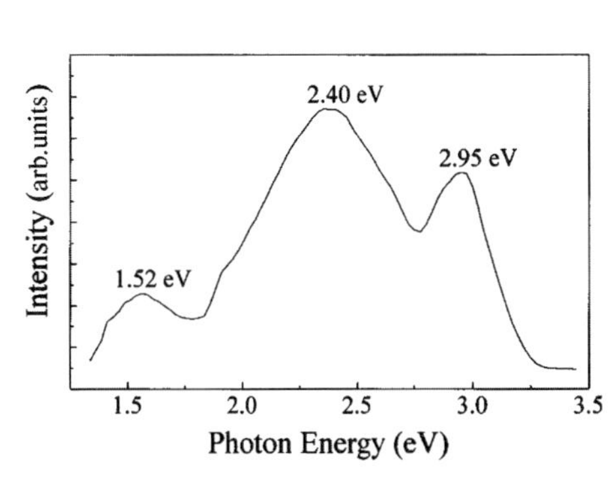

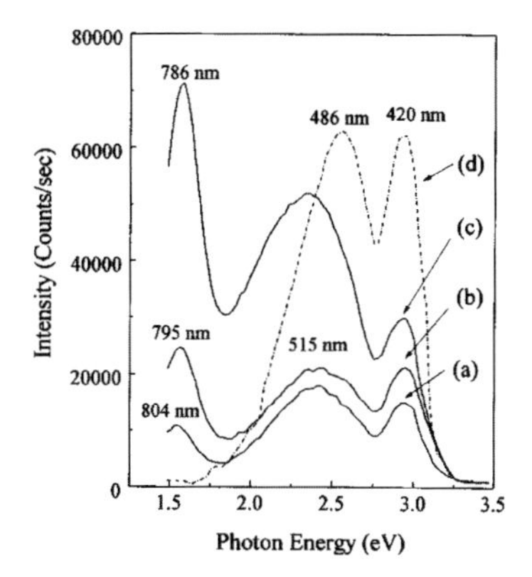

Fig. 4.27. *Left*: PL spectra of the as-grown Si NW. *Right*: PL spectra of the oxidized Si NW at 700 °C: (**a**) 5 min; (**b**) 10 min; (**c**) 15 min. (**d**) PL spectrum of a Si NW sample oxidized for 30 min at 900 °C. After [69]

Oxidation for 15 min (wires of about 4 nm in diameter) results in a rapid increase of the red peak, now centered at 786 nm. If the Si wire is oxidized for 30 min at 900 °C, a completely amorphous silicon oxide forms and the corresponding PL shows that the red band disappears, whereas the green and blue ones increase in intensity. These characteristics have been interpreted by assigning the red band to quantum confinement effects and possibly interface effects and the green and blue ones to radiative recombinations from defect centers in the silicon oxide [69].

The PL properties of Si NWs with a conical shape fabricated by reactive ion etching (RIE) in silicon-on-insulator substrates have been studied as a function of the oxidation and annealing temperature [80]. Nanocones whose tips are as large as 5 nm were obtained and show, for oxidation at 900 °C for 1 h, intense PL in the yellow-green at about 530 nm. When they are oxidized at 1000 °C, PL centered at about 730 nm or at 650 nm, depending on the etching time, is observed. The yellow-green PL has been interpreted in terms of defect states, the red one in term of a combination of defect states and quantum confinement effects.

The most complete study of the electronic properties of the Si NW has been done by using a combination of XPS, XAFS both in TEY and FLY mode, electron energy loss spectroscopy (EELS) and X-ray excited optical luminescence [98, 105, 106]. The Si NWs were prepared by the laser ablation technique and show diameters of the order of 10–20 nm. The results were compared with those for Si(100) and porous Si in order to understand the role of structure and dimensionality. The TEY measurements show that the Si NWs are crystalline and covered by silicon oxide. They also exhibit a smaller plasmon frequency than bulk Si, as theoretically predicted [107]. The comparison between XANES data for two Si NWs of different diameters with those of Si(100) and porous Si shows that, even though porous Si shows the usual blue-shift of the edge, the same is not true for the NW (see top left of Fig. 4.28). However the XEOL of as-prepared Si NWs (diameter 26 nm) shows PL peaked at 295, 460 and 530 nm (Fig. 4.28, top right).

Upon HF dipping, the PL at shorter wavelengths disappears, whereas the PL in the visible remains. This difference is due to the fact that XANES measures the average properties of a large number of wires of different diam-

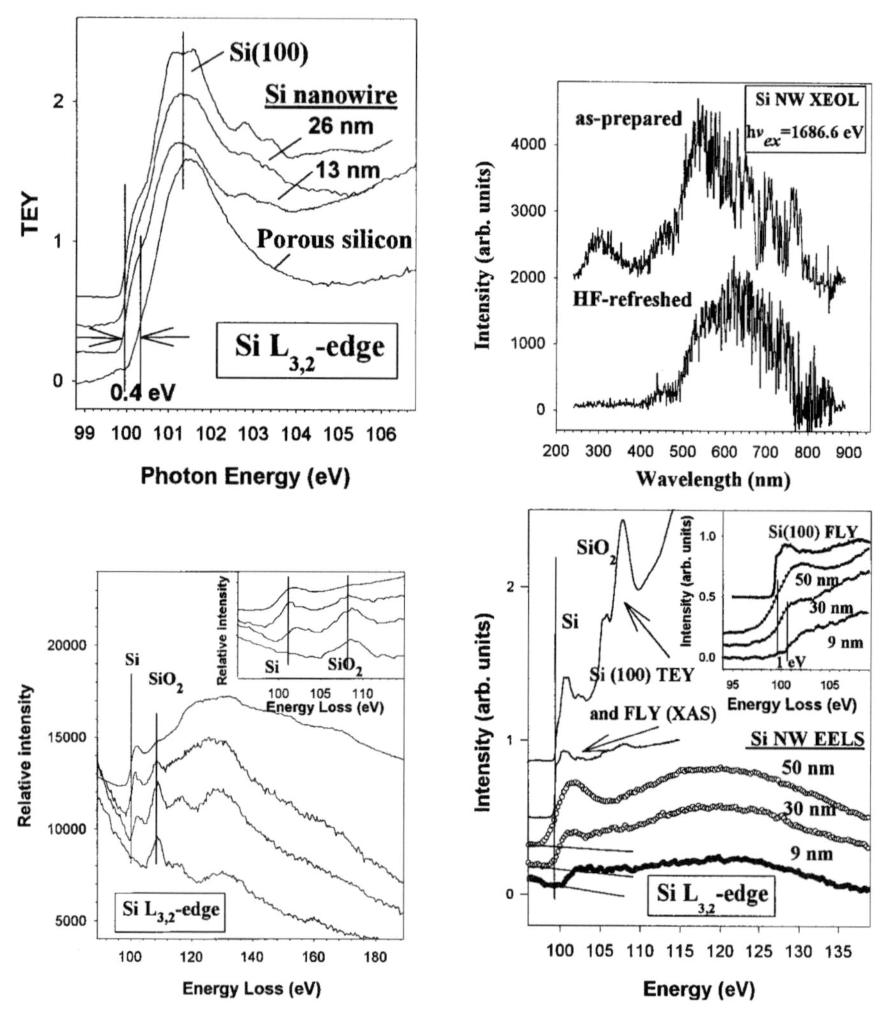

Fig. 4.28. *Top left*: XAS for Si NW, Si(100) and porous Si. The *vertical lines* mark the threshold. *Top right*: XEOL of a 20–26 nm wire. *Bottom left*: site-selected EELS from a HF refreshed (*top spectrum*) and several regions of an ambient 11–15 nm wire. *Vertical lines* mark edges. The *inset* shows the alignment. *Bottom right*: site-selected EELS of three HF etched wires and comparison with Si(100) TEY and FLY. The *inset* shows a more detailed comparison. After [106]

eters, whereas XEOL probes the luminescent grains in the specimen. This is confirmed by the use of site-selective EELS. Using a 50 nm beam from a third-generation source and probing different regions of a Si NW it is possible to use the SiO_2 resonance maximum to align the results. By proper alignment of the energy scale, a 1 eV quantum confinement originated shift for the 9 nm wide NW can be determined [106]. Such a value is probably related to small Si nanocrystals formed in the NW.

Solid state devices based on Si NWs have been fabricated and tested. They are based on a thin metal film, an insulating layer in which Si NWs are embedded, an n-type Si substrate and an ohmic contact. The wires are in contact with the Si substrate from one side and with the metal from the other. EL is observed in the visible range, slightly shifted to longer wavelength with respect to the PL, when the voltage exceeds 12–14 V [108].

Electron field emission has been observed in doped and undoped NW. The emission increases with decreasing diameter and with doping [109, 110]. The surface reactivity of the NW toward reductive metal ions deposition has been investigated and a periodic array of intramolecular junctions of Si NWs has been developed [111, 112]. Si nanowire devices have been fabricated, whose conductance can be improved by four orders of magnitude by doping and thermal treatment [67]. Finally Si nanopillar-based photonic crystals have been reported (see Chap. 6) [113].

4.3 Silicon Nanocrystals

4.3.1 Fabrication Techniques

Silicon nanocrystals, also referred to as silicon quantum dots, have been widely studied due to their interesting luminescent properties [53, 114, 115, 116, 117, 118, 119, 120, 121, 122, 123, 124, 125, 126, 127, 128, 129, 130, 131, 132, 133, 134, 135, 136, 137, 138, 139, 140, 141, 142, 143, 144, 145, 146, 147, 148, 149, 150, 151, 152, 153, 154, 155, 156, 157, 158, 159, 160, 161, 162, 163, 164, 165, 166, 167, 168, 169]. They can be produced by several different techniques. In some cases the techniques are compatible with VLSI technology, making them particularly appealing for applications. More important is a distinction of methods in terms of the matrix surrounding the fabricated nanocrystals or, when formed in a gas phase, the surface termination and passivation properties. These, in fact, may have a strong impact on the luminescence properties. A schematic picture of a few Si nanocrystal fabrication techniques is displayed in Fig. 4.29.

In the upper row deposition techniques are shown. They can vary from silane decomposition in the gas phase [126] to deposition through laser ablation [122, 123, 124, 125] or aerosol techniques [127, 128]. In these methods accurate size selection can be obtained by various methods before deposition. Both the density and size of the nanoparticles can be controlled, leading also

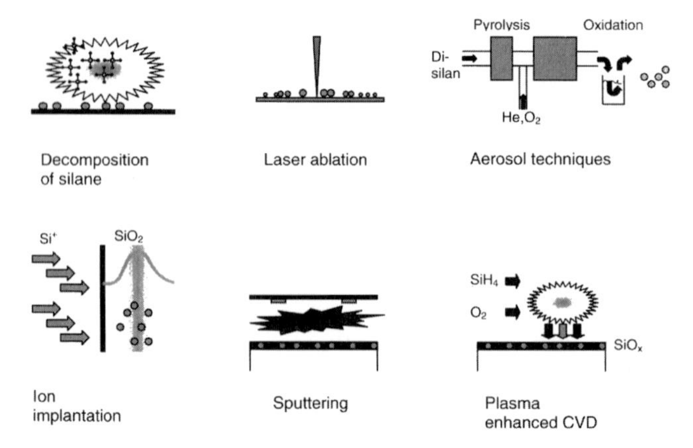

Fig. 4.29. Si nanocrystal fabrication techniques. The lower techniques produce nanocrystals in a SiO_2 matrix. After [114]

to extremely low concentrations. These methods, however, are more easily found in research laboratories and are difficult to integrate with current silicon technology. In the lower part of Fig. 4.29 are grouped VLSI-compatible techniques (also available as standard processes in submicron circuit facilities) producing nanocrystals embedded in an oxide matrix. Starting with ion implantation, Si ions of moderate keV energies are implanted at high doses (of the order of $10^{16}/cm^2$) into a thermal oxide followed by a high-temperature anneal to form the nanocrystals [115, 116, 117, 118, 119, 120, 121]. The annealing also provides good surface passivation due to the excellent interface properties of Si/SiO_2. The high dose required to have sufficient excess Si in the oxide coupled with the necessity of multi-energy implants to obtain flat profiles makes this technique not very flexible when thick layers are required. In contrast, sputtering [129, 130] or chemical vapour deposition (CVD) [131, 132] techniques are more favorable although somewhat less controllable. To control the excess Si concentration during sputtering, Si samples of corresponding areal density are positioned on the SiO_2 target. In CVD techniques, the Si concentration is controlled by varying the gas flow rate, e.g. the quotient between the silane and oxygen flow rates.

4.3.2 Structural Properties

In the present and the following paragraphs we will mainly refer to plasma-enhanced chemical vapour deposition (PECVD) formed nanocrystals (nc) as a key example to study the properties of Si nc [131, 132]. The source gases are SiH_4 and N_2O and a variation of the ratio N_2O/SiH_4 varies the composition of the deposited SiO_x film (with $x < 2$). After high temperature annealing ($T > 1000°C$) separation of the two phases and formation of Si nc is achieved.

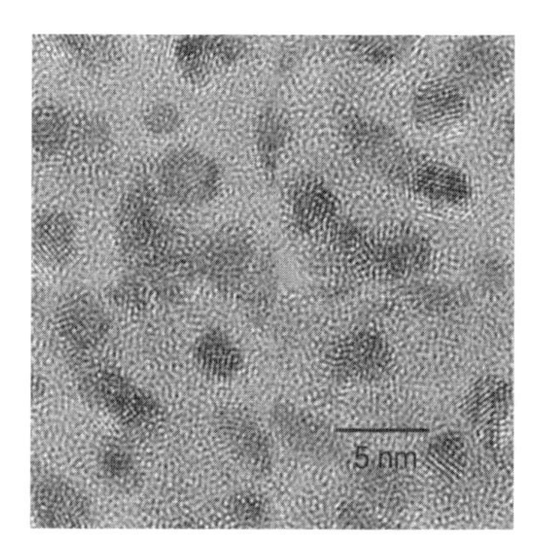

Fig. 4.30. High-resolution TEM micrograph of Si nanocrystals embedded in a SiO_2 matrix. Courtesy of C. Spinella

This can be seen in Fig. 4.30 which shows a high-resolution TEM image of Si nc embedded within SiO_2.

Figure 4.31 shows dark field plan-view TEM micrographs relative to a sample with 44 at.% Si after thermal annealing at 1100, 1200 and 1250 °C.

All micrographs show that the amorphous SiO_x matrix contains a high density of small clusters, that, on the basis of the electron diffraction analysis, can be identified as silicon nanocrystals. Their formation is due to thermal annealing, because they are completely absent in the as-deposited sample. The statistical analysis of the crystal size distribution (reported in Fig. 4.31 as a histogram) obtained from the TEM micrographs indicates that the thermal process at 1100 °C (Fig. 4.31a) induces the formation of crystals having a mean radius of 1.0 nm, as deduced by fitting the size distribution with a Gaussian curve. The distribution has a standard deviation (σ) of 0.2 nm, accounting for all the different crystal sizes detected in the micrograph. From the analysis of Fig. 4.31 it is evident that, for the same silicon composition, the mean radius of the silicon crystals strongly depends on the annealing temperature; indeed, from the analysis of the micrographs relative to samples annealed at higher temperatures, an increase of the mean crystal size up to 1.3 nm at 1200 °C (Fig. 4.31b) and 2.1 nm at 1250 °C (Fig. 4.31c) can be clearly noted. This phenomenon implies also an increase of the width of the distribution; indeed, the value of σ becomes 0.4 nm at 1200 °C and 0.5 nm at 1250 °C.

The same effect of increase of the grain size distribution with increasing temperature has been observed for the other silicon concentrations; for instance, a sample with 42 at.% Si does not show the presence of any detectable nanocrystal at 1100 °C, while very small clusters become well distinguishable at 1200 °C (mean radius of 0.7 nm, $\sigma = 0.2$ nm), and their size remarkably increases at 1250 °C (mean radius of 1.7 nm, $\sigma = 0.5$ nm).

Figure 4.32 shows plan-view TEM micrographs of samples with 37 at.% Si, 39 at.% Si and 42 at. % Si, after thermal annealing at 1250 °C, together with the relative nanocrystal size distributions, fitted with a Gaussian curve. Fig-

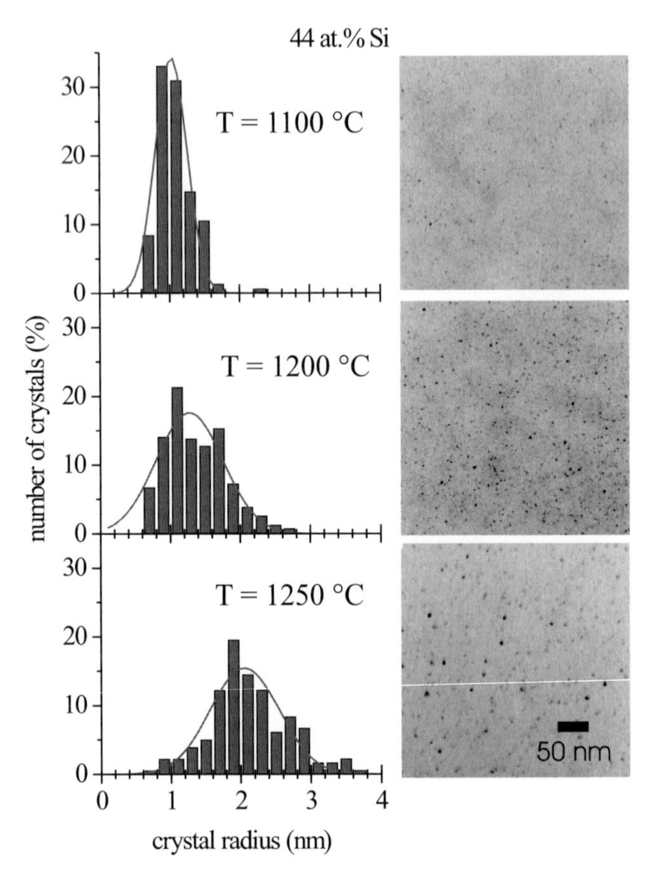

Fig. 4.31. Plan view TEM micrographs and relative Si nanocrystal size distribution for a SiO_x film having a Si concentration of 44 at.%, after annealing at (**a**) 1100 °C, (**b**) 1200 °C, and (**c**) 1250 °C. After [132]

ure 4.32 clearly demonstrates that, for the same annealing temperature, the crystal mean radius increases with the silicon concentration of the films. In summary by properly varying deposition parameters and annealing temperature the size distribution of nc can be varied in a wide range. It should be added that quite isolated Si nc can also be obtained by PECVD deposition of alternate layers of thin Si (or SiO) films and thicker SiO_2 films. These superlattices (SLs) after high temperature annealing may result in the balling up of the ultrathin Si-rich layer with the formation of layers of Si nanocrystals. The size of these nanocrystals can also be controlled through a proper choice of film thickness and annealing temperature. Their property, however, is to be much more isolated than in the previous case.

In addition, recently Zacharias et al. [170, 171] have demonstrated that phase separation and thermal crystallization of SiO/SiO_2 superlattices re-

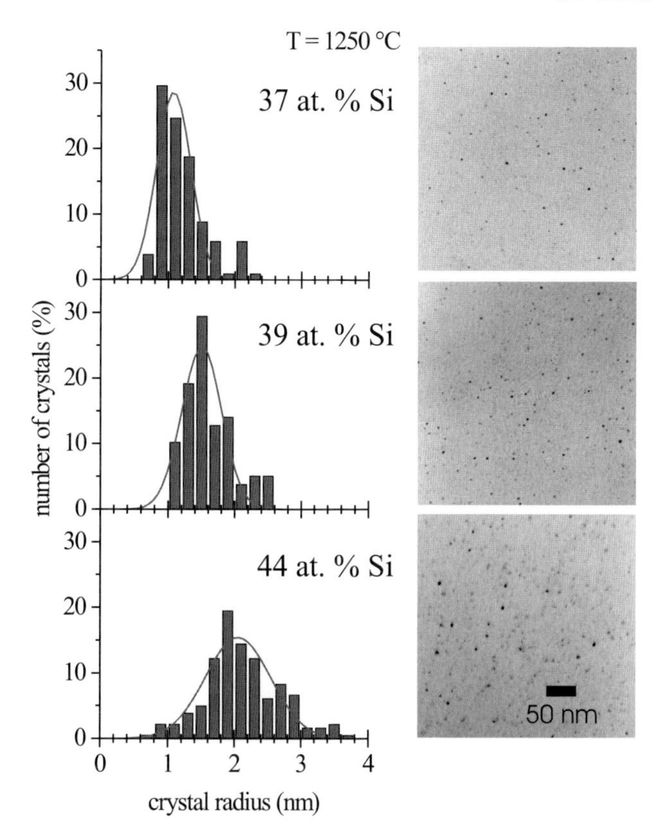

Fig. 4.32. Plan-view TEM micrographs and relative Si nanocrystal size distribution for SiO_x films having Si concentration of (**a**) 37 at.%, (**b**) 39 at.%, and (**c**) 44 at.%, after annealing at 1250 °C. After [132]

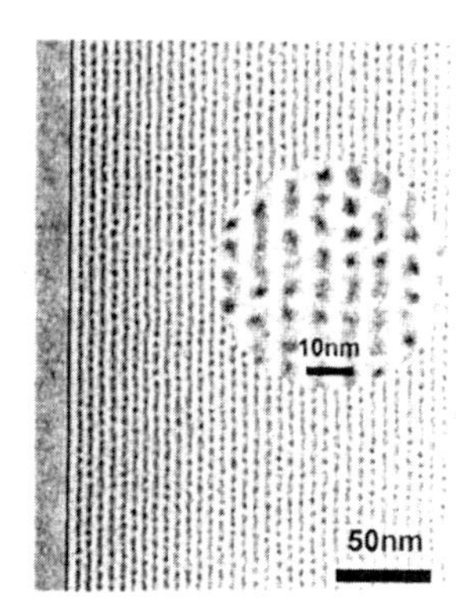

Fig. 4.33. Si/SiO_2 superlattice after annealing. The formation of an ordered array of Si quantum dots is clearly visible. After [170]

sults in ordered arranged silicon nanocrystals with extremely narrow size distributions. The preparation method, which is fully compatible with VLSI technology, enables independent control of particle size as well as of particle density and spatial positioning. An example is shown in Fig. 4.33 where an array of dots after phase separation is shown. This method appears to be extremely powerful and flexible.

4.3.3 Photoluminescence

All of the produced Si nanocrystals emit light at room temperature in the range 700–1100 nm.

As an example in Fig. 4.34 the normalized PL spectra of Si nc obtained by annealing at 1250 °C SiO_x samples having different Si contents are reported. The luminescence signal clearly shows a marked blue shift with decreasing Si content as a result of the smaller size of the Si nc. Indeed, the average nc radius (as observed by TEM) increases from 1.1 to 2.1 nm by increasing the Si content from 37 at.% to 44 at.%.

These data are consistent with a large number of observations supporting the quantum confinement model. The decay time of the emitted radiation is of particular interest since it reflects the confinement properties of the nc. Figure 4.35 shows the decay time of the PL intensity at two fixed detection wavelengths, 700 nm and 950 nm, for different samples after shutting off a 10 mW 488 nm laser beam at room temperature. It is quite interesting to note that the decay time at 700 nm (Fig. 4.35) (i) increases with decreasing Si content, (ii) is characterized by a stretched exponential shape which becomes more and more similar to a single exponential with decreasing Si content.

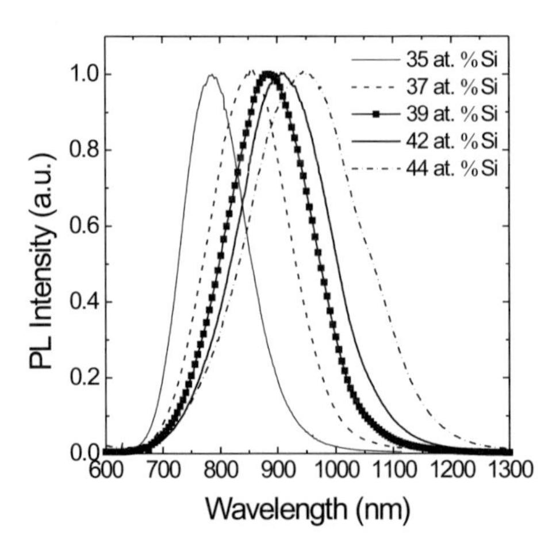

Fig. 4.34. Normalized PL spectra of SiO_x thin films with different silicon concentrations annealed at 1250 °C for 1 h. Spectra were measured at room temperature, with a laser pump power of 10 mW. After [131]

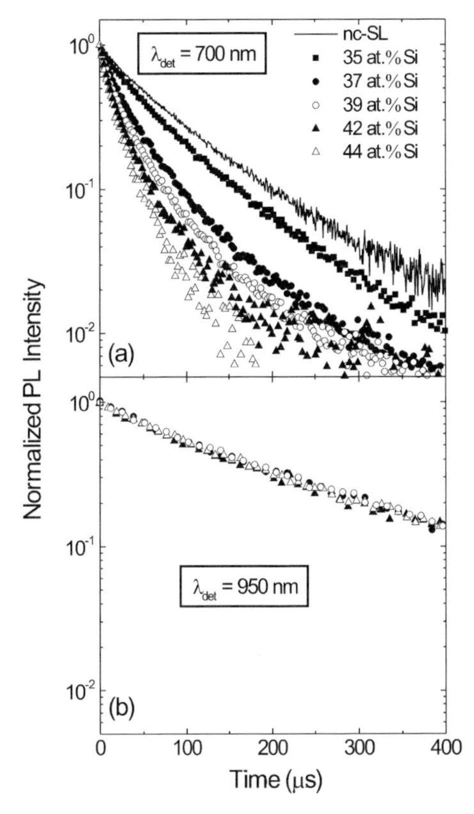

Fig. 4.35. Measurements of the time decay of the PL signal at 700 nm (**a**) and at 950 nm (**b**) for a Si/SiO_2 superlattice (with a Si layer thickness of 0.9 nm) after annealing at 1200 °C for 1 h (continuous line) and for SiO_x samples with different Si contents annealed at 1250 °C for 1 h. Data were taken at room temperature and with a laser pump power of 10 mW. After [131]

Extremely interesting is the behavior of the 950 nm signal decay time, which is characterized by almost single exponentials with the same lifetime of 175 μs for all the silicon contents.

Stretched exponential functions have been widely observed in the literature in the decay time of both porous Si and Si nc [133, 134]. In a stretched exponential the decay line shape is given by:

$$I(t) = I_0 \exp\left[-{}^{t}\!/_{\tau}\right]^{\beta}, \tag{4.2}$$

where $I(t)$ and I_0 are the intensity as a function of time and at $t = 0$, τ is the decay time and β is here a dispersion factor. In general $\beta \leq 1$. The smaller is β the more "stretched" is the exponential. Figure 4.36 reports the quantitative results of the fits to the data in Fig. 4.35a with equation (1). The factor β decreases from 0.85 to 0.63 and τ from 65 μs to 10 μs on increasing

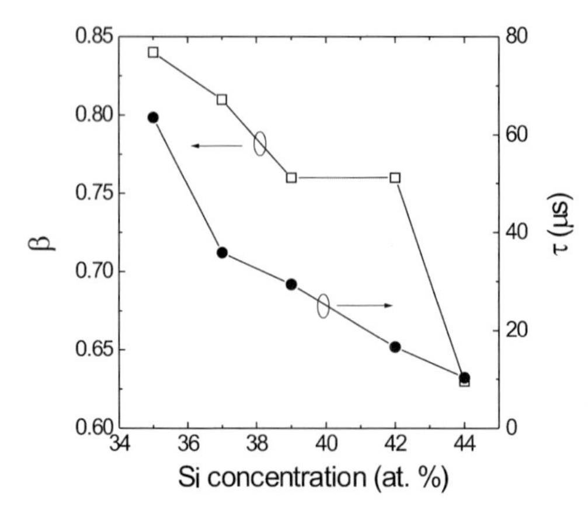

Fig. 4.36. Values of the decay lifetime τ and of the parameter β (as obtained from the fit to the 700 nm luminescence time-decay data reported in Fig. 4.30a) as a function of the Si concentration for SiO_x films annealed at 1250 °C for 1 h. After [131]

the Si content from 35 at.% to 44 at.%. A decrease in β has been associated [133, 134] with a redistribution of the energy within the sample with a transfer from smaller nc (having larger gaps) to larger nc (having smaller gaps). This picture is consistent with the data, showing smaller β and τ values in Si-rich samples in which the nc concentration is higher and hence the energy transfer is more probable. It should be stressed that, since all measurements are performed at the same wavelength they reveal the properties of the same class of nc (i.e. having the same size) embedded within different samples.

The markedly different behavior observed demonstrates that the environment of the nc plays a quite important role in determining its decay time. The more the nc are isolated (larger β) the larger is the decay time τ (since energy transfer becomes less probable). Moreover, larger nc (those emitting at 950 nm, see Fig. 4.35b) cannot transfer their energy to the surrounding nc since their energy is not sufficient due to their smaller band gaps. Therefore they act as "isolated" nc in all systems in the sense that once excited they will re-emit the energy only radiatively. This explains the identical lifetime with all surroundings and the almost single exponential behavior.

An important issue to be investigated is the excitation cross-section of these nc and its dependence on the nc density and on the excitation and detection wavelength. The PL intensity is in general given by:

$$I \propto \frac{N^*}{\tau_{\mathrm{rad}}}, \tag{4.3}$$

where N^* is the concentration of the excited nc and τ_{rad} the radiative lifetime. The rate equation for nc excitation is

$$\frac{dN^*}{dt} = \sigma\phi(N - N^*) - \frac{N^*}{\tau}, \tag{4.4}$$

where σ is the excitation cross-section, ϕ the photon flux, N the total number of nc and τ the decay time, taking into account both radiative and non-radiative processes. If a pumping laser pulse is turned on at $t = 0$, the PL intensity, according to the former equations, will increase with the following law:

$$I(t) = I_0 \left\{ 1 - \exp\left[-\left(\sigma\phi + \frac{1}{\tau} \right) t \right] \right\}, \tag{4.5}$$

with I_0 being the steady state PL intensity. The risetime τ_{on} will hence follow the relationship:

$$\frac{1}{\tau_{on}} = \sigma\phi + \frac{1}{\tau}. \tag{4.6}$$

A measure of the risetime as a function of photon flux ϕ will therefore give direct information on the excitation cross-section. As an example in Fig. 4.37 the PL risetime for Si nc within a superlattice pumped at 488 nm, with pump powers from 1 mW to 60 mW, at 300 K and detected at 776 nm is reported. As predicted by (4.4) and (4.5) the risetime becomes shorter and shorter by increasing pump power. By fitting these risetime curves with (4.4) the values of τ_{on} at the different pump powers are obtained. The reciprocal of τ_{on} is reported in Fig. 4.38 as a function of the photon flux. The data follow a straight line according to (4.5) with a slope $\sigma \sim 1.8 \times 10^{-16}$ cm^2.

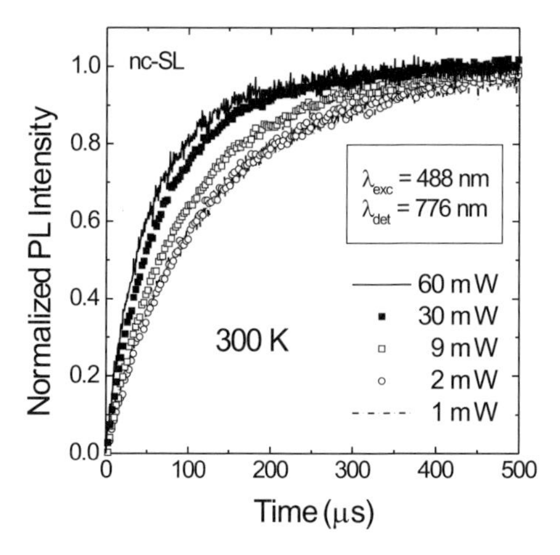

Fig. 4.37. Time resolved PL intensity at 776 nm when switching on the pumping laser at $t = 0$ for a Si nc/SiO_2 superlattice having a Si layer thickness of 0.9 nm and annealed at 1200 °C for 1 h. The excitation wavelength was 488 nm. Data were taken at room temperature and at different pump powers and are normalized to the maximum intensity. After [131]

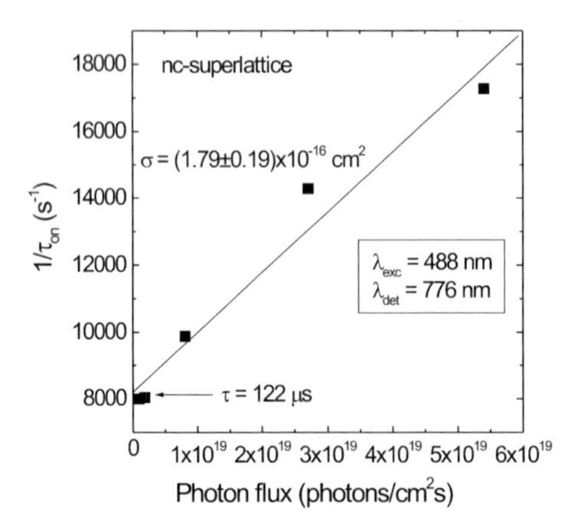

Fig. 4.38. Reciprocal of the τ_{on} time as obtained from the data in Fig. 4.37 as a function of the pump laser photon flux. The slope gives the cross-section for Si nc excitation. After [131]

The intercept of the fitted straight line with the vertical axis gives the lifetime of the Si nc in the system at the measured wavelength. The obtained value (122 μs) is in agreement with decay time measurements at 776 nm on the same sample. Such measurements were performed for several samples and the cross-section values as a function of the detection energy are summarized in Fig. 4.39. Several features are noteworthy. First of all the cross-section is almost constant with detection energy while it shows a quite strong increase with increasing excitation energy. In fact, by increasing the excitation energy to 2.71 eV the cross-section increases to 3.5×10^{-16} cm^2. This is probably due to the increase in the density of states within the nc with increasing energy. Moreover, systems in which Si nc are almost isolated (as in nanocrystalline SL and in the sample with 35 at.% Si) are characterized by very similar cross-section values.

Different behavior is observed in nc in which a quite substantial energy transfer is present. As an example Fig. 4.39 reports the data for a SiO_x sample with 42 at.% silicon. The effective excitation cross-section is much higher than in the previous case, being 8×10^{-16} cm^2 at an excitation energy of 2.54 eV. This strong increase is probably due to the fact that, in the presence of energy transfer, the effective excitation cross-section of each nc is increased since excitation can occur not only through direct photon absorption but also through energy transfer from a nearby nc. A single photon absorption can now excite several nc since the energy is transferred from one nc to the other. On the other hand only one nc per incoming photon is excited at a time since energy transfer requires nonradiative de-excitation from the nc. In fact, a factor of 4 increase in the effective cross-section corresponds to a factor of 4 decrease in the lifetime. This is a clear indication that the increase in

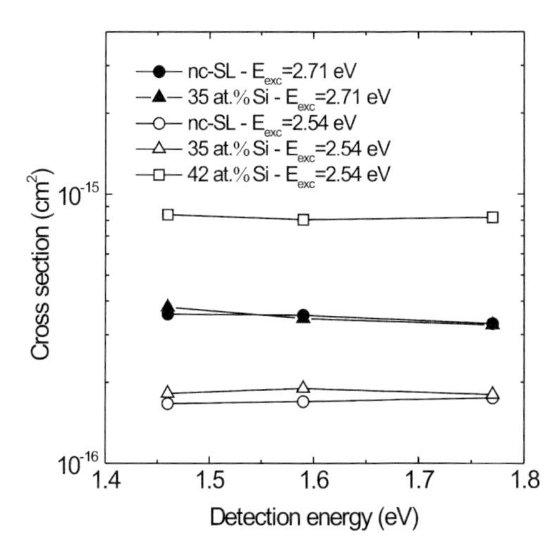

Fig. 4.39. Values of the cross-section for Si nc excitation measured at room temperature as a function of the energy of the detected luminescence signal. Data are reported for both a nc-Si/SiO_2 superlattice with a Si layer thickness of 0.9 nm (*circles*) and for SiO_x films with 35 at. % (*triangles*) and 42 at.% Si (*squares*). The excitation energy was 2.54 eV (*open symbols*) or 2.71 eV (*closed symbols*). After [131]

cross-section and decrease in time decay are different aspects of the same physical process.

Recently, Linnros' group at the Royal Institute of Technology in Stockholm performed single silicon quantum dot spectroscopy [158]. These results are extremely important since one might expect to observe a huge luminescence line narrowing from a single dot. Indeed Fig. 4.40 shows the spectra of three dots. The detection time was 30 min at the excitation intensity of 0.5 W/cm^2 and the spectral resolution was about 10 nm.

The PL spectrum is formed by a single band, which can be well fitted by a single Gaussian peak lying in the range 1.58–1.88 eV (600–785 nm). The full width at half maximum (FWHM) values are in the range 120–210 meV. The bottom part of Fig. 4.40 shows a sum of nine spectra of individual dots measured under identical conditions in one sample. As expected, this ensemble spectrum is significantly broader than individual spectra. The relatively large bandwidth of the PL spectrum (120–210 meV) is common for single nc at room temperature. It might be due to participation of one, two or more phonons in optical transitions (the energy of LO phonon in the Γ point of bulk Si at RT is 64.4 meV). Other effects like the existence of several localized states in one Si nc [172], stress in a Si core, vibrations on the Si-SiO_2 interface, etc., could also contribute to the broadening of the PL spectrum.

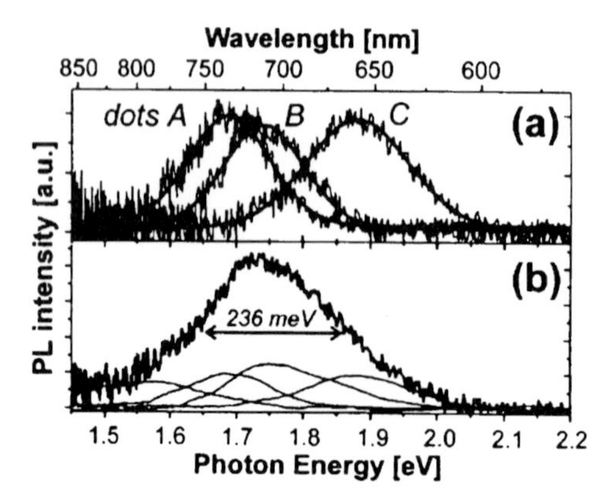

Fig. 4.40. PL spectra of three different Si nc under 325 nm excitation (0.5 W/cm^2) at room temperature (**a**). The *bold smooth lines* are Gaussian fits (FWHM is 122, 120 and 152 meV for dots A, B and C, respectively). Panel (**b**) shows a sum of nine spectra of different single nc from one sample. Four of these individual spectra (smoothed out) are shown under the sum spectrum. After [158]

4.3.4 Temperature Dependence

In the low pump power regime the PL intensity increases linearly with pump power and it is proportional to the ratio τ/τ_{rad}, being given by:

$$I \propto \sigma\phi\frac{\tau}{\tau_{\mathrm{rad}}}N\,, \tag{4.7}$$

One way to understand the de-excitation processes in detail is by studying both time decay and luminescence intensity as a function of temperature. Since N is temperature independent and σ is expected to be also independent of temperature, the only temperature dependencies are due to τ and τ_{rad}. If τ coincides with τ_{rad} one expects the PL intensity to be temperature independent. In contrast, in the presence of nonradiative processes the temperature dependence of the radiative rate ($R = 1/\tau_{\mathrm{rad}}$) can be extracted by the ratio between the PL intensity and the decay time.

The temperature dependence of both PL intensity and decay time are reported in Fig. 4.41 and Fig. 4.42, respectively, for a sample of isolated Si nc obtained through annealing of a SL. Figure 4.41 reports the integrated PL intensity in the temperature range 15–300 K. Data were taken at a low pump power (5 mW) for which the PL intensity increases linearly with power density at all temperatures.

The PL intensity increases with increasing temperature, reaches its maximum value at around 150 K and then slightly decreases. The variation is however weak and the maximum intensity change is $\sim$ 50%. The PL decay data of Fig. 4.42 are also taken at 5 mW and for a fixed wavelength of 790 nm. At all temperatures the decay curves maintain their single exponential feature with the time decay decreasing with increasing temperature. For

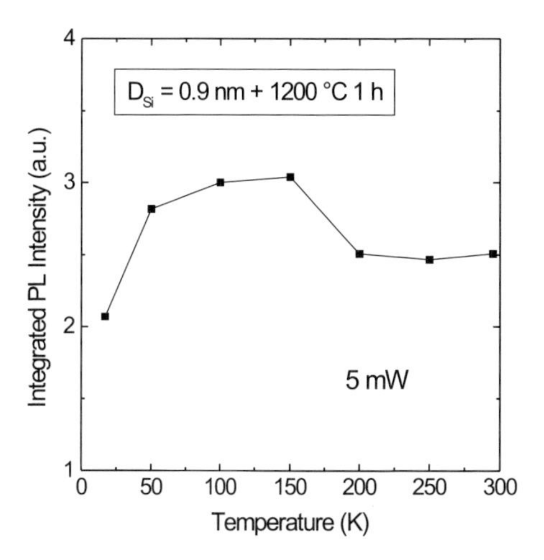

Fig. 4.41. Integrated PL intensity as a function of the temperature for Si/SiO_2 SLs with $D_{Si} = 0.9$ nm after annealing at 1200 °C for 1 h. The laser pump power was 5 mW. After [53]

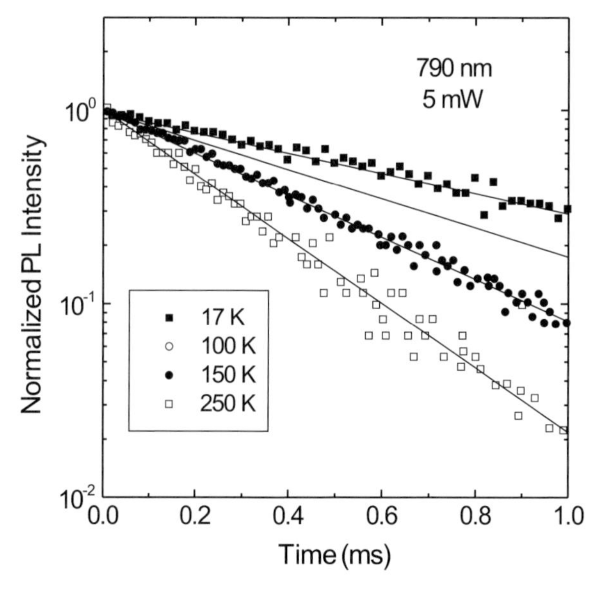

Fig. 4.42. Measurements of the time decay of the PL signal (at 790 nm) for Si/SiO_2 SLs with $D_{Si} = 0.9$ nm after annealing at 1200 °C for 1 h. Data were taken at temperatures between 17 and 250 K and with a laser pump power of 5 mW. The *continuous lines* are single exponential fits to the experimental data with lifetimes of 0.8 ms at 17 K, 0.55 ms at 100 K, 0.4 ms at 150 K and 0.25 ms at 250 K. After [53]

instance, the decay time passes from 0.8 ms at 17 K to 0.25 ms at 250 K, with a variation by only a factor of 3. We can obtain the radiative rate R at each temperature by dividing the PL intensity by the decay time.

The result of such an exercise is shown in Fig. 4.43 with the radiative rate reported in arbitrary units. The radiative rate increases by a factor of 4 on going from 17 K to 300 K. This increase in radiative rate is partially counterbalanced by a simultaneous slight increase in the efficiency of the nonradiative processes as testified by the time decay curves of Fig. 4.42.

The main point at this stage is to identify the physical reasons for the increase in both radiative and nonradiative rates. As far as radiative rates are concerned the behavior can be explained with a model proposed by Calcott et al. [173] for porous silicon and applied by Brongersma [139] to silicon nanocrystals. According to this model the exchange electron–hole interaction splits the excitonic levels by an energy Δ. The lowest level in this splitting is a triplet state and the upper level is a singlet state (inset in Fig. 4.43). The triplet state (threefold degenerate) has a radiative decay rate R_{T} much smaller than the radiative decay rate R_{S} of the singlet. Once excited the excitonic population will be distributed according to a thermal

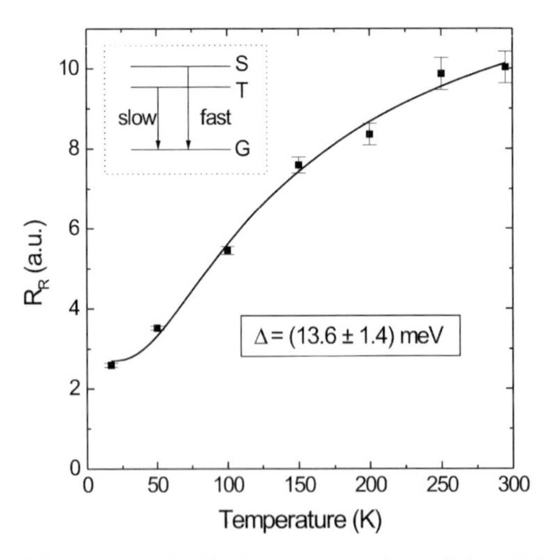

Fig. 4.43. Radiative rate, evaluated by dividing the PL intensity (taken from Fig. 4.41) by the decay time (taken from the fit of Fig. 4.42), as a function of temperature, for Si/SiO_2 SLs with $D_{\mathrm{Si}} = 0.9$ nm after annealing at 1200 °C for 1 h. The experimental data have been fitted according to the model proposed on the different transition rates from the singlet or triplet states to the ground state (see *inset*). The resulting splitting between the two excited levels is 13.6 ± 1.4 meV. After [53]

equilibrium law. Hence at a temperature T the radiative decay rate will be:

$$R_{\mathrm{R}} = \frac{3R_{\mathrm{T}} + R_{\mathrm{S}} \exp\left(-\frac{\Delta}{kT}\right)}{3 + \exp\left(-\frac{\Delta}{kT}\right)} . \tag{4.8}$$

This rate just tells us that, by increasing temperature, the relative population of the singlet state will increase and, since the radiative rate of the singlet state is much higher than that of the triplet state, also the total radiative rate will consequently increase. Indeed the continuous line in Fig. 4.43 is a fit to the data with $\Delta = 13.6 \pm 1.4$ meV, and $R_{\mathrm{T}}/R_{\mathrm{S}} = 0.056 \pm 0.008$ (the absolute values of the two rates cannot be deduced since the experimental values of R_{R} are found in arbitrary units).

A further remark needs to be made. Several studies have recently pointed out that light emission in silicon nanocrystals is not the result of a free exciton recombination, but rather of the recombination of an exciton trapped in surface states at the Si = O interfacial bond [147]. According to this picture excitons generated within a nc are very rapidly (in the ns range) trapped at interfacial states where they decay radiatively in times of the order of some hundreds of microseconds. This picture explains the Stokes shift between emission (at around 800 nm) and absorption (at around 400 nm) in these structures. It is also fully compatible with the temperature dependence results by assuming that the whole discussion refers indeed to trapped excitons.

4.3.5 Optical Gain

The discovery of optical gain in Si nc [165] opened the way towards the possible future fabrication of a silicon laser.

Following the initial observation of optical gain in Si-nc prepared by ion-implantation [165], other works [167, 168, 169] have demonstrated the possibility of stimulated emission in Si-nc in spite of the severe competition with fast nonradiative processes (such as Auger). Although a clear understanding of the microscopic gain mechanism is still under debate, it has been realized that interface radiative states associated to oxygen atoms can play a crucial role in determining the emission properties of the Si-nc systems.

Light amplification has been demonstrated by using the variable stripe length (VSL) method. In the VSL method the sample is optically excited by a laser beam in a stripe-like geometry as shown in the insert of Fig. 4.44.

The amplified spontaneous emission (ASE) signal $\mathrm{I}_{\mathrm{ASE}}$ is collected as a function of the excitation illuminated length l from the edge of the sample, in a 90° configuration with respect to the excitation. As a result of stimulated emission, the spontaneous emitted light is amplified along the amplification axis of the sample (waveguide axis). Referring to a simple one-dimensional amplification model, it is possible to relate the amplified spontaneous emission intensity with the small-signal-modal gain coefficient $\mathrm{g}_{\mathrm{mod}}$ and with the length l of the excited region:

$$I_{\mathrm{ASE}} = \frac{J_{\mathrm{sp}}(\Omega)}{g_{\mathrm{mod}}}(\exp(g_{\mathrm{mod}}l) - 1) \tag{4.9}$$

where J_{sp} is the spontaneous emitted power corresponding to an appropriate emission solid angle Ω. The modal gain coefficient is related to the material gain coefficient g_m through the optical confinement factor Γ of the waveguide.

From a fit of the experimental data with (4.9), the net modal gain coefficient $g_{mod} = \Gamma\, g_m - \alpha$ (where α are the propagation losses) can be deduced for every wavelength within the emission spectrum. Figure 4.44 shows, as an example, the amplified spontaneous emission intensity versus excitation stripe length at a recording wavelength of 800 nm in Si nc.

Clear exponential behavior according to (4.9) is observed demonstrating the presence of optical gain. Indeed, several other observations demonstrate the existence of net gain. For instance Fig. 4.45 shows the observation of strong emission lineshape narrowing either when the pump power density is increased with fixed excitation length l (Fig. 4.45a), or when the excitation length l is increased with a fixed pump power (Fig. 4.45b). When l and the pump power are fixed and the observation angle ϕ is changed (Fig. 4.45c), one measures a significant broadening of the amplified emission spectrum as soon as one departs from the strict one-dimensional amplifier configuration, i.e. when $\phi > 0°$.

The onset of stimulated emission should be accompanied by a reduction in lifetime. Luminescence decays at low pumping fluence J_p are stretched exponentials with observation energy-dependent lifetimes in the microsecond

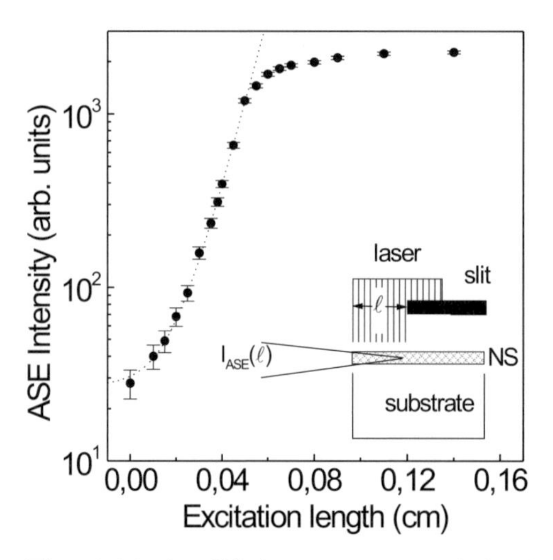

Fig. 4.44. Amplified spontaneous emission intensity at a wavelength of 800 nm versus excitation stripe length (l) for Si nanocrystals embedded in a quartz matrix. A fit to the data with (4.9) is shown as the *dashed line*. The *inset* shows the experimental method. After [132]

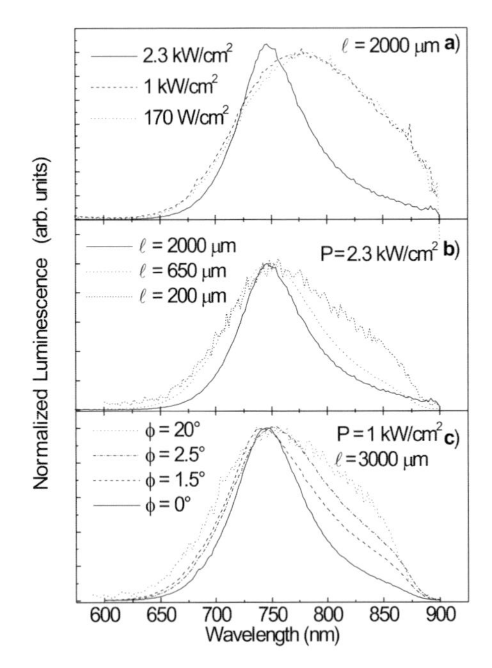

Fig. 4.45. Amplified spontaneous emission (ASE) spectra for different measurement conditions. (**a**) ASE spectra for a constant excitation length $l = 2000$ μm and various power densities P; (**b**) ASE spectra for constant $P = 2.3$ kW/cm^2 and various l; (**c**) ASE spectra for constant $P = 1$ kW/cm^2 and $l = 3000$ μm and various observation angles ϕ. After [165]

range. However at high J_p (Fig. 4.46a) time-resolved measurements in the VSL configuration show a new *fast* decay component (full width at half maximum, FWHM, of about 10 ns) which is superimposed on the usual slow component. Some very peculiar characteristics of the fast component are: (i) it disappears when either the excitation length l is decreased at a fixed J_p (Fig. 4.46b) or when J_p is decreased for a fixed l (Fig. 4.46a); (ii) it shows a superlinear increase vs l for high J_p which can be fitted with the usual one-dimensional amplifier equation yielding a net optical gain; (iii) it shows a threshold behavior in intensity: at low J_p the emission is sublinear to the power 0.5, while for higher J_p population inversion is achieved and a superlinear increase to a power about 3 is observed, suggesting the onset of the stimulated regime; (iv) the lifetime of the fast component signal decreases significantly when the stimulated regime is entered.

The mechanisms underlying optical gain in silicon nanocrystals are still under debate and a definite picture has not yet been achieved. A possible scheme is however depicted in Fig. 4.47.

Ab initio DFT calculations [174] suggest that Si nc in SiO_2 are coated by a 1 nm thick stressed silica shell. This stressed SiO_2 could enhance the for-

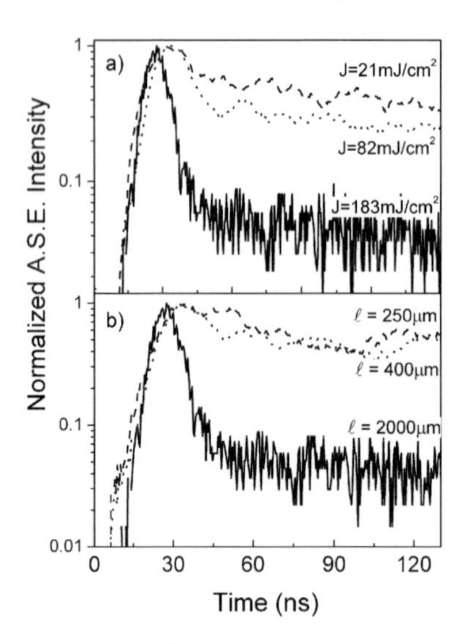

Fig. 4.46. (**a**) Normalized amplified spontaneous emission measured under VSL geometry with a pumping length $l = 2$ mm at different pumping fluences. The measured sample is 42 at% of Si annealed at 1250 °C for 1 h. The excitation wavelength was 355 nm. (**b**) Here the effect of the pumping length l on the fast ASE dynamics is shown. The pumping fluence is fixed at 183 mJ/cm^2 and only the pumping length is varied. After [166]

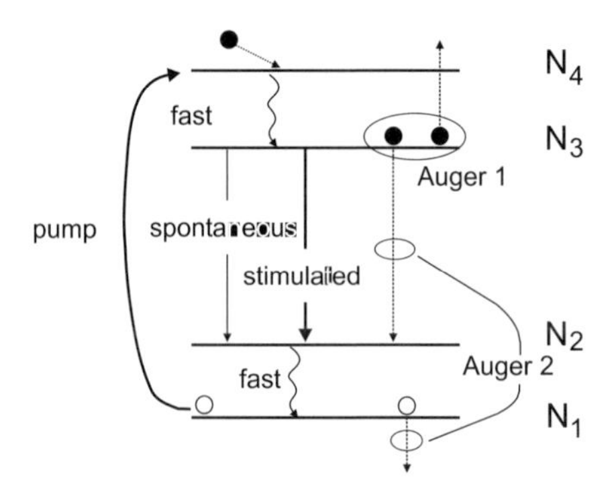

Fig. 4.47. A possible scheme for the mechanisms underlying optical gain in silicon nanocrystals

mation of oxygen-related states, like silanone bonds at the interface between Si nc and SiO_2. The energetics of the silanone-like bond as a function of the Si = O interatomic distances are, for Si nc, typical of a four-level scheme, with different local surface atom rearrangements for the Si-nc ground and excited state configurations, causing a significant Stokes shift between absorption and luminescence (see Chap. 1). Within this scheme, photon excitation induces a strong structural relaxation of small H-saturated and O-saturated nanocrystals leading to new transitions involving surface localized states. In this picture levels 1 and 4 are associated with absorption transitions in the Si-nc ground-state configuration while levels 2 and 3 are associated to the localized states in the excited-state configuration. It remains clear however that accurate theories must still be developed for a detailed explanation of the relevant Si-nc gain physics.

4.3.6 Electroluminescence

The achievement of electroluminescence (EL) in Si quantum dots is somewhat difficult since it requires the right balance between good optical properties and good electrical conduction.

Nevertheless several reports of room temperature EL have been made in MOS-like structures in which within the oxide a high density of Si (or Ge) quantum dots is present [159, 160, 161, 162, 163, 164]. As an example Fig. 4.48 reports the cross-sectional TEM image of a typical device. The layer sequence is clearly visible: from the top, the front metallization layer, the poly-Si film, the SiO_x active region and the Si substrate are distinguishable. The Si nc present in the SiO_x layer are clearly shown as white dots on the dark SiO_2 background.

In order to investigate the charge transport in these systems, the current density-voltage (J-V) characteristics of the devices should be measured. Fig. 4.49 reports the J-V curves of the samples under both forward and reverse bias conditions. In the device with a silicon content of 39 at.% in the SiO_x film (continuous line) a current density of 1.6 $\mu A/cm^2$ is observed for a forward bias voltage of 50 V. By increasing the Si concentration in the SiO_x

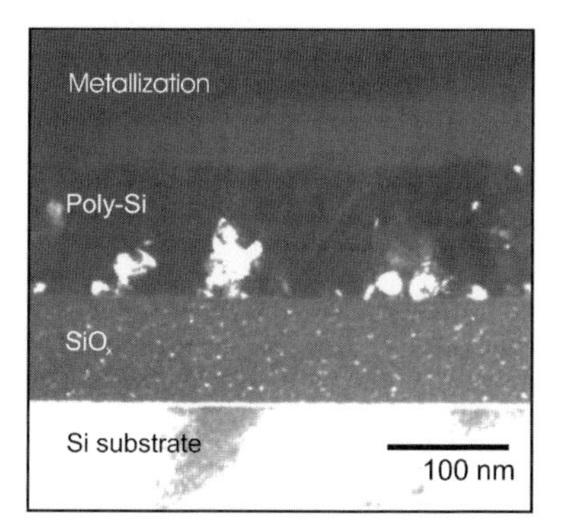

Fig. 4.48. Dark field cross-sectional TEM micrograph of the device having a Si content of 46 at.% in the SiO_x layer. Starting from the *bottom* the Si substrate, the SiO_x active region, the poly-silicon film and the metallization layer are clearly visible. In the SiO_x layer Si nc are clearly shown as bright spots on a dark background. After [159]

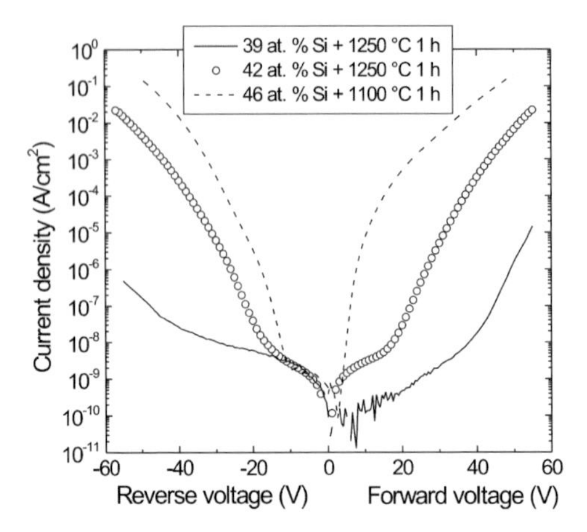

Fig. 4.49. Current density as a function of voltage for Si nc devices. After [161]

layer, a clear increase in the current density that flows through the device is observed. In fact, in the sample with a silicon content of 42 at.% a higher current of 7 mA/cm^2 is measured for the same applied voltage.

Finally, in the device with a silicon content of 46 at.% (dash-dotted line) even higher current densities are observed. For example, with an applied voltage of 50 V a current density of 0.2 A/cm^2 is measured. Lower current densities, although of the same order of magnitude as under forward bias conditions, are measured under reverse bias. From the comparison of the J-V curves we can conclude that by increasing the Si concentration in the SiO_x layer there is a very strong increase in the current that can pass through the device for a fixed applied voltage.

The mechanism of carrier injection through the device may involve two contributions: direct tunneling and Fowler–Nordheim tunneling. For all of the samples direct tunneling is observed at low voltages with current densities that depend on the Si concentration in the SiO_x layer. It is reasonable to think that this mechanism is due to the direct tunneling between nanocrystals that are separated from each other by about 3 nm of SiO_2. By increasing the voltage, the current density increases with the typical exponential behavior of Fowler–Nordheim tunneling. Figure 4.50 shows the room temperature photoluminescence and electroluminescence spectra measured in the device with a silicon content of 42 at. % in the SiO_x layer. The PL spectrum was taken by illuminating the device with a laser pump power of 10 mW and presents a peak centered at 890 nm. By comparing this PL spectrum with that measured in the same annealed SiO_x layer before the poly-Si deposition, a small shift towards higher wavelengths is found.

This is probably due to the absorbance of the emitted light by the polysilicon layer which is stronger for the shorter wavelengths. The electroluminescence spectrum was measured by biasing the device with a voltage of

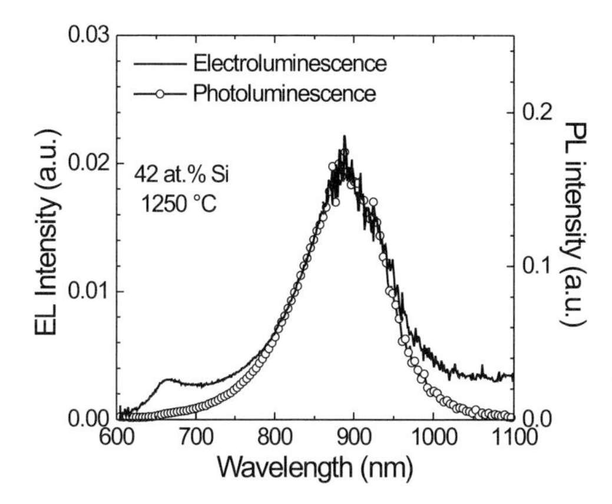

Fig. 4.50. Comparison between the photoluminescence and the electroluminescence spectra of the device with a Si content of 42 at. % in the SiO_x layer. The PL spectrum was measured with a laser pump power of 10 mW. The EL spectrum was measured with a voltage of 48 V and a current density of 4 mA/cm^2. After [159]

48 V. Under these conditions the current density passing through the device is $\sim$ 4 mA/cm^2 and the EL spectrum presents two peaks: a weak one at 660 nm and a most intense one at 890 nm. The EL peak at 660 nm, both for the position and the shape, can be attributed to the presence of defects in the oxide matrix. It is interesting to note that the electroluminescence peak at 890 nm is very similar both in position and shape to the photoluminescence peak measured in the very same sample. It is therefore straightforward to attribute this emission to electron–hole pair recombination in the silicon nanocrystals dispersed in the oxide layer. Indeed, it has been demonstrated that also the temperature dependence in electroluminescence has the same bell shape as in photoluminescence. This is further proof that both EL and PL come from the very same recombination in Si nc. On the other hand the excitation mechanism is quite different. It has been demonstrated [160] that the excitation cross-section of Si nc under EL conditions is about 10^{-14} cm^2. This should compare with a value of about 10^{-16} cm^2 in photoluminescence. Indeed, excitation of Si nc is much more efficient under electroluminescence. It is likely that excitation in EL is due to the impact of energetic electrons with consequent generation of an electron–hole pair within the nc. Though this mechanism appears to be quite efficient the main problems with these devices are: (i) the low current densities passing through the devices, (ii) the high voltages usually required, and (iii) the deterioration of the devices upon operation. Recent advances have in part solved these problems and devices operating at 4 V have been reported [160]. The reliability of these MOS devices is however still an open issue.

It should be added that quite efficient blue-emitting devices have also been fabricated by Ge ion implantation (and subsequent Ge nc formation) in silicon oxide [161]. In these devices emission is indeed coming not from the nc themselves but from defects present within the oxide and excited through electrical pumping. These devices can show external quantum efficiencies up to 0.1%. As in the previous case reliability issues are still under intense investigation.

References

1. J. Kanicki (ed.): *Amorphous and Microcrystalline Semiconductor Devices*. Artech House (1991)
2. D. Kruangam, W. Boonkosum, S. Panyakeow: J. Non-Cryst. Solids **164–166**, 809 (1993)
3. F. Giorgis, C.F. Pirri, C. Vinegoni, L. Pavesi: J. Lumin. **80**, 423 (1999)
4. R. Tsu, A. Filios, C. Lofgren, J. VanNostrand, C.G. Wang: Solid-State Electron. **40**, 221 (1996)
5. R. Tsu, A. Filios, C. Lofgren, K. Dovidenko, C.G. Wang: Electrochem. and Sol. St. Lett. **1**, 80 (1998)
6. Q. Zhang, A. Filios, C. Lofgren, R. Tsu: Physica E **8**, 365 (2000)
7. Y.-J. Seo, R. Tsu: Jpn. J. Appl. Phys. **40**, 4799 (2001)
8. Y.-J. Seo, J.C. Lofgrene, R. Tsu: Appl. Phys. Lett. **79**, 788 (2001)
9. S. Ossicini, A. Fasolino, F. Bernardini: 'The electronic properties of low-dimensional Si structures'. In: *Optical Properties of Low Dimensional Silicon Structure* ed. by D.C. Bensahel, L.T. Canham, S. Ossicini, NATO ASI Series, Vol. 244 (Kluwer Academic Publishers, Dordrecht 1993) pp. 219–228
10. S. Ossicini, A. Fasolino, F. Bernardini: Phys. Rev. Lett. **72**, 1044 (1994)
11. S. Ossicini, A. Fasolino, F. Bernardini: phys. stat. sol (b) **190**, 117 (1995)
12. E. Degoli, S. Ossicini: Phys. Rev. B **57**, 14776 (1998)
13. F. Arnaud d'Avitaya, L. Vervoort, F. Bassani, S. Ossicini, A. Fasolino, F. Bernardini: Europhys. Lett. **31**, 25 (1995)
14. F. Bassani, L. Vervoort, I. Mihalcescu, J.C. Vial, F. Arnaud d'Avitaya: J. Appl. Phys. **79**, 4066 (1996)
15. F. Bassani, I. Mihalcescu, J.C. Vial, F. Arnaud d'Avitaya: Appl. Surf. Sci. **117/118**, 670 (1997)
16. F. Bassani, L. Vervoort, F. Arnaud d'Avitaya: Thin Solid Films **297**,179 (1997)
17. J.C. Vial et al.: Phys. Rev. B **45**, 14171 (1992)
18. F. Bassani, S. Menard, I. Berbezier, F. Arnaud d'Avitaya, I. Mihalcescu: Mat. Sci. Eng. B **69–70**, 340 (2000)
19. M. Watanabe, T. Maruyama, S. Ikeda: J. Lumin. **80**, 253 (1999)
20. A.G. Nassiopoulou et al.: J. Lumin. **80**, 81 (1999)
21. S. Ménard et al.: Mat. Sci. Eng. B **69–70**, 464 (2000)
22. V. Ioannou-Sougleridis, T. Ouisse, A.G. Nassiopoulos, F. Bassani, F. Arnaud d'Avitaya: J. Appl. Phys. **89**, 610 (2001)
23. V. Ioannou-Sougleridis, A.G. Nassiopoulos, T. Ouisse, F. Bassani, F. Arnaud d'Avitaya: Appl. Phys. Lett. **79**, 2076 (2001)
24. A.N. Kholod et al.: J. Appl. Phys. **85**, 7219 (1999)
25. A.N. Kholod, V.E. Borisenko, A. Zaslavski, F. Arnaud d'Avitaya: Phys. Rev. B **60**, 15975 (1999)

26. M. Liniger et al.: Journ. Appl. Phys. **89**, 6281 (2001)
27. Y. Takahashi, T. Furuta, Y. Ono, T. Ishiyama, M. Tabe: Jpn. J. Appl. Phys. **34**, 950 (1995)
28. Y. Kanemitsu, S. Okamoto: Phys. Rev. B **56**, R15561 (1997)
29. S. Nihinyanagi, K. Nishimoto, Y. Kanemitsu: J. Non-Cryst. Sol. **299–302**, 1095 (2002)
30. S. Okamoto, Y. Kanemitsu: Sol. Stat. Comm. **103**, 573 (1997)
31. Z.H. Lu, D.J. Lockwood, J.-M. Baribeau: Nature **378**, 258 (1995)
32. D.J. Lockwood, Z.H. Lu, J.-M. Baribeau: Phys. Rev. Lett. **76**, 539 (1996)
33. S.V. Novikov, J. Sinkkonen, O. Kilpelä, S.V. Gastev: J. Vac. Sci. Technol. B **15**, 1471 (1997)
34. J. Keränen, T. Lepistö, L. Ryen, S.V. Novikov, E. Olsson: J. Appl. Phys. **84**, 6827 (1998)
35. Y. Kanemitsu, Y. Fukumishi, M. Iiboshi, S. Okamoto, T. Kushida: Physica E **7**, 456 (2000)
36. Y. Kanemitsu, M. Iiboshi, T. Kushida: Appl. Phys. Lett. **76**, 2200 (2000)
37. Y. Kanemitsu, T. Kushida: Appl. Phys. Lett. **77**, 3550 (2000)
38. B.T. Sullivan, D.J. Lockwood, H.J. Labbé, Z.H. Lu: Appl. Phys. Lett. **69**, 3149 (1996)
39. D.J. Lockwood, J.-M. Baribeau, B.T. Sullivan: J. Vac. Sci. Technol. B **16**, 1707 (1998)
40. B. Averboukh et al.: J. Appl. Phys. **92**, 3564 (2002)
41. L. Tsybeskov et al.: Appl. Phys. Lett. **72**, 43 (1998)
42. L. Tsybeskov et al.: Appl. Phys. Lett. **75**, 2265 (1999)
43. L. Khriachtchev, S. Novikov, O. Kilpelä: J. Appl. Phys. **87**, 7805 (2000)
44. L. Khriachtchev, O. Kilpelä, S. Karirinne, J. Keränen, T. Lepistö: Appl. Phys. Lett. **78**, 323 (2001)
45. Z.H. Lu, D. Grozea: Appl. Phys. Lett. **80**, 255 (2002)
46. V. Mulloni, R. Chierchia, C. Mazzoleni, G. Pucker, L. Pavesi: Phil. Mag. B **80**, 705 (2000)
47. G. Pucker et al.: J. Appl. Phys. **88**, 6044 (2000)
48. G. Pucker, P. Bellutti, M. Cazzanelli, Z. Gaburro, L. Pavesi: Opt. Mat. **17**, 27 (2001)
49. G. Pucker, P. Bellutti, L. Pavesi: Spectrochem. Acta Part A **57**, 2019 (2001)
50. P. Photopoulos, A.G. Nassiopoulos, D.N. Kouvatsos, A. Travlos: Appl. Phys. Lett. **76**, 3588 (2000)
51. P. Photopoulos, A.G. Nassiopoulos: Appl. Phys. Lett. **77**, 1816 (2000)
52. B.V. Kamenev, A.G. Nassiopoulos: J. Appl. Phys. **90**, 5735 (2001)
53. V. Vinciguerra, G. Franzò, F. Priolo, F. Iacona, C. Spinella: J. Appl. Phys. **87**, 8165 (2000)
54. M. Benyoucef, M. Kuball, J.M. Sun, G.Z. Zhong, X.W. Fan: J. Appl. Phys. **89**, 7903 (2001)
55. H. Kageshima, K. Shiraishi: Phys. Rev. Lett. **81**, 5936 (1998)
56. H. Kageshima, K. Shiraishi: Surf. Sci. **438**, 102 (1999)
57. E. Degoli, S. Ossicini: Surf. Sci. **470**, 32 (2000)
58. E. Degoli, S. Ossicini: Opt. Mat. **17**, 95 (2001)
59. P. Carrier, L.J. Lewis, M.W.C. Dharma-Wardana: Phys. Rev. B **64**, 195330 (2001)
60. Z. Gaburro, G. Pucker, P. Bellutti, L. Pavesi: Solid St. Comm. **114**, 33 (2000)

61. L. Heikkilä, T. Kuusela, H.P. Hedman: J. Appl. Phys. **89**, 2179 (2001)
62. A. Irrera et al.: Phys. E, (2003)
63. L. Tsybeskov et al.: Europhys. Lett. **55**, 552 (2001)
64. T. Ouisse, V. Ioannou-Sougleridis, D. Kouvatsos, A.G. Nassiopoulos: J. Phys. D **33**, 2691 (2000)
65. M. Rosini, C. Jacoboni, S. Ossicini: Phys. Rev B **66**, 155332 (2002); Physica E **16**, 455 (2003)
66. Y. Cui, X. Duan, J. Hu, C.M. Lieber: J. Phys. Chem. B **104**, 5213 (2000)
67. S.W. Chung, J.Y. Yu, J.R. Heat: Appl. Phys. Lett. **76**, 2068 (2000)
68. Y. Cui, C.M. Lieber: Science **291**, 851 (2001)
69. D.P. Yu et al.: Phys. Rev. B **59**, R2498 (1999)
70. H.I. Liu, N.I. Maluf, R.F. W. Pease: J. Vac. Sci. Technol. B **10**, 2846 (1992)
71. Y. Wada et al.: J. Vac. Sci. Technol. B **12**, 48 (1994)
72. A.G. Nassiopoulos, S. Grigoropoulos, D. Papadimitriu, E. Gogolides: phys. stat. sol. (b) **190**, 91 (1995)
73. H. Namatsu et al.: J. Vac. Sci. Technol. B **15**, 1688 (1997)
74. A.G. Nassiopoulos, S. Grigoropoulos, E. Gogolides, D. Papadimitriu: Appl. Phys. Lett. **66**, 1114 (1995)
75. W. Chen, H. Ahmed: Appl. Phys. Lett. **63**, 1116 (1993)
76. T. Tada et al.: Microelectr. Eng. **41/42**, 539 (1998); J. Vac. Sci. Technol. B **16**, 3934 (1998)
77. A.P.G. Robinson, R.E. Palmer, T. Tada, T. Kanayama, J.E. Preece: Appl. Phys. Lett. **72**, 1302 (1998)
78. K. Seeger, R.E. Palmer: J.Phys. D **32**, L129 (1999)
79. K. Seeger, R.E. Palmer: Appl. Phys. Lett. **74**, 1627 (1999)
80. A. Wellner, R.E. Palmer, J.G. Zheng, C.J. Kiely, K.W. Kolasinski: J. Appl. Phys. **91**, 3294 (2002)
81. R.P. Wang et al.: Phys. Rev. B **61**, 16827 (2000)
82. T. Ono, H. Saitoh, M. Esashi: Appl. Phys. Lett. **70**, 1852 (1997)
83. R. Hasunuma, T. Komeda, H. Mukaida, H. Tokumoto: J. Vac. Sci. Technol. B **15**, 1437 (1997)
84. A.M. Morales, C.M. Lieber: Science **279**, 208 (1998)
85. Y. Cui, L.J. Lauhon, M.S. Gudiksen, J. Wang, C.M. Lieber: Appl. Phys. Lett. **78**, 2214 (2001)
86. D.P. Yu et al.: Solid State Comm. **105**, 403 (1998)
87. G.W. Zhou, Z. Zhang, Z.G. Bai, S.Q. Feng, D.P. Yu: Appl. Phys. Lett. **73**, 677 (1998)
88. D.P. Yu et al.: Appl. Phys. Lett. **72**, 3458 (1998)
89. H.Z. Zhang et al.: Appl. Phys. Lett. **73**, 3396 (1998)
90. Y.F. Zhang et al.: Appl. Phys. Lett. **72**, 1835 (1998)
91. Y.H. Tang et al.: J. Appl. Phys. **85**, 7981 (1999)
92. Y.H. Tang et al.: J. Vac. Sci. Technol. B **19**, 317 (2001)
93. N. Wang, Y.F. Zhang, Y.H. Tang, C.S. Lee, S.T. Lee: Appl. Phys. Lett. **73**, 3902 (1998)
94. Y.F. Zhang et al.: Appl. Phys. Lett. **75**, 1842 (1999)
95. N. Wang, Y.H. Tang, Y.F. Zhang, C.S. Lee, S.T. Lee: Phys. Rev. B **58**, R16024 (1998)
96. Y.H. Tang et al.: Appl. Phys. Lett. **79**, 1673 (2001)
97. D.D.D. Ma, C.S. Lee, S.T. Lee: Appl. Phys. Lett. **79**, 2468 (2001)

98. Y.H. Tang et al.: Appl. Phys. Lett. **80**, 3709 (2002)
99. J.Z. He et al.: Appl. Phys. Lett. **80**, 1812 (2002)
100. J.L. Liu et al.: J. Vac. Sci. Technol. B **13**, 2137 (1995)
101. Q. Gu, H. Dang, J. Cao, J. Zhao, S. Fan: Appl. Phys. Lett. **76**, 3020 (2000)
102. V. Ovchinnikov, A. Malinin, S. Novikov, C. Tuoviven: Mat. Sci. Eng. B **69–70**, 459 (2000)
103. M.K. Sunkara, S. Shama, R. Miranda, G. Lian, E.C. Dickey: Appl. Phys. Lett. **79**, 1546 (2001)
104. D, Papadimitriu, A.G. Nassiopoulos: J. Appl. Phys. **84**, 1059 (1998)
105. Y.F. Zhang et al.: Phys. Rev. B **61**, 8298 (2000)
106. X.H. Sun et al.: J. Appl. Phys. **90**, 6379 (2001)
107. O. Bisi, C. Bertarini: 'Plasmons in quantum wires'. In: *Silicon-Based Microphotonics: from Basics to Applications.* ed. by O. Bisi, S.U. Campisano, L. Pavesi and F. Priolo (IOS Press, Amsterdam 1999) pp. 261–277
108. A.G. Nassiopoulos, S. Grigoropoulos, D. Papadimitriu: Appl. Phys. Lett. **69**, 2267 (1996); Thin Solid Films **297**, 176 (1997)
109. F.C.K. Au et al.: Appl. Phys. Lett. **75**, 1700 (1999)
110. Y.H. Tang et al.: Appl. Phys. Lett. **79**, 1673 (2001)
111. X.H. Sun et al.: J. Appl. Phys. **89**, 6396 (2001)
112. D.D.D. Ma et al.: Appl. Phys. Lett. **81**, 3233 (2002)
113. V.V. Poborchil, T. Tada, T. Kanayama: J. Appl. Phys. **91**, 3299 (2002)
114. J. Linnros: 'Silicon nanostructures'. In: *Silicon-Based Microphotonics: from Basics to Applications.* ed. by O. Bisi, S.U. Campisano, L. Pavesi and F. Priolo (IOS Press, Amsterdam 1999) pp. 47–86
115. T. Shimizu-Iwayama, K. Fujita, S. Nakao, K. Saitoh, T. Fujita, N. Itoh: J. Appl. Phys. **75**, 7779 (1994)
116. T. Shimizu-Iwayama, S. Nakao, K. Saitoh: Appl. Phys. Lett. **65**, 1814 (1994)
117. P. Mutti, G. Ghislotti, S. Bretoni, L. Bonoldi, G.F. Cerofolini, L. Meda, E. Grilli, M. Guzzi: Appl. Phys. Lett. **66**, 851 (1994)
118. H. Cheong, W. Paul, S.P. Withrow, J.G. Zhu, J.D. Budai, C.W. White, D.M. Hembree: Appl. Phys. Lett. **68**, 87 (1996)
119. G. Ghislotti, B. Nielsen, P. Asoka-Kumar, K.G. Lynn, A. Gambhir, L.F. Di Mauro, C.E. Bottani: J. Appl. Phys. **79**, 8660 (1996)
120. J.G. Zhu, C.W. White, J.D. Budai, S.P. Withrow, Y. Chen: J. Appl. Phys. **78**, 4386 (1995)
121. K.S. Min, K.V. Shcheglov, C.M. Yang, H.A. Atwater, M.L. Brongersma, A. Polman: Appl. Phys. Lett. **69**, 2033 (1996)
122. E. Werwa, A.A. Seraphin, L.A. Chin, Chuxin Zhou, K.D. Kolenbrander: Appl. Phys. Lett. **64**, 1821 (1994)
123. L.N. Dinh, L.L. Chase, M. Balooch, L.J. Terminello, F. Wooten: Appl. Phys. Lett. **65**, 3111 (1994)
124. L. Patrone, D. Nelson, V. Safarov, M. Sentis, W. Marine: J. Lumin. **80**, 217 (1999)
125. G. Ledoux, J. Gong, F. Huisken, O. Guillois, C. Reynaud: Appl. Phys. Lett. **80**, 4834 (2002)
126. H. Morisaki, F.W. Ping, H. Ono, K. Yazawa: J. Appl. Phys. **70**, 1869 (1991)
127. W.L. Wilson, P.F. Szajowski, L.E. Brus: Science **262**, 1242 (1993)
128. K.A. Littau, P.F. Szajowski, A.J. Muller, A. Kortan, L.E. Brus: J. Phys. Chem. **97**, 1224 (1993)

129. S. Hayashi, T. Nagareda, Y. Kanzawa, K. Yamamoto: Jpn. J. Appl. Phys. **32**, 3840 (1993)
130. Y. Kanzawa, T. Kageyama, S. Takeoka, M. Fujii, S. Hayashi, K. Yamamoto: Solid State Comm. **102**, 533 (1997)
131. F. Priolo, G. Franzó, D. Pacifici, V. Vinciguerra, F. Iacona, A. Irrera: J. Appl. Phys. **89**, 264 (2001)
132. F. Iacona, G. Franzó, C. Spinella: J. Appl. Phys. **87**, 1295 (2000)
133. J. Linnros, N. Lalic, A. Galeckas, V. Grivckas: J. Appl. Phys. **291**, 6128 (1999)
134. J. Linnros, A. Galeckas, N. Lalic, V. Grivickas: Thin Solid Films **297**, 167 (1997)
135. H. Takagi, H. Ogawa, Y. Yamazaki, A. Ishizaki, T. Nakagiri: Appl. Phys. Lett. **56**, 2379 (1990)
136. S. Schuppler, S.L. Friedman, M.A. Marcus, D.L. Adler, Y.-H. Xie, F.M. Ross, Y.J. Chabal, T.D. Harris, L.E. Brus, W.L. Brown, E.E. Chaban, P.F. Szajowski, S.B. Christman, P.H. Citrin: Phys. Rev. B **52**, 4910 (1995)
137. L.N. Dinh, L.L. Chase, M. Balooch, W.J. Siekhaus, F. Wooten: Phys. Rev. B **54**, 5029 (1996)
138. H.Z. Song, X.M. Bao: Phys. Rev. B **55**, 6988 (1997)
139. M.L. Brongersma, A. Polman, K.S. Min, E. Boer, T. Tambo, H.A. Atwater: Appl. Phys. Lett. **72**, 2577 (1998)
140. T. Shimizu-Iwayama, N. Kurumado, D.E. Hole, P.D. Townsend: J. Appl. Phys. **83**, 6018 (1998)
141. T. van Buuren, L.N. Dinh, L.L. Chase, W.J. Siekhaus, L.J. Terminello: Phys. Rev. Lett. **80**, 3803 (1998)
142. T. Takagahara, K. Takeda: Phys. Rev. B **46**, 15578 (1992)
143. B. Delley, E.F. Steigmeier: Phys. Rev. B **47**, 1397 (1993)
144. Y. Kanemitsu, T. Ogawa, K. Shiraishi, K. Takega: Phys. Rev. B **48**, 4883 (1993)
145. T. Inohuma, Y. Wakayama, T. Muramoto, R. Aoki, Y. Kurata, S. Hasegawa: J. Appl. Phys. **83**, 2228 (1998)
146. P. Deak, M. Rosenbauer, M. Stutzmann, J. Weber, M.S. Brandt: Phys. Rev. Lett. **69**, 2531 (1992)
147. M.V. Wolkin, J. Jorne, P.M. Fauchet, G. Allan, C. Delerue: Phys. Rev. Lett. **82**, 197 (1999)
148. D. Kovalev, H. Heckler, G. Polisski, F. Koch: Phys. Stat. Sol. (b) **251**, 871 (1999)
149. Y. Kanemitsu, S. Okamoto: Phys. Rev. B **58**, 9652 (1998)
150. V.I. Klimov, Ch. Schwarz, D.W. McBranch, C.W. White: Appl. Phys. Lett. **73**, 2603 (1998)
151. D. Kovalev, J. Diener, H. Heckler, G. Polisski, N. Kunzner, F. Koch: Phys. Rev. B **61**, 4485 (2000)
152. V.R. Nikitenko, Y.-H. Tak, H. Bässler: J. Appl. Phys. **84**, 2334 (1998)
153. V.R. Nikitenko, V.I. Arkhipov, Y.-H. Tak, J. Pommerehne, H. Bässler, H.-H. Hornold: J. Appl. Phys. **81**, 7514 (1997)
154. K.S. Zhuravlev, A. Yu. Kobitski: Semiconductors **34**, 1203 (2000).
155. A. Yu. Kobitski, K.S. Zhuravlev, H.P. Wagner, D.R.T. Zahn: Phys. Rev. B **63**, 115423 (2001)
156. H. Song, X. Bao: Phys. Rev. B **55**, 6988 (1997)
157. F. Gourbilleau, X. Portier, C. Ternon, P. Voivenel, R. Madelon, R. Rizk: Appl. Phys. Lett. **78**, 3058 (2001)

158. J. Valenta, R. Jushasz, J. Linnros: Appl. Phys. Lett. **80**, 1072 (2002)
159. G. Franzó, A. Irrera, E.C. Moreira, M. Miritello, F. Iacona, D. Sanfilippo, G. Di Stefano, P.G. Fallica, F. Priolo: Appl. Phys. A **74**, 1 (2002)
160. A. Irrera, D. Pacifici, M. Miritello, G. Franzó, F. Priolo, F. Iacona, D. Sanfilippo, G. Di Stefano, P. G. Fallica: Appl. Phys. Lett. **81**, 1866 (2002)
161. L. Rebohle, J. Von Borany, R.A. Yankov, W. Skorupa, I.E. Tyschenko, H. Frob, K. Leo: Appl. Phys. Lett. **71**, 2809 (1994)
162. S. Fujita, N. Sugiyama: Appl. Phys. Lett. **74**, 308 (1999)
163. N. Lalic, J. Linnros: J. Lumin. **80**, 75 (1999)
164. P. Photopoulos, A.G. Nassiopoulou: Appl. Phys. Lett. **77**, 1816 (2000)
165. L. Pavesi, L. Dal Negro, C. Mazzoleni, G. Franzó, F. Priolo: Nature **408**, 440 (2000)
166. L. Dal Negro, M. Cazzanelli, N. Daldosso, Z. Gaburro, L. Pavesi, F. Priolo, D. Pacifici, G. Franzó, F. Iacona: Physica E **16**, 297 (2003); Appl. Phys. Lett. **82**, 4636 (2003)
167. L. Khriachtchev, M. Rasanen, S. Novikov, J. Sinkkonen: Appl. Phys. Lett. **79**, 12 (2001)
168. M. Nayfeh, S. Rao, N. Barry, A. Smith, S. Chaieb: Appl. Phys. Lett. **80**, 13 (2002)
169. K. Luterova et al.: J. Appl. Phys. **91**, (2002)
170. M. Zacharias, J. Heitmann, R. Scholz, K. Kahler, M. Schmidt, J. Blasing: Appl. Phys. Lett. **80**, 661 (2002)
171. L.X. Yi, J. Heitman, R. Scholz, M. Zacharias: Appl. Phys. Lett. **81**, 4248 (2002)
172. M. Luppi, S. Ossicini: J. Appl. Phys. **94**, 2130 (2003)
173. P.D.J. Calcott, K.J. Nash, L.T. Canham, M.J. Kane, D. Brumhead: J. Phys. Condens. Matter **5**, L91 (1993)
174. M. Luppi, S. Ossicini: SPIE **4808**, 73 (2002); phys. stat. sol. (a) **197**, 251 (2003)

5 Light Emission of Er^{3+} in Silicon

Erbium doping of Si has recently emerged as a quite promising route towards Si-based optoelectronics [1, 2, 3, 4, 5, 6, 7, 8, 9, 10, 11, 12, 13, 14, 15, 16, 17, 18, 19, 20, 21, 22, 23, 24, 25, 26, 27, 28, 29, 30, 31, 32, 33, 34, 35, 36, 37, 38, 39, 40, 41, 42, 43, 44, 45, 46, 47, 48, 49, 50, 51, 52, 53, 54, 55]. Erbium is a rare earth which, in its 3+ state, is characterized by a radiative intra $4f$ shell atomic-like transition emitting photons at 1.54 μm. The energy level scheme of the Er^{3+} ion is depicted in Fig. 5.1.

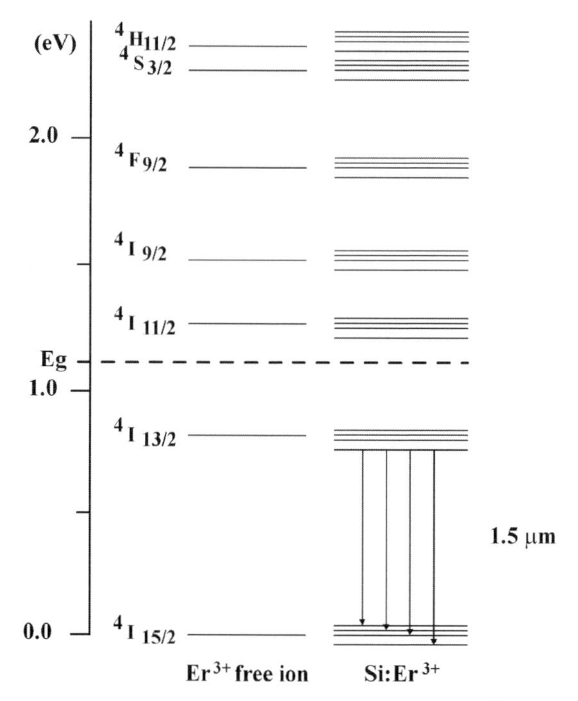

Fig. 5.1. Energy diagram of $4f$ levels in Er^{3+}. On the *left-hand side* the diagram for the free ion is shown, while on the *right-hand side* the splitting due to the interaction with the crystalline field in a solid is evidenced. Transitions at around 1.54 μm between the levels in the first excited manifold and those in the ground state are shown. Also reported as a *dashed line* is the energy of the silicon band gap

The 4*f* levels in Er^{3+} are degenerate in the free ion and radiative transitions between them are forbidden by the electric dipole selection rule. As soon as Er is introduced within a solid matrix the interaction with the neighboring host atoms produces a crystalline field which partially removes the degeneration by means of the Stark effect. This process also produces a partial mixing among the 4*f* states of different parity and inter shell transitions can now occur. In particular, the transition between states of the first excited multiplet ($^4I_{13/2}$) and the ground state ($^4I_{15/2}$) occurs at a wavelength of about 1.54 μm. This wavelength is almost independent of the host matrix since in the electronic configuration of Er^{3+} ($[Xe]4f^{11}5s^25p^6$) the outer electrons screen the 4*f* shell from the external world. This internal 4*f* shell transition has therefore all the characteristics of an atomic-like transition: almost independent of host material and temperature, and an extremely sharp linewidth (0.01 nm). Moreover, the emission wavelength is strategic in telecommunication technology since it matches a window of maximum transmission for silica optical fibers (see Fig. 5.2).

The introduction of Er impurities in Si would in principle allow electronic excitation of the 4*f* transition through a carrier-mediated process with a subsequent radiative de-excitation. Indeed, the first Er-doped light emitting diode (LED) operating at 1.54 μm and at 77 K was demonstrated in 1985 [9]. However, the achievement of room temperature operation has been hampered for a long time by a strong temperature quenching of the Er luminescence. Important potential applications have driven worldwide an increasing interest in the understanding of the Si:Er system and we have now gained a huge amount of information leading to the fabrication of room temperature operating LEDs [37, 38, 39, 40, 41, 42, 43, 44, 45, 46, 47].

The process of Er luminescence in Si is a phenomenon which involves different but equally important steps. First Er has to be incorporated at high concentrations in Si without the formation of precipitates. Second, it has to

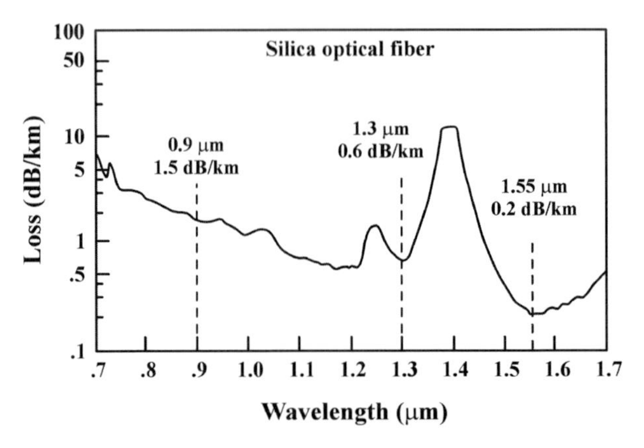

Fig. 5.2. Transmission loss spectrum of silica optical fibers

be incorporated in its optically active 3+ state. Then the electronic system of Si should be in contact with the Er $4f$ shell. This is needed to produce a rate of excitation through electron–hole mediated processes of the levels involved in the transition that is as efficient as possible. In particular the excitation process should be more efficient than competing routes of electron–hole recombination such as Auger recombination and recombination at deep levels. Finally, once excited, Er should decay radiatively. This radiative decay will be in competition with nonradiative de-excitation processes which can be extremely severe due to the long radiative lifetime of Er in Si (2 ms). A diagram of the several processes leading to Er luminescence in Si is reported in Fig. 5.3 and it will be used as a scheme throughout this chapter.

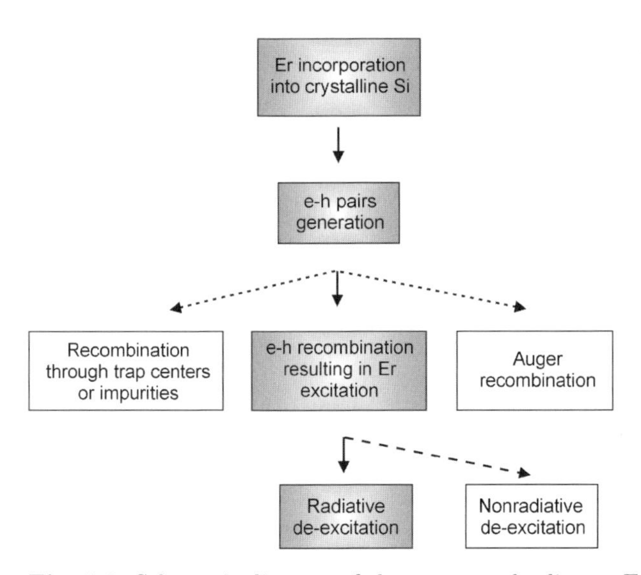

Fig. 5.3. Schematic diagram of the processes leading to Er luminescence in Si

5.1 The Incorporation of Er in Si

Several methods can be used to incorporate Er in Si. Among them we can mention ion implantation, molecular beam epitaxy, chemical vapor deposition, ion beam assisted deposition, and liquid phase epitaxy. Ion implantation is one of the most widespread methods and it is fully compatible with VLSI technology. Once Er has been implanted, however, high-temperature annealing is needed in order to restore crystallinity and activate the Er. This annealing might produce the precipitation of Er into the optically inactive silicide phase [11]. In Fig. 5.4 the precipitate density as measured by transmission electron microscopy (TEM) is reported versus the Er concentration. It is clear that at Er concentrations above $1\times10^{18}/cm^3$ precipitation occurs. Increasing the Er content increases the precipitate density. Therefore the low solid solubility of Er in Si represents a fundamental limit towards high-intensity Er luminescence.

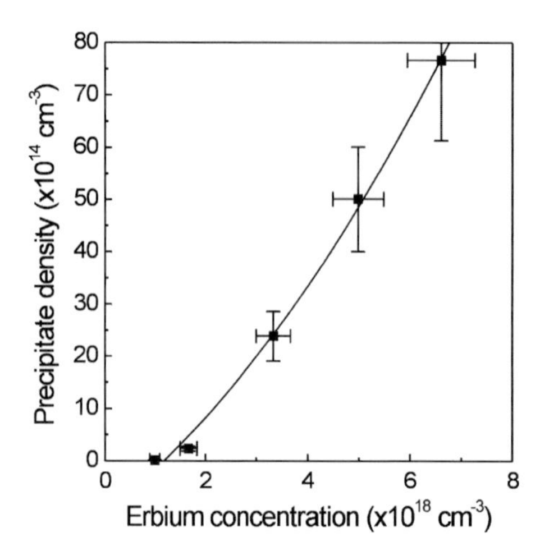

Fig. 5.4. Precipitate density as a function of the erbium concentration. After [11]

A method to incorporate Er in Si is low temperature ion implantation of Er in Si where a thick amorphous Si (a-Si) layer is produced [17]. Then a layer-by-layer epitaxial crystallization is stimulated by thermal treatment at 600°C. As a result of the low solid solubility of Er in Si, however, Er segregation occurs at the recrystallizing interface and Er is partially swept towards the surface during recrystallization. This process is shown in Fig. 5.5 where Er profiles, obtained by Rutherford backscattering spectrometry, are shown at different stages of the annealing process [17]. The *arrows* indicate the position of the crystal–amorphous (c–a) interface for the different annealing times.

In order to achieve the incorporation of a high concentration of electrically and optically active Er atoms in Si a change of the chemical environment around the Er impurity is needed. To this end, it is necessary to co-implant Er and O and obtain almost constant profiles at a ratio 1:10 [19]. Figure 5.6 shows the Er and O chemical profiles as obtained by secondary ion mass spectroscopy (SIMS) taken before and after the annealing process (620 °C, 3 hr) for an O-doped sample. During annealing the O profile remains unmodified at a constant value of $10^{20}/cm^3$ between 0.3 and 1.8 μm. On the other hand the Er profile changes upon annealing. A peak of Er is segregated at $\sim$ 2.3 μm, which is where the end of range defects are probably left. More-

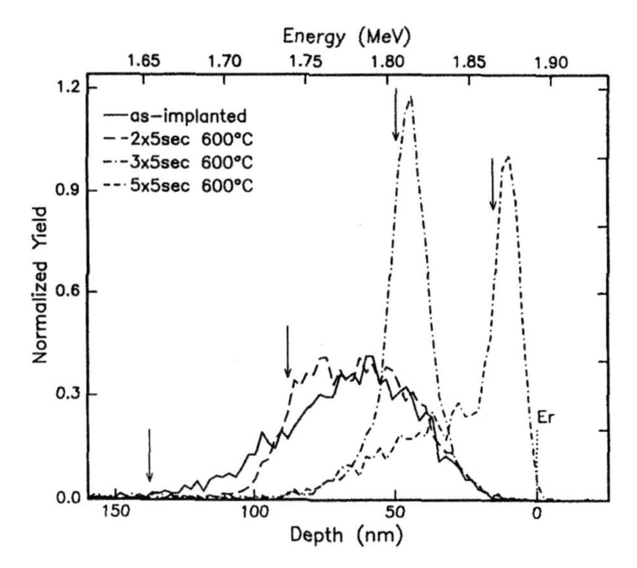

Fig. 5.5. Erbium profiles, as obtained by Rutherford backscattering, prior to and at various stages during annealing at 600 °C. The *arrows* indicate the position of the moving crystal–amorphous interface at the various stages. The segregation process is clearly evident. After [17]

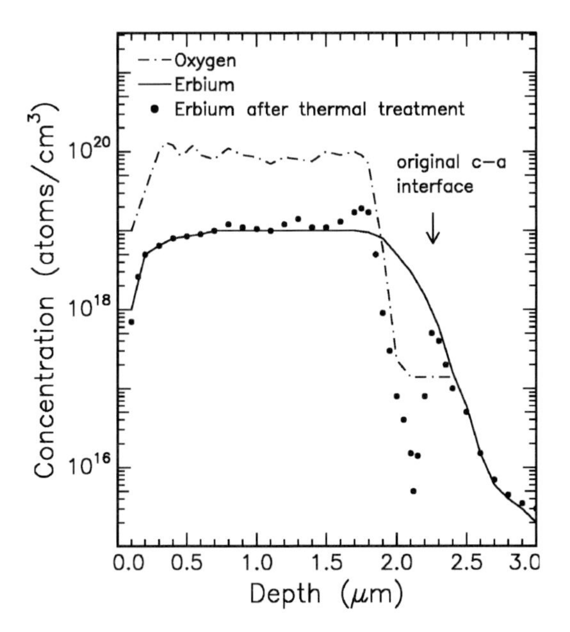

Fig. 5.6. Er and O profiles before and after annealing at 620 °C: 3 hrs for multiple implants in Si. The O profiles are identical both before and after annealing. The position of the original crystal–amorphous interface is also indicated in the figure. After [19]

over, in the region between 2.3 and 1.8 μm Er has been swept by the moving c–a interface. As soon as Er enters the O-doped region this redistribution is however stopped. A clear Er peak is indeed observed at 1.8 μm (which is due to the Er coming from the deeper region) and after that the Er profile is almost unchanged. This redistribution is attributed to segregation processes occurring at the moving c–a interface [17]. While the interface is passing through, Er is partially swept into the amorphous phase and partially incorporated into the crystal, with an Er peak growing at the moving interface. This process depends on both the interface velocity and Er diffusivity. Oxygen is known to retard the c–a interface velocity [56] and to reduce Er diffusivity through the formation of Er-O complexes [23]. Therefore the segregation process abruptly changes when the c–a interface enters the O-doped region and Er remains trapped in the regrowing crystal. More information on the Er-O interaction has been obtained by analyzing the samples prior to and after recrystallization by extended X-ray absorption fine structure (EXAFS) analysis. This analysis provides important information on the chemical environment around Er [14, 34].

An important conclusion arises. In spite of high Er and O contents, no Er-O coordination is observed after implantation and low temperature annealing (450 °C). During epitaxial regrowth of the implanted layer a strong interaction occurs between Er and O with the formation of an O-rich first shell similar to Er_2O_3. On the basis of the atomic diffusivities of O and Er in bulk amorphous and crystalline Si, this interaction would be kinetically inhibited also at 620 °C. It can be speculated therefore that the high atomic mobility at the amorphous–crystal interface during the solid phase epitaxy at 620 °C is needed to start the formation of an Er-O coordination which is then accomplished for all of the Er sites after annealing at 900 °C. Therefore we can state that it is the formation of Er-O complexes (with one Er atom surrounded by six O atoms in a fashion similar to Er_2O_3) that enhances the effective solid solubility of Er in Si. These conclusions have also been confirmed by electron paramagnetic resonance (EPR) analysis [33].

O is not the only co-implantation species able to enhance the effective solid solubility of Er in Si. Similar results have also been reported for F co-implantation with the formation of Er-F complexes. Moreover, processes analogous to those reported above for implanted samples have also been observed with other incorporation techniques. For instance, in the case of molecular beam epitaxy it was shown that Er is segregated towards the surface and cannot be incorporated in the crystal in the absence of O [54]. In contrast, when the layer is grown in the presence of an O over-pressure both Er and O can be incorporated within crystalline Si.

The low solid solubility of Er in Si is a problem for incorporation and can be solved by the use of co-dopants (such as O or F) which, through the formation of Er-impurity complexes, avoid Er precipitation into the optically inactive silicide phase and enhance the effective solid solubility of Er in crystalline Si.

5.2 Electrical Properties of Er in Si

Several studies [13, 14, 23, 28, 55] have shown that Er acts as a donor in Si and its electrical activity depends on the O content. In other words it is the Er-O complexes that introduce donor levels in the Si band-gap. Therefore the electrical activity of Er should critically depend on the O content of the Si substrate and should differ in float zone (FZ) Si (containing 1×10^{16} O/cm^3) and Czochralski (CZ) Si (containing 1×10^{18} O/cm^3). For example in Fig. 5.7 the donor concentration, as obtained by spreading resistance profiling (SRP), for 5 MeV, 1×10^{15} Er/cm^2 implanted in FZ, CZ and oxygen doped Si (at a concentration of 8×10^{19}/cm^3) are compared.

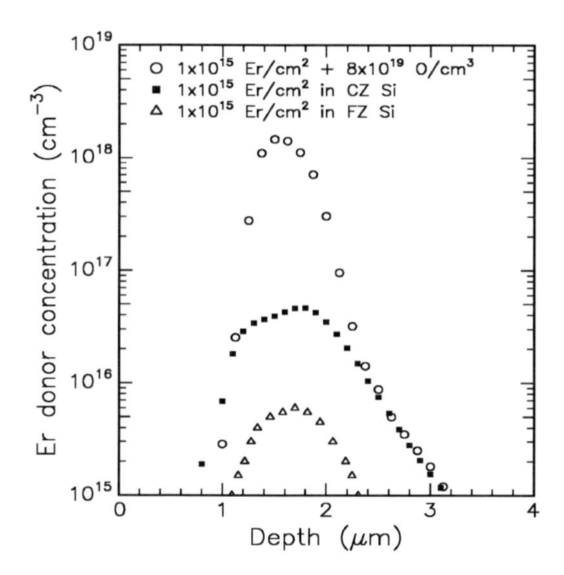

Fig. 5.7. Spreading resistance profiles showing the Er donor concentration for an Er implant of 5 MeV, 1×10^{15}/cm^2 after annealing at 900 °C for 30 s. The profiles refer to FZ Si (*triangles*), CZ Si (*squares*) and FZ Si implanted with 8×10^{19} O/cm^3 (*circles*). After [23]

The maximum donor concentration introduced by Er is 6×10^{15}/cm^3 in FZ Si and increases to 5×10^{16}/cm^3 in CZ Si. A further increase of the O concentration (to 8×10^{19}/cm^3) produces a further strong enhancement in the electrical activation which reaches a peak of 1.5×10^{18}/cm^3. In this case 15% of the implanted Er is electrically active.

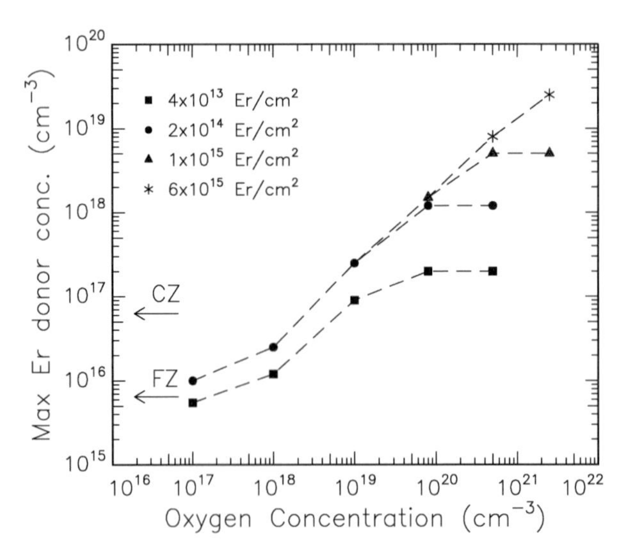

Fig. 5.8. Maximum Er donor concentration, as extracted from spreading resistance profiles, versus oxygen concentration. The data refer to 5 MeV Er implants at different doses. Data for samples implanted at doses of 4×10^{13} Er/cm^2 (*squares*), 2×10^{14} Er/cm^2 (*circles*), 1×10^{15} Er/cm^2 (*triangles*) and 6×10^{15} Er/cm^2 (*stars*) are reported. Data for implants in FZ and CZ Si (without any further intentional O doping) are also reported. After [23]

O plays a key role on the electrical activation of Er. Figure 5.8 summarizes the results of a study aimed at understanding the dependence of the peak donor concentration introduced by the Er atoms as a function of the O concentration. Samples of Er implanted in FZ and CZ Si have donor concentrations of 6×10^{15}/cm^3 and 2–5 $\times 10^{16}$/cm^3 which are nearly independent of the Er dose (above 4×10^{13} Er/cm^2). This means that Er activation is limited by the lack of oxygen atoms. At 4×10^{13} Er/cm^2 by increasing the O content also the donor concentration increases and saturates at a value of 2×10^{17}/cm^3 at an O concentration of 1×10^{20}/cm^3. At this point the Er activation is 50%. A further increase in the O content does not produce any further electrical activation. This means that at this stage electrical activation is limited by the presence of Er atoms. By increasing the Er dose to 2×10^{14} Er/cm^2 the donor concentration increases to 1×10^{18}/cm^3 and saturates again at an O concentration of 5×10^{20}/cm^3. A further increase of the Er dose produces a consequent increase in the donor concentration which reaches a maximum value of 2×10^{19}/cm^3 for an Er dose of 6×10^{15}/cm^2 and an O concentration of 2.5×10^{21}/cm^3. This electrical activation of Er in Si demonstrates that huge amounts of electrically active Er atoms can be incorporated in Si crystals provided that the O concentration is increased as well.

It should be noted, from Fig. 5.8, that in the case of 2×10^{19} Er/cm^3 almost 100 O atoms per Er ion are needed to electrically activate all of the Er. This is not always so. Under optimized conditions [19] it is possible to electrically activate all of the Er with only six O atoms per Er ion. In these conditions, therefore, all of the O atoms are used to link with Er (in a fashion similar to erbium oxide) and no extra O is wandering around in the Si sample.

5.3 Photoluminescence

A typical photoluminescence (PL) spectrum of Er in Si is shown in Fig. 5.9.

A clear peak is present at 1.54 μm with several other small peaks at longer wavelengths. This behavior is typical of transitions between the lowest level in the first excited manifold and the different levels in the ground state manifold. The overall spectrum is almost host independent. However, since the Stark splitting which determines the formation of the manifold is generated by the atoms surrounding Er, high-resolution PL spectra show a small change in shape reflecting a change in chemical surroundings. This is clarified in Fig. 5.10 which shows PL spectra of Er in Er and O co-implanted samples for two different O contents. The shape of the spectra is modified as the O content changes reflecting a change in the Er surroundings. In fact, while at low O contents the peak at 1.538 μm dominates and a peak at 1.556 μm is present; as the O is increased a new peak at 1.534 μm takes over and the peak at 1.556 μm disappears. These changes are typical and reflect the symmetry of the crystalline field around Er.

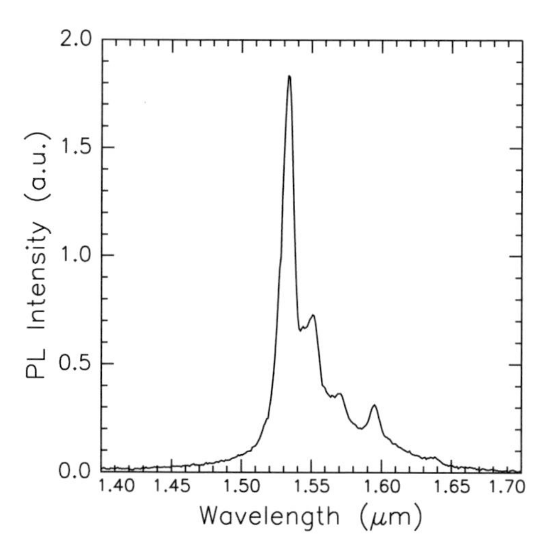

Fig. 5.9. Typical photoluminescence spectrum of Er in crystalline Si

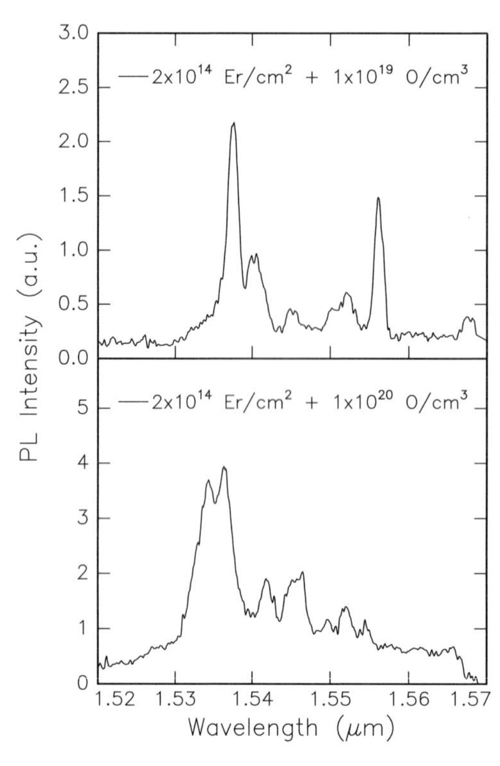

Fig. 5.10. High-resolution photoluminescence spectra of Er implanted in CZ Si at 5 MeV and at a dose of $2 \times 10^{14}/\text{cm}^2$. The spectra are taken at 3 K and at a constant pump power of 120 mW. The two spectra refer to samples having different O contents

5.3.1 Excitation Mechanisms

Excitation spectroscopy data demonstrate that the excitation of Er in Si occurs through carrier mediated processes (Fig. 5.11). A Si sample and a SiO_2 sample have both been doped with Er by ion implantation. The data in Fig. 5.11 show that for the Er doped SiO_2 sample the luminescence intensity at 1.54 μm has a maximum at a pump wavelength of 980 nm and strongly decreases for both shorter and longer wavelengths. This is a clear indication that excitation of Er^{3+} in SiO_2 occurs through the transition from the $^4I_{15/2}$ state to the $^4I_{11/2}$ manifold of Er^{3+}, i.e. a direct transition which exactly matches the 0.98 μm wavelength of the incident photons (see the insert). Subsequently the ion decays nonradiatively to the first excited state ($^4I_{13/2}$) and then to the ground state ($^4I_{15/2}$), emitting a photon at 1.54 μm. As soon as the pumping wavelength is changed out from the resonance condition, the 1.54 μm luminescence strongly reduces. In contrast, the 1.54 μm luminescence of the Er-doped Si sample is almost independent of pump wavelength. This is a clear indication that Er^{3+} is excited through an indirect process:

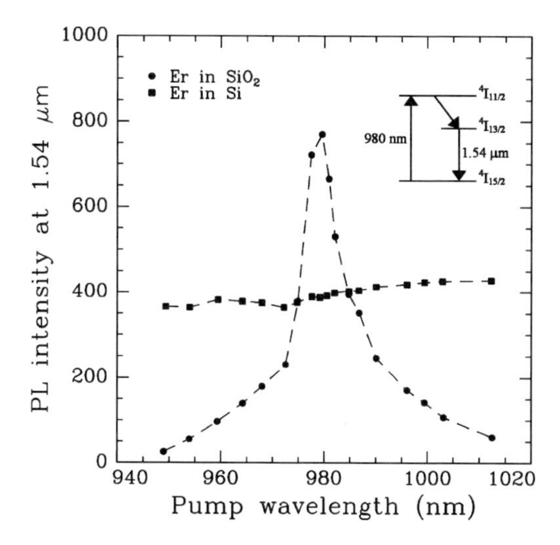

Fig. 5.11. Photoluminescence intensity at 1.54 μm as a function of the wavelength of the pumping laser for Er in SiO_2 (*circles*) and Er in Si (*filled squares*). The Er implantation dose was $\sim 1 \times 10^{15}/\mathrm{cm}^2$. The measurements were done at a fixed temperature of 77 K at a fixed pump power of 70 mW

electron–hole (e–h) pairs are photogenerated in Si and Er excitation is stimulated by an e–h mediated process. This makes the Er:Si system particularly interesting since one can think of applications where exciting carriers are not generated by the light but are injected at a pn junction producing electroluminescence. However, the main limiting factor towards these applications is the strong temperature quenching [12] which reduces the signal by three orders of magnitude on going from 77 K to 300 K. Indeed impurities play a major role in reducing this quenching [24, 29].

The effect of O on temperature quenching is reported in Fig. 5.12. In this figure the PL intensity at 1.54 μm for Er in Si containing different oxygen concentrations is reported as a function of the reciprocal temperature. The data show that the increase of O content produces a large reduction in the temperature quenching. In addition and independently of the O concentration, the PL temperature (T) dependence is characterized by two well-defined regimes. At low T the PL decreases with a small activation energy while at higher T the slope changes and the PL decreases with an activation energy of ~0.15 eV. Oxygen produces a shift towards higher values of T at which the 0.15 eV activated regime begins.

These results can be understood in terms of the major changes that the Er-O interaction produces in the local environment around the Er atoms. In particular, EXAFS [14, 34] and EPR [33] analysis have shown that O is changing the Er site in Si. This change also produces important modifications to the electronic properties of Er-doped Si. The influence of Er+O co-doping

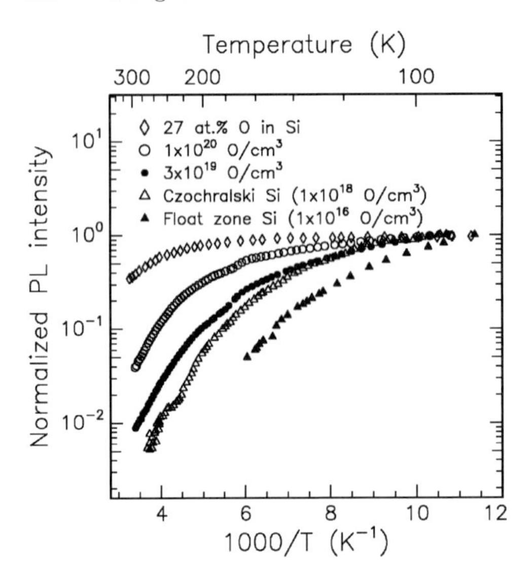

Fig. 5.12. Photoluminescence intensity at 1.54 μm versus reciprocal temperature for Er in Si containing different O contents. The total amount of Er is in all cases $\sim 1 \times 10^{15}/\text{cm}^2$. The O content was varied by implanting Er in float zone (FZ) Si ($\sim 1 \times 10^{16}$ O/cm^3), in Czochralski (CZ) silicon ($\sim 1 \times 10^{18}$ O/cm^3) or by co-implanting O in order to reach a uniform O concentration of $3 \times 10^{19}/\text{cm}^3$ or $1 \times 10^{20}/\text{cm}^3$. Alternatively, Er was implanted in O-rich semi-insulating polycrystalline Si (SIPOS) having an O content of 27 at.%. The data are normalized to the photoluminescence intensity at 77 K

on the electronic properties of Si can be explored by deep level transient spectroscopy (DLTS) analysis [13, 28].

In Fig. 5.13 the DLTS spectra for a solely Er-implanted sample and for Er and O co-implanted samples are reported. The DLTS spectrum of Er alone reveals the presence of four well-separated peaks on a continuous background. These peaks are localized at $E_C - 0.51$ eV (E_1), $E_C - 0.34$ eV (E_2), $E_C - 0.26$ eV (E_3) and $E_C - 0.20$ eV (E_4), respectively. The four different peaks are a signature of different Er configurations and/or Er-defect complexes. O produces major modifications in the deep-level spectra. For the lowest O concentration, the intensities of the deepest levels are slightly reduced while a large change occurs in the shallower level E_4. In particular this level shifts to lower temperatures where a more complex structure starts to develop. These features are more strikingly evident for highest O concentrations where the concentrations of the deepest levels are reduced by about one order of magnitude. Moreover, in this sample, a new well-defined peak at $E_C - 0.15$ eV (E_5) develops and dominates the DLTS spectrum. It should be noted that this level is not due solely to the O implants. In fact the spectrum for a sample only implanted with O (without any Er) is also reported

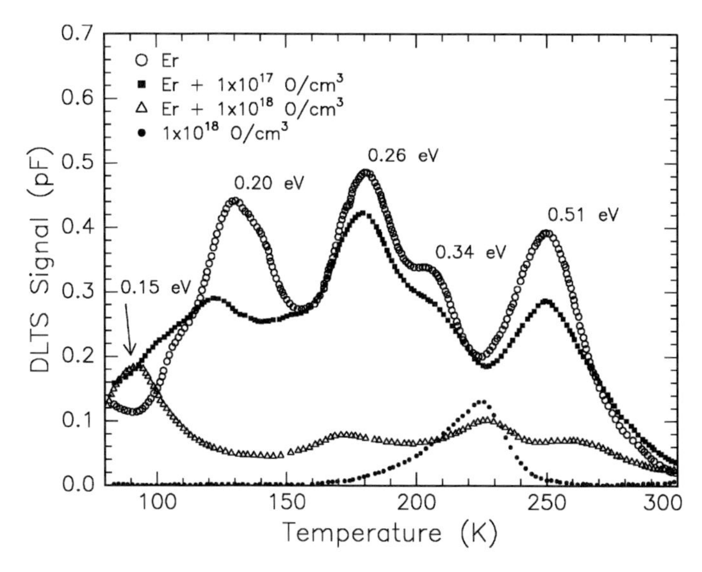

Fig. 5.13. DLTS spectra for the solely Er implanted epitaxial Si (*solid line*) and for samples co-implanted with multiple O implants in order to achieve a concentration of 10^{17} (*dashed-dot line*) or $10^{18}/\mathrm{cm}^3$ (*dashed line*). The spectrum for a sample solely implanted with O (to a concentration of $10^{18}/\mathrm{cm}^3$) is also reported (•). After [28]

and reveals a single peak at $E_\mathrm{C} - 0.34$ eV. Therefore the E_5 peak represents a clear signature of Er+O complexes.

These results have important implications on the excitation of Er^{3+} in Si. In fact it has been shown [26, 27] that efficient pumping of the Er^{3+} can only be achieved if the e-h recombination occurs through an Er-related level in the Si band gap. Indeed, according to the above results the beneficial effect of O co-doping is twofold. First of all, Er-O complexes introduce a well-defined level in the Si band gap which can represent a pathway for e-h recombinations ending up in Er excitation. In particular it should be noted that 0.15 eV (the energy of the level) corresponds to the activation energy for luminescence temperature quenching. It is then plausible to think that carrier thermalization represents the limiting process towards higher luminescence intensities. Moreover, O co-doping reduces detrimental levels in the Si band gap (those reducing carrier lifetime without producing Er excitation) and thus improves luminescence efficiency.

One important issue concerns the dependence of the deep level properties of Er upon thermal annealing. Indeed this has been studied in detail [28] and it has been shown that the two levels which might provide a pathway to Er excitation (the 0.15 eV in O-codoped samples and the 0.20 eV level in pure Si samples, see Fig. 5.13) have very different thermal evolution.

This is illustrated in Fig. 5.14 where the intensity of the 0.20 eV level (in pure Si) and of the 0.15 eV level (in Er+O coimplanted layers) are reported as a function of the annealing temperature [28].

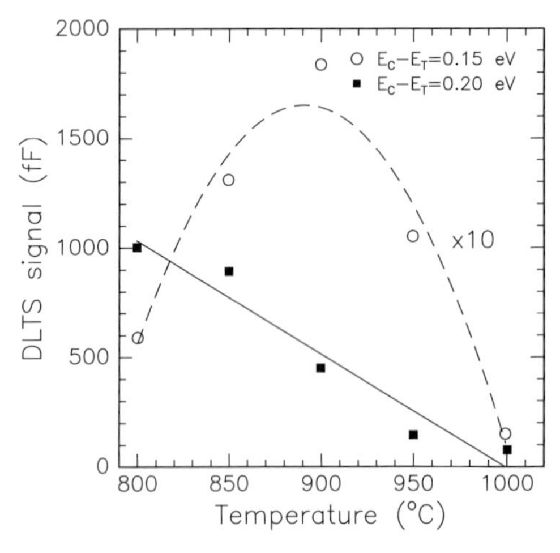

Fig. 5.14. DLTS signal as a function of annealing temperature for the peak at $E_C - 0.20$ eV (Er implanted Si) and for the peak at $E_C - 0.15$ eV (Er and O co-implanted Si). After [28]

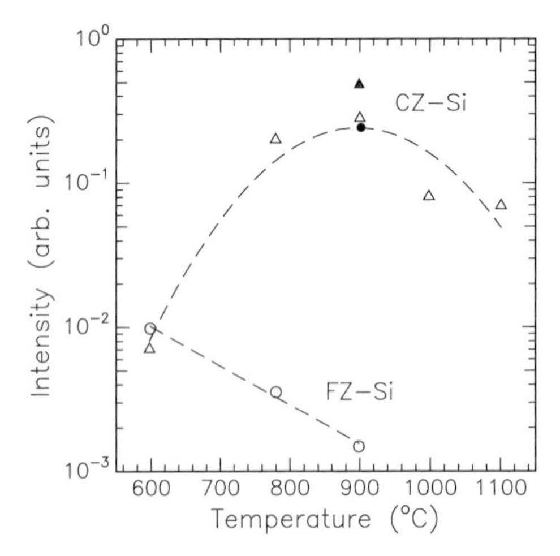

Fig. 5.15. Annealing temperature dependence of the Er luminescence intensity at 4.2 K for FZ and CZ Si. Er was implanted at room temperature at 5.25 MeV. The Er peak concentration was $1 \times 10^{18}/\mathrm{cm}^3$. The annealing treatments were 30 min long. The *filled data points* are from FZ samples with additional oxygen implantation ($1 \times 10^{18}/\mathrm{cm}^3$). After [12]

It is seen that the intensity of the 0.20 eV level monotonically decreases as a function of the annealing temperature (probably due to the annealing of the Er-defect complexes responsible for it). On the other hand the 0.15 eV level first increases, reaches a maximum at 900 °C, and then decreases at higher temperatures. This behavior is probably due to the kinetics of formation and dissociation of Er-O complexes. If these levels are a gateway for the energy transfer from Si to Er similar trends should be observed in the PL intensity as a function of annealing T. This is indeed the case as shown in Fig. 5.15 [12].

5.3.2 Auger Processes

The donor behavior of Er often leads to the introduction of free carriers in the conduction band of Si. Langer and coworkers [57, 58, 59] predicted that an excited impurity within a semiconductor matrix can decay nonradiatively by an Auger process to free carriers. This process is schematically depicted in Fig. 5.16. The energy released by the Er de-excitation is given to the Si matrix to excite a free electron (a) or a free hole (b). When this process is operative, the Er decay lifetime is given by:

$$\frac{1}{\tau} = C_{\mathrm{A}}\, n \tag{5.1}$$

where C_{A} is the Auger coefficient and n the carrier concentration. According to the Langer theory the Auger coefficient in (5.1) has the following expression:

$$C_{\mathrm{A},e} = \frac{1}{\tau_{\mathrm{rad}}} \frac{1}{n_0} \tag{5.2}$$

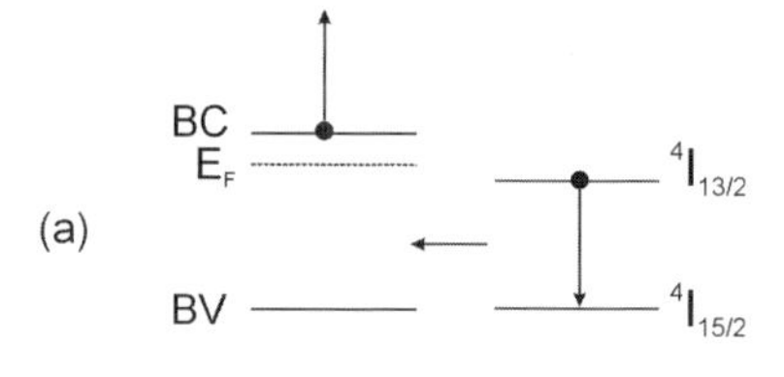

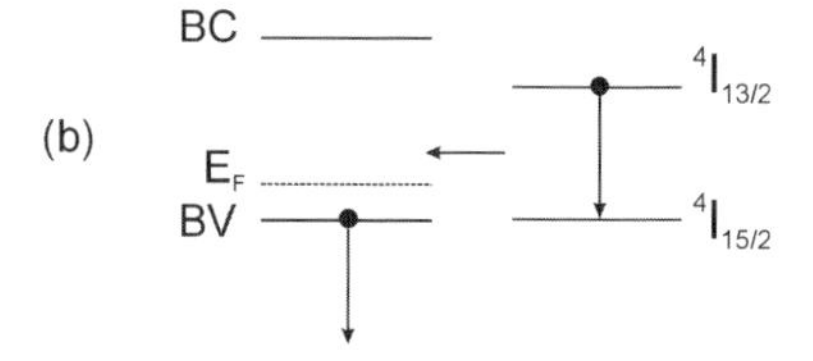

Fig. 5.16. Schematic representation of the Er Auger nonradiative de-excitation process with free electrons (**a**) or free holes (**b**)

where τ_{rad} is the radiative lifetime of the impurity under investigation and n_0 a critical concentration given by:

$$n_0 = 4\pi^{5/2} n_r^5 \left[\frac{m_0}{m^*} 137 a_0\right] \lambda_0^{7/2} \tag{5.3}$$

where n_r is the refractive index of the host matrix, m_0 and m^* are the electron mass and the effective mass, respectively, a_0 is the Bohr radius and λ_0 is the wavelength of the impurity emission. In the case of Er in Si this theory yields $n_0 = 7 \times 10^{14}/\mathrm{cm}^{-3}$ and therefore $C_{\mathrm{A,}e} \sim 1\times10^{-12}\mathrm{cm}^3/\mathrm{s}$. These theoretical predictions have been experimentally tested by directly measuring the decay time of the 1.54 μm luminescence in samples having different carrier concentrations [46].

Following (5.1) Fig. 5.17 reports the experimentally determined reciprocal of the decay time as a function of free carrier concentration for both p-type and n-type Er doped Si. It is interesting to note that all of the data fall into a unique straight line in agreement with (5.1) with a slope C_{A} given by $4.4\times10^{-13}\mathrm{cm}^3/\mathrm{s}$.

These data have several important implications. For example, in a region doped with 10^{19} electrons/cm^3 the lifetime for the Auger process will be only 200 ns to be compared with $\sim$ 2 ms of the radiative lifetime. This means that only 1 in 10 000 excited Er ions will eventually emit a photon. Auger quenching with free carriers therefore represents a severe limit towards the achievement of high efficiencies in the Er:Si system.

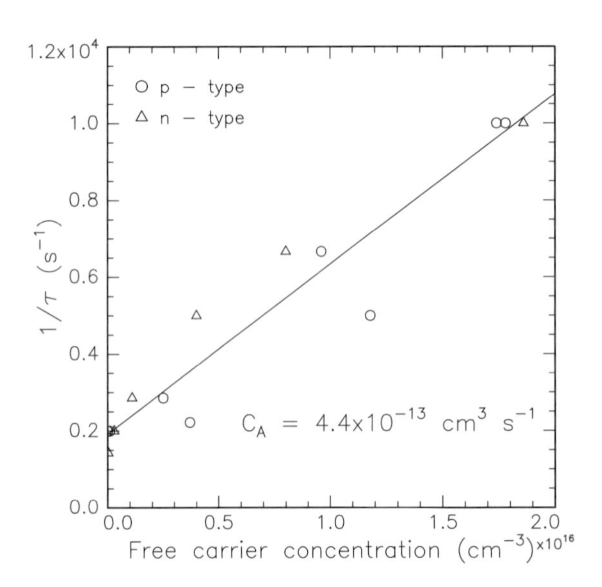

Fig. 5.17. Reciprocal of the time decay of the 1.54 μm signal as a function of free carrier concentration for both p-type (○) and n-type (Δ) Si. Data were taken at 15 K with a laser pump power of 1 mW. From a fit to the data it is possible to estimate the Auger coefficient as $C_{\mathrm{A}} = 4.4\times10^{-13}\mathrm{cm}^3/\mathrm{s}$. After [46]

Another observation is needed. Since Er (particularly in the presence of O) behaves as a donor in Si, the free electron concentration (and hence according to (5.1), the decay lifetime) varies enormously along the Er profile. This implies that in the peak of the Er profile (where the electron concentration is also at its maximum) de-excitation processes play a stronger role than in the tails. Therefore, in a typical Er-doped sample, most of the luminescence signal is likely to come only from those Er atoms present in the tails of the profile (where de-excitation processes are less efficient).

Auger processes demonstrate that there is a strong coupling between free carriers and Er. This on the one hand produces Auger de-excitation of excited Er^{3+}, while on the other suggests that also the reverse process, i.e. impact excitation of the Er from energetic carriers, should be highly probable. Indeed, impact excitation of Er in Si has been observed to occur within reverse biased Er doped LEDs [37, 47] and will be discussed in detail later in this chapter. We can calculate the cross-section for Auger de-excitation as $\sigma_{\mathrm{A}} = C_{\mathrm{A},e}/v_{\mathrm{th}}$, v_{th} being the thermal velocity. We obtain $\sigma_{\mathrm{A}} \sim 5 \times 10^{-20}\mathrm{cm}^2$. It is interesting to note that this value is three orders of magnitude smaller with respect to the measured cross-section for impact excitation of Er in Si ($\sigma_{\mathrm{imp}} \sim 6 \times 10^{-17}\mathrm{cm}^2$ [41]). This is unexpected since the two processes seem just the reverse of one another. On the other hand while de-excitation always involves a transition from the $^4\mathrm{I}_{13/2}$ to the $^4\mathrm{I}_{15/2}$ state, impact excitation with sufficiently energetic electrons can involve transitions from the ground state ($^4\mathrm{I}_{15/2}$) to any of the higher states ($^4\mathrm{I}_{13/2}$, $^4\mathrm{I}_{11/2}$, $^4\mathrm{I}_{9/2}$, etc.), provided that carriers are hot enough. A fast nonradiative de-excitation from the excited state to the first excited state will then occur. The cross-section for impact excitation then represents the integral of the cross-sections of several different transitions. Together with this observation it should be noted that the cross-sections for impact excitation and Auger de-excitation are also proportional to the density of states in the correspondent final carrier configuration. Hence impact excitation is proportional to the density of states close to the bottom of the conduction band while Auger de-excitation is proportional to the density of states ~ 0.8 eV above the bottom of the conduction band. The former is higher than the latter. These observations can explain why $\sigma_{\mathrm{imp}} > \sigma_{\mathrm{A}}$. The implications of this result are enormous. In fact, $\sigma_{\mathrm{imp}} = \sigma_{\mathrm{A}}$ would have implied that population inversion by impact excitation is impossible since very high excitation rates necessarily imply also very high de-excitation rates and, at most, only half of the Er could be excited, thus not allowing (also in principle) laser operation. The fact that $\sigma_{\mathrm{imp}} \gg \sigma_{\mathrm{A}}$, instead, demonstrates that population inversion is achievable and a laser is, in principle, feasible.

5.3.3 Nonradiative De-Excitation Processes

Figure 5.18 shows the PL intensity taken at a pump power of 1 mW as a function of the reciprocal temperature for Er in both n-type Si (4×10^{13} P/cm^3) and p-type Si (3×10^{17} B/cm^3). In the same plot, as closed circles,

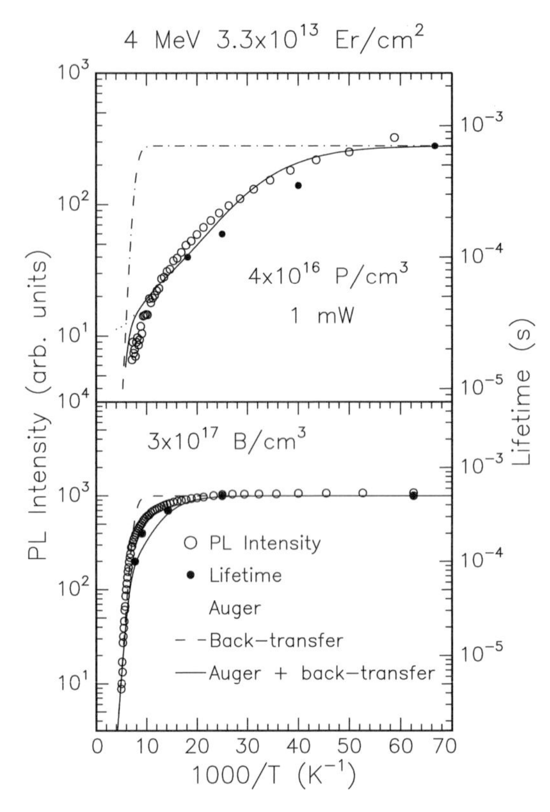

Fig. 5.18. Temperature dependence of the PL intensity at 1.54 µm (○) and of the lifetime (•) for the samples co-implanted with 4 MeV, 3.3×10^{13} Er/cm^2 and 4×10^{16} P/cm^3 (**a**) or 3×10^{17} B/cm^3 (**b**). Data were taken at a pump power of 1 mW. Fits to the data by taking into account nonradiative Auger process (*dotted line*), back-transfer process (*dash-dotted line*) and Auger + back-transfer (*continuous line*) processes are also shown. After [46]

the measured lifetime for both samples (right-hand scale) is reported. In both cases the PL intensity and the associated lifetime are closely related. In n-type Si the PL intensity (and the lifetime) continuously decreases as T is increased from 15 K towards higher values. The lifetime, however, can be measured up to 55 K since above it becomes resolution limited.

In p-type Si the Er luminescence and lifetime are both independent of T in the range from 15 to 40 K. However above 40 K an initially smooth and suddenly strong (above 130 K) quenching occurs. In this latter case both luminescence and lifetime are seen to rapidly decrease with an activation energy of 0.15 eV. This regime is well known and has been attributed to energy back-transfer [24, 36, 45, 46]. In this process the energy is transferred back from excited Er (which decays nonradiatively) to the electronic system of Si. In particular, since Er introduces a level in the Si bandgap at $E_{\mathrm{T}} = 0.15$ eV from the bottom of the conduction band, the back-transfer will consist in bringing an electron from the valence band to this level. However, extra energy is necessary to complete the process, i.e.

$$\Delta E = (E_{\mathrm{G}} - E_{\mathrm{T}}) - E_{\mathrm{Er}} = 0.15\,\mathrm{eV} \tag{5.4}$$

where E_{G} is the Si band gap and $E_{\mathrm{Er}} = 0.8$ eV is the energy of the excited Er atom. Hence, the energy required for the back-transfer process corresponds to the activation energy of the observed quenching. This process severely affects the luminescence yield up to room temperature.

In order to understand the behavior of the data in Fig. 5.18 it should be noted that all of the de-excitation processes directly affect the T dependence of the PL intensity. In fact, the PL intensity is given by:

$$I_{\mathrm{PL}} \propto \frac{N^*_{\mathrm{Er}}}{\tau_{\mathrm{rad}}} \tag{5.5}$$

where N^*_{Er} is the excited E_{r} concentration and τ_{rad} the radiative lifetime.

The excited E_{r} concentration is, in general, obtained from the following rate equation:

$$\frac{\mathrm{d}N^*_{\mathrm{Er}}}{\mathrm{d}t} = \sigma_{\mathrm{eh}}\phi(N_{\mathrm{Er}} - N^*_{\mathrm{Er}}) - N^*_{\mathrm{Er}}\left(\frac{1}{\tau_{\mathrm{bt}}} + C_{\mathrm{A}}\ n + \frac{1}{\tau_0}\right) \tag{5.6}$$

where σ_{eh} is the excitation cross-section through e-h recombinations, ϕ the carrier flux, N_{Er} the total Er concentration, τ_{bt} the lifetime for the back-transfer process, n the electron concentration and τ_0 the lifetime below 15 K due to Auger de-excitations with bound carriers. If the electron concentration introduced by the pumping laser is small, n in (5.6), and hence the Er lifetime, will be independent of ϕ. In this regime we can define:

$$\frac{1}{\tau} = \frac{1}{\tau_{\mathrm{bt}}} + C_{\mathrm{A}}\ n + \frac{1}{\tau_0}\ . \tag{5.7}$$

Under steady state conditions the rate equation (5.6) yields:

$$N^*_{\mathrm{Er}} = \frac{\sigma_{\mathrm{eh}}\phi\tau}{\sigma_{\mathrm{eh}}\phi\tau + 1} N_{\mathrm{Er}} \tag{5.8}$$

and, for small pump powers ($\sigma_{\mathrm{eh}}\phi \ll 1/\tau$):

$$N^*_{\mathrm{Er}} \sim \sigma_{\mathrm{eh}}\phi\tau N_{\mathrm{Er}}\ . \tag{5.9}$$

Therefore, under these conditions, the luminescence intensity is:

$$I_{\mathrm{PL}} \propto \frac{N^*_{\mathrm{Er}}}{\tau_{\mathrm{rad}}} = \sigma_{\mathrm{eh}}\phi N_{\mathrm{Er}}\frac{\tau}{\tau_{\mathrm{rad}}} \tag{5.10}$$

i.e. I_{PL} is proportional to the decay time. Any change in the decay time should then be reflected in the luminescence intensity.

The T dependencies for both p- and n-type Si can therefore be fitted inserting in (5.7) the measured Auger coefficient C_A, the electron (hole) concentration vs T, the measured low T lifetime τ_0 and describing the back-transfer as:

$$\frac{1}{\tau_{bt}} = W_0 e^{\frac{-0.15\text{eV}}{kT}} \tag{5.11}$$

where W_0 (the rate of the back-transfer process) is the only fitting parameter. Fits to the data are also reported in Fig. 5.18 as a continuous line and show very good agreement with the experimental data. The value of W_0 has been determined to be 1×10^9 s^{-1}. In the same figure, for clarity, the effect of only the back-transfer (dot-dashed line) and the Auger (dotted line) processes are reported. It is very clear that, in both p-type and n-type samples, back-transfer alone cannot take into account the low T quenching since it determines a quite sharp decrease starting from T above 130 K. On the other hand, Auger processes, while describing nicely the low T data, result in a saturation at 150 K when most of the shallow levels are ionized and therefore cannot explain the higher T quenching. Only a combination of both processes gives a nice explanation of the data.

A question naturally arises. What is going to happen if we increase the pump power. In this case we may enter a regime in which $\sigma_{eh}\phi > 1/\tau$ and hence we will have:

$$I_{PL} \propto \frac{N^*_{Er}}{\tau_{rad}} \sim \frac{N_{Er}}{\tau_{rad}} . \tag{5.12}$$

Therefore within this regime the PL intensity is not proportional to the Er lifetime and it assumes a constant value. However, since with increasing T, τ decreases, in a measurement of the PL intensity vs $1/T$ at a fixed pump power, one can pass from a regime in which $\sigma_{eh}\phi > 1/\tau$ (at low T) to a regime where $\sigma_{eh}\phi < 1/\tau$ (at high T). The PL intensity vs $1/T$ then reflects two different regimes and special care should be taken in the interpretation of the data. This effect is clearly shown in Fig. 5.19. The two temperature dependencies show completely different behavior. In fact, at 1 mW there is a continuous decrease in intensity as a result of donor ionization and nonradiative Auger de-excitation with free electrons. The PL intensity is then seen to vary by two orders of magnitude between 15 K and 100 K in a fashion similar to the lifetime. In contrast, at 200 mW the PL intensity decreases only by less than a factor of 3 in the same T range. At higher T, however, a strong decrease in the PL intensity occurs (with an activation energy of ~ 0.15 eV) and the intensity is seen to decrease by two orders of magnitude between 100 K and room temperature. This decrease resembles that observed in p-type samples (Fig. 5.18), it has been reported several times in the literature [12, 24, 29, 32, 45, 46] and it seems to be a quite common feature to all Er-doped Si samples, independently of the method of preparation.

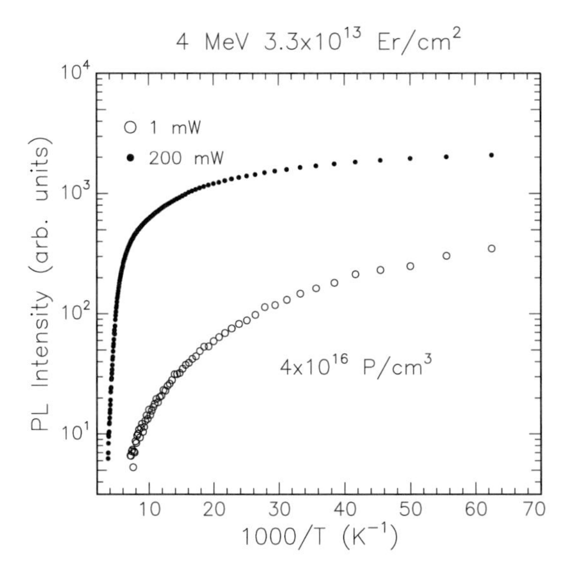

Fig. 5.19. Temperature dependence of the PL intensity at 1.54 μm for the sample co-implanted with 4 MeV, 3.3×10^{13} Er/cm^2 and 4×10^{16} P/cm^3 at two different pump powers: 1 mW (∘) and 200 mW (•). After [46]

The mechanism producing this quenching has been attributed either to a decrease in the excitation cross-section σ_{eh} or in the lifetime τ. Indeed, I_{PL} is predicted to depend linearly on both σ_{eh} and τ (5.10). The following conclusions can be drawn. In p-type samples, where the 0.15 eV Er-related level is empty, τ is responsible for the temperature quenching. The nonradiative de-excitation process is back-transfer. In n-type Si, however, the situation is rather different. In principle back-transfer is expected to be unlikely. In fact in n-type Si the level at 0.15 eV from the conduction band will mostly be full, and hence cannot accommodate an electron coming from the valence band. It has then been proposed [19] that, alternatively, the luminescence quenching with 0.15 eV activation energy might be due to a reduction in σ_{eh} as a result of the thermalization of an electron from the Er-related 0.15 eV level in the bandgap to the conduction band. This process clearly affects the excitation efficiency (being excitation related to the recombination of an electron in this level with a hole in the valence band). Moreover, this process is thermally activated with an activation energy of 0.15 eV.

These two different points of view represent indeed two aspects of a single process. In p-type Si the 0.15 eV level is empty and thus the back-transfer process can occur. However, if the electron remains trapped in the Er-related level, Er excitation will occur again and the back-transfer process would have had no effect. In order for the back-transfer to be completed, thermalization of the electron in the trap level to the conduction band should necessarily occur.

In this case the electron will become de-localized and the excited Er atom will be lost forever. The entire process is depicted in Fig. 5.20.

Indeed, the direct evidence of the back-transfer is all based on the generation of a photocurrent in Er-doped reverse-biased diodes irradiated at 1.54 μm and in Er-doped Si solar cells [36], demonstrating that the back-transfer is always coupled to thermalization. This is best illustrated by the results reported in Fig. 5.21 [36].

Here the spectral response of an Er-doped solar cell is reported as a function of wavelength. Superimposed on a decreasing background which is attributed to absorption by implantation induced defects, a pronounced room temperature erbium absorption spectrum is observed at around 1.54 μm. This represents direct evidence for a back-transfer process in which the photon is absorbed by Er and the energy is subsequently used to generate free carriers within silicon.

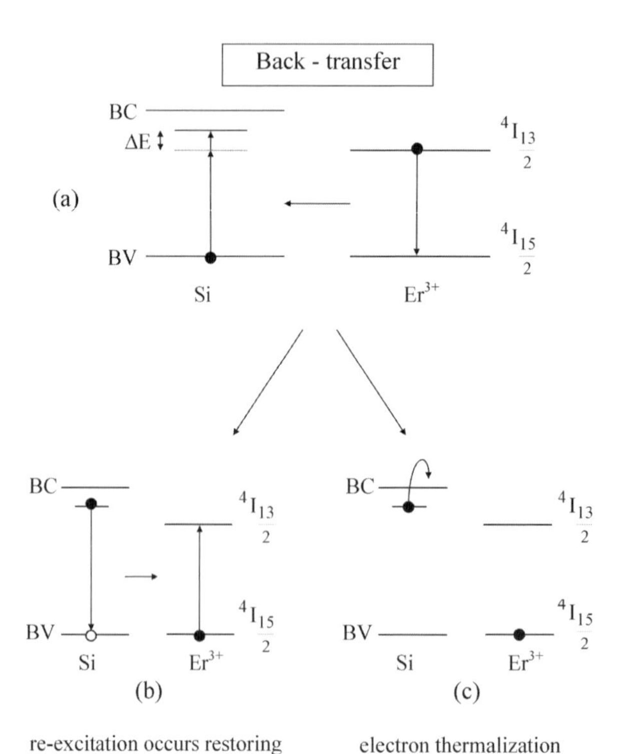

Fig. 5.20. Schematic picture of the back-transfer process (**a**) and of the two possible processes that may occur after Er gives back the energy to the Si matrix: (**b**) the electron in the Er-related level recombines with a hole in the valence band causing the re-excitation of Er; (**c**) the electron in the Er-related level thermalizes in the conduction band and cannot excite Er

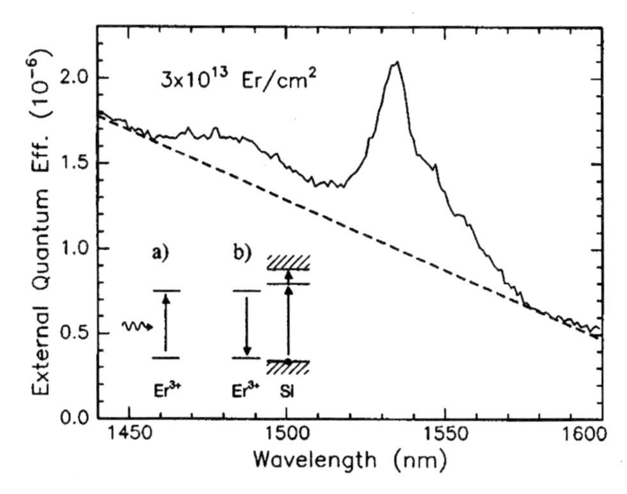

Fig. 5.21. Spectral response of an erbium implanted PERL (passivated emitter, rear locally diffused) solar cell. The *dashed line* is a linear interpolation of the background signal. Schematically indicated are the processes of Er excitation (**a**) and energy back-transfer followed by exciton dissociation (**b**). After [36]

5.3.4 Quantum Efficiency

All of the nonradiative de-excitation processes that have been illustrated in the previous sections tend to decrease the luminescence efficiency for Er in Si. Efficiency is also affected by those processes producing e-h recombinations not ending up in Er excitation. We have seen that increasing T nonradiative de-excitations increase (Auger and back-transfer) and also excitation is affected (electron thermalization). The question is what is the efficiency of the overall process under the best conditions (i.e. low T). We can define three different efficiencies. The first one is the excitation efficiency which is the ratio between the Er excitations occurring per unit area and time and the e-h pairs generated per unit area and time. Therefore this is a monitor of the probability of exciting Er. The second one is the internal quantum efficiency defined as the photons emitted by Er per unit area and time divided by the e-h pairs generated per unit area and time. Therefore this efficiency also takes into account the probability that an Er atom, once excited, decays radiatively. Finally we have the external quantum efficiency taking into account that, due to the difference in refractive indices between Si and air, only a fraction of the emitted photons is able to exit from the sample.

In order to obtain this information the pump power dependence of luminescence yield and time-decay can be measured. The data of luminescence yield versus pump power are reported in Fig. 5.22a for Er in Si at 15 K.

From these measurements the different efficiencies can be derived by making comparison with an Er-implanted SiO_2 sample in which excitation by direct photon absorption occurs with a well-known cross-section of σ_{ph} =

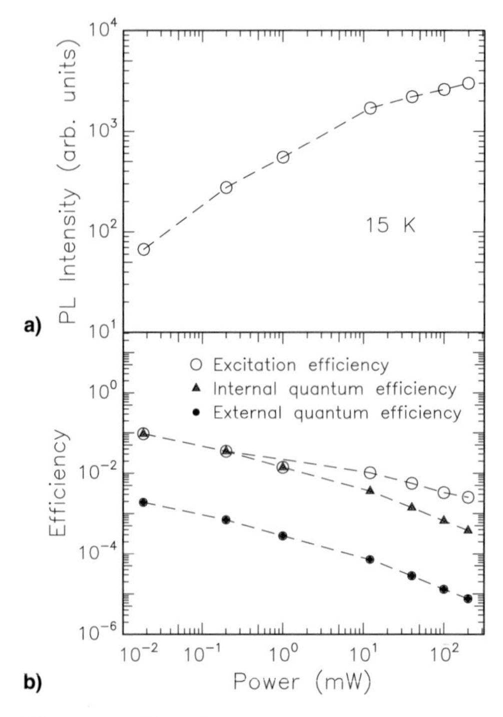

Fig. 5.22. Photoluminescence intensity at 1.54 µm measured at 15 K as a function of pump power for Er doped Si (**a**). The pump power dependence of the excitation efficiency (○), internal quantum efficiency (◇) and external quantum efficiency (•) as extracted from (**a**) are shown in (**b**). After [46]

8×10^{-21}cm^2 [60]. The efficiencies are reported in Fig. 5.22b. It is interesting to note that at low pump powers the excitation efficiency for Er in Si is 10%. This is evidence that at low T and low pump powers competitive phenomena are not so important and a large fraction of e-h pairs produces Er excitation. Moreover, at low pump powers nonradiative decay processes are strongly reduced. If the measured 2 ms decay time corresponds indeed to the radiative lifetime (as we are assuming), then nonradiative processes are totally inhibited and also the internal quantum efficiency is 10%. This is direct proof that Er luminescence in Si can be a very efficient process under appropriate circumstances. Note that this process in Si is orders of magnitude more efficient than in SiO_2. The internal quantum efficiency for Er in SiO_2 at an equivalent Er dose (3.3×10^{13} Er/cm^2) would have been only 2.5×10^{-7}. The very high efficiency of the luminescence of Er in Si is the result of an extremely efficient excitation. Indeed from these data we can estimate the cross-section for excitation through the formula:

$$\eta_{\mathrm{i}} = \sigma_{\mathrm{eh}}\phi_{\mathrm{Er}}\frac{\tau}{\tau_{\mathrm{rad}}} \tag{5.13}$$

η_i being the internal quantum efficiency, σ_{eh} the cross-section for excitation through e-h recombinations mediated by an Er-related level in the gap and ϕ_{Er} the areal density of Er. We obtain $\sigma_{eh} = 3 \times 10^{-15}/cm^2$. The excitation cross-section for Er^{3+} in Si through carrier-mediated recombinations is about seven orders of magnitude higher than that for Er^{3+} in SiO_2 through direct photon absorption.

Figure 5.22 also shows that the efficiencies decrease with increasing pump power. This is not surprising since several competing phenomena set in as the power is increased. First of all, due to the limited amount of Er atoms, e-h recombination routes not producing Er excitations become increasingly more important. Furthermore, as the density of injected carriers increases, non-radiative Auger de-excitations will take place, further decreasing the overall efficiency.

5.4 Electroluminescence

Light emitting devices (LEDs) have been fabricated by taking into account the main problems of Auger and back-transfer nonradiative de-excitations. Though the first Er-doped LED was fabricated in 1985 [9] it took ten years of international effort to achieve room temperature electroluminescence which has now been proven by several groups [37, 38, 39, 40, 41, 42, 43, 44, 47]. State-of-the-art devices have room temperature quantum efficiencies of 5×10^{-4} and can operate at modulation frequencies more than 10 MHz.

A schematic drawing of a typical light emitting Er-doped silicon diode is reported in Fig. 5.23. The device consists of a highly doped pn junction with Er in the active region of the device. Current–voltage characteristics in the T range 100–300 K show that this diode is, under forward bias, always in the recombination regime. Moreover, the reverse bias characteristics are almost temperature independent and present a breakdown voltage of ~ -5 V (due to the Zener effect).

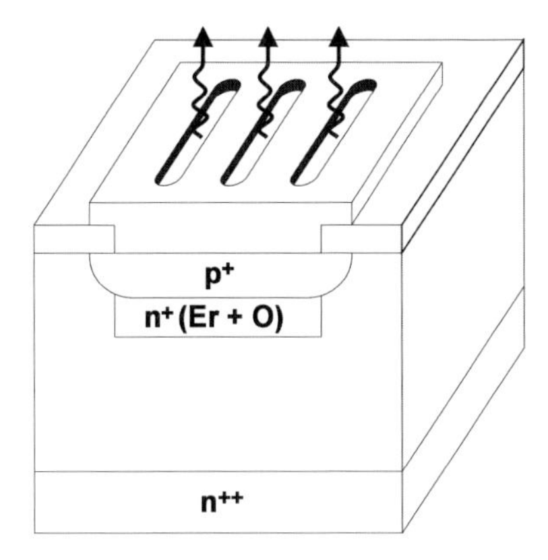

Fig. 5.23. Schematic cross-section of the light emitting silicon diode

Room temperature electroluminescence (EL) spectra of Er+O and Er+F doped silicon diodes are reported in Fig. 5.24.

Under forward bias a weak, but clearly visible, peak at about 1.54 μm is observed both in O- and F-doped samples (with the peak slightly higher in the oxygen case). Interestingly, under reverse bias conditions in the breakdown regime a much more intense EL peak is detected. This peak is about 20 times higher than under forward bias with an identical current density. Moreover, it is five times higher in the O-doped sample with respect to the F-doped one. The shape of the spectra in the two cases (O and F co-doping) is also slightly different suggesting that light emission arises from different Er sites.

Further insight into the mechanisms behind this process can be gained by exploring the EL intensity as a function of T (Fig. 5.25). Several interesting features can be noticed.

First of all the T dependence of the PL intensity and of the forward-bias EL intensity appears interestingly similar. In both cases (and for both Er+O

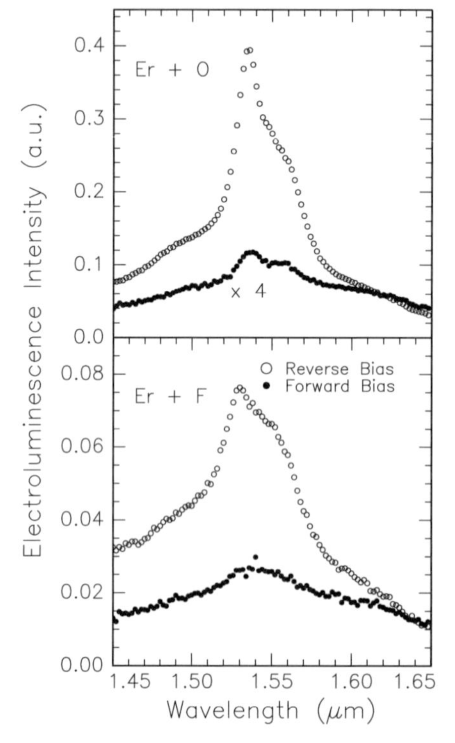

Fig. 5.24. Room temperature electroluminescence spectra under forward bias (•) and reverse bias conditions (○) for the Er+O doped device (*upper part*) and for the Er+F doped device (*lower part*). The spectrum of the Er+O diode under forward bias conditions is multiplied by a factor of 4. The spectra are taken at a constant current density of 2.5 A/cm^2. After [37]

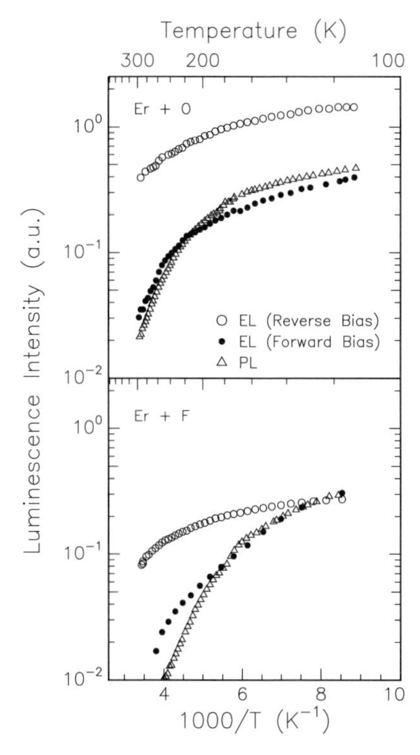

Fig. 5.25. Temperature dependence of the EL intensity under both forward (•) and reverse bias (∘) conditions. Data are taken for both the Er+O doped diode (*upper part*) and for the Er+F doped diode (*lower part*). A constant current density of 2.5 A/cm^2 was used. For comparison the temperature dependence of the PL intensity (D) in the two cases is also reported and is normalized to the forward bias EL value at 110 K. After [37]

and Er+F samples) the luminescence decreases on going from 100 K to 300 K. This quenching is reduced under reverse bias pumping. In this case (and for both O and F co-doping) the EL intensity decreases only by a factor of 4 on going from 100 to 300 K. Thus a strong EL peak is visible even at RT. In particular, in the Er+F diode, the reverse-bias EL is equal to its forward bias counterpart at 100 K but, due to its reduced thermal quenching, at RT the EL is five times higher under reverse bias than under forward bias.

These data give us a hint to the understanding of the Er luminescence in Si. In particular, the similarity in the temperature dependence between PL and forward-bias EL demonstrates that similar excitation mechanisms are operative in these two cases. A different excitation mechanism might be operative for reverse-bias EL. Due to the large doping concentration on the two sides of the junction, the diode under reverse bias has a very thin depletion layer ($\sim$ 70 nm) in which an intense electric field is present ($\sim 1.5 \times 10^6$ V/cm at −5 V). Under these conditions the carriers produced by band-to-band tunneling can be accelerated by the electric field and produce impact excitation of the Er^{3+} ions from the $^4I_{15/2}$ to the $^4I_{13/2}$ level. The energy required for this process is ∼ 0.8 eV. Indeed direct evidence on impact excitation has been provided [47]. It should be noted that the reverse bias operation is particularly efficient since nonradiative Auger de-excitation with free carriers is inhibited within the depletion region. Moreover, since a different excitation mechanism is now operative also Er ions which do not have the 0.15 eV level in the Si bandgap can be excited. For these Er ions energy back-transfer cannot occur and therefore nonradiative de-excitations become less probable. This is the main advantage of reverse bias operation through impact excitation.

In Fig. 5.26 we report the EL yield as a function of the current density. Both under forward and reverse bias the EL intensity first increases linearly with current and then saturates. The data in Fig. 5.26 are normalized to the corresponding saturation value EL_{max} obtained at high current densities, which is assumed to be due to complete excitation of all of the excitable Er ions.

When comparing the non-normalized EL yield in the linear region, reverse bias signal is ∼ 10 times higher than the forward bias counterpart. Moreover the saturation current density under forward bias is much higher and the saturation value is ∼ 1/2 of that observed under reverse bias. These data

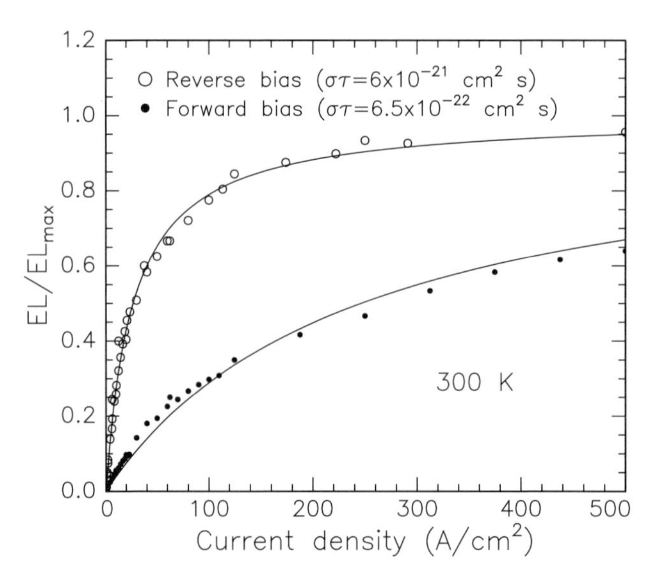

Fig. 5.26. Electroluminescence intensity at 1.54 μm versus current density under forward (•) and reverse (○) bias operation. Data were taken at room temperature and are normalized to the corresponding saturation intensity (EL_{max}). EL_{max} under forward bias is twice as low as under reverse bias. *Continuous lines* are fits to the data. After [41]

demonstrate that forward bias electroluminescence at RT is in general much less efficient than reverse bias EL (the slope in the linear region is smaller) and that the total number of excitable sites is smaller (a lower saturation value is reached).

In order to understand the reasons for this behavior the data should be analyzed in detail. The data in Fig. 5.26 can be described by solving the rate equation which accounts for the excitation and de-excitation processes of Er^{3+}:

$$\frac{dN^*_{Er}}{dt} = \sigma \frac{J}{q} \left(N_{Er} - N^*_{Er}\right) - \frac{N^*_{Er}}{\tau} \tag{5.14}$$

where N_{Er} is the total amount of excitable Er, N^*_{Er} is the excited Er population, σ is the cross-section for excitation, J is the current density passing through the device, q is the electron charge and τ is the lifetime of the excited state $^4I_{13/2}$ taking into account both radiative and nonradiative processes.

At steady state, solving (5.14) and taking into account that the EL intensity is proportional to N^*_{Er}/τ_{rad} where τ_{rad} is the radiative lifetime of Er^{3+}, we obtain:

$$\frac{EL}{EL_{max}} = \frac{\sigma\tau\frac{J}{q}}{\sigma\tau\frac{J}{q} + 1} \tag{5.15}$$

with $EL_{max} \propto N_{Er}/\tau_{rad}$.

From a fit to the data of Fig. 5.26 using (5.15) we obtain a $\sigma\tau$ value of 6×10^{-21} and 6.5×10^{-22} cm^2 s for reverse and forward bias, respectively. The values of $\sigma\tau$ clearly confirm the higher efficiency of the reverse bias condition. However it is not clear whether this is due to a higher efficiency in the excitation process (high σ) or to a reduced probability of nonradiative de-excitation paths (a high τ).

In order to separate the contributions of the cross-section σ and the lifetime τ, it is necessary to have an independent measurement of one of these two parameters. This can be achieved by studying the time evolution of the EL signal at 1.54 μm during diode turn-on and turn-off. In fact the time dependence of the EL yield during diode turn-on is given by:

$$EL(t) = \frac{\sigma\tau\frac{J}{q}}{\sigma\tau\frac{J}{q} + 1} EL_{max} \left\{1 - e^{-(\sigma\frac{J}{q} + \frac{1}{\tau})t}\right\} \tag{5.16}$$

while, during turn-off, is:

$$EL(t) = \frac{\sigma\tau\frac{J}{q}}{\sigma\tau\frac{J}{q} + 1} EL_{max} \left(e^{-\frac{t}{\tau}}\right) . \tag{5.17}$$

Therefore the EL signal from the diode is turning off with a characteristic time τ_{off} that is equal to the decay lifetime τ, while at turn-on the EL signal approaches the steady state with a characteristic time τ_{on} given by:

$$\frac{1}{\tau_{\text{on}}} = \sigma \frac{J}{q} + \frac{1}{\tau} . \tag{5.18}$$

In Fig. 5.27 we report the room temperature modulation of the EL signal at 1.54 μm obtained under reverse and forward bias, by using a current signal that turns the diode on and off at a frequency of 5 kHz.

Under reverse bias, the turn-on of the emission is much slower than the turn-off. In fact a fit to the data can be done by using a τ_{on} of 90 μs and a τ_{off} of 12 μs (the time response of the detector used). In contrast, under forward bias both τ_{on} and τ_{off} are detector limited.

Let us start by analyzing reverse bias operation. As predicted by (5.18), information on σ and τ can be obtained by measuring τ_{on} for different current densities. This can be done under reverse bias since τ_{on} is not detector limited. The RT results are reported in Fig. 5.28 and can be fitted, according to (5.18), using a cross-section of 6×10^{-17}cm^2 and a lifetime of 100 μs at room temperature.

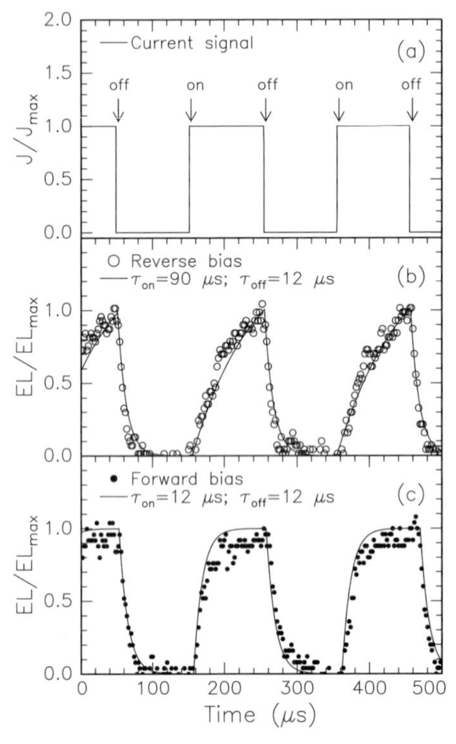

Fig. 5.27. Modulation of the EL intensity with a current signal (**a**) at 5 kHz. The current density is 5.6 A/cm^2 in both cases. Data were taken at room temperature and are shown for both reverse (**b**) and forward (**c**) bias. The *continuous lines* are fits to the data. After [41]

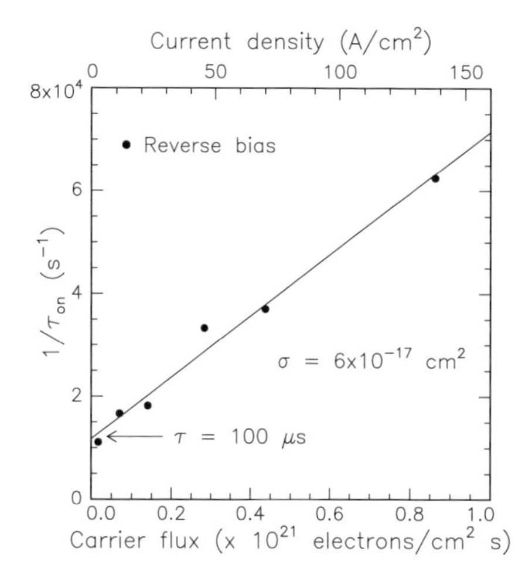

Fig. 5.28. Reciprocal of the risetime ($1/\tau_{\text{on}}$) of the EL signal at diode turn-on as a function of the current density. After [39]

The large value of τ measured during pumping (100 μs at RT) poses the question of why τ_{off} is so short (< 12 μs, see Fig. 5.27). Since on the basis of (5.18) exactly the opposite behavior is expected (i.e. $\tau_{\text{on}} \leq \tau_{\text{off}}$), the results strongly suggest that the Er lifetime is quite different during pumping and at the turn-off.

The explanation for this behavior is schematically depicted in Fig. 5.29. When the diode is under reverse bias in the breakdown regime (-5.5 V), e-h pairs are preferentially formed by band to band tunneling in the center of the depletion region where the electric field is at its maximum ($\sim 1 \times 10^6$ V/cm). Electrons and holes will be accelerated in opposite directions and travel through only half of the depletion layer, gaining energy during this trip. Of course only close to the edge of the depletion region, where the energy of hot carriers is at its maximum, are carriers sufficiently energetic (> 0.8 eV) to excite Er by impact. Therefore most of the Er ions are excited within the depletion region but, preferentially, at the edges of it. When the diode is turned off the depletion region shrinks and the excited Er ions are suddenly embedded in a heavily doped ($\sim 1 \times 10^{19}$ carriers/cm^3) region.

In the presence of this large concentration of free carriers, fast nonradiative Auger de-excitation of the Er^{3+}, with the energy released to free electrons, sets in and produces a drastic reduction in the decay time of the Er ions. Using the value of the Auger coefficient for this process (1×10^{-12}cm^3/s), a decay lifetime of 100 ns is expected in a region which contains 1×10^{19}/cm^3 free electrons. Of course this nonradiative process is fully inhibited during pumping which occurs in the depletion region. This mechanism allows one to achieve modulation of the diodes at high frequency (see Fig. 5.30). At low cur-

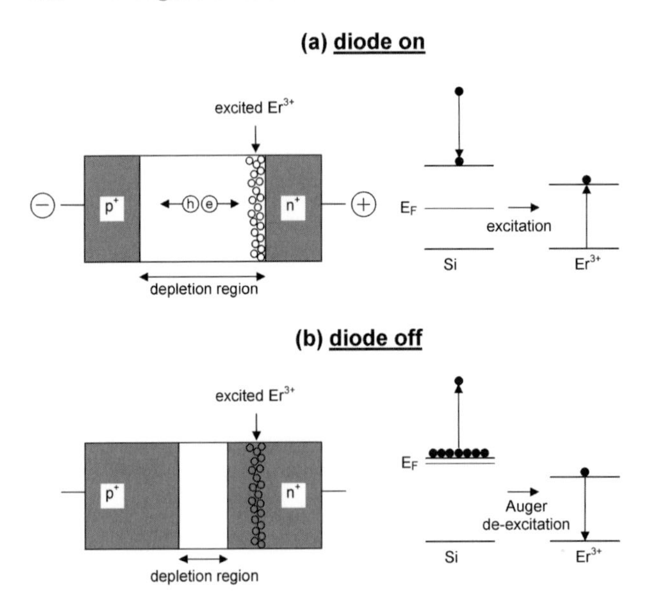

Fig. 5.29. Mechanism of reverse bias operation. When the diode is on (**a**) Er is excited by impact of the hot carriers produced by tunneling in the center of the depletion region. However, only at the edges of the depletion layer are carriers energetic enough to excite Er. At the diode turn off (**b**) the depletion layer shrinks and the excited Er atoms are embedded in a heavily doped region. Fast nonradiative Auger de-excitation to free carriers occurs. After [41]

rent densities the EL signal decreases faster as the frequency is increased due to a long τ_{on} which, at 5.6 A/cm^2, is 90 ms. However, as soon as the current density is increased, τ_{on} decreases because $\sigma J/q$ increases (Fig. 5.28) and fast modulation is achievable. For instance at 300 A/cm^2 the cut-off frequency is 80 kHz and it should be noted that this is totally due to the time resolution of the detector, while the LED itself is much faster: the lifetime should be 200 ns with 1×10^{19} carriers/cm^3 and 20 ns with 1×10^{20} carriers/cm^3. This results in modulation frequencies of 5 MHz and 50 MHz respectively. In fact, modulation at frequencies over 10 MHz has been demonstrated.

It is interesting to note that the different decay times during pumping and at the turn-off solves one of the major problems towards the application of Er-doped LEDs: since the internal quantum efficiency of the luminescence is given by $\eta_{\text{i}} = \sigma\phi_{\text{Er}}\tau/\tau_{\text{rad}}$ (where ϕ_{Er} is the Er areal density), the simultaneous achievement of high efficiency (requiring a long τ, close to τ_{rad} which is $\sim$ 2 ms) and fast modulation (requiring a short τ) has been in general considered impossible. The results presented in this section clearly show that using a specific diode structure it is possible to solve this problem since a long lifetime during pumping and a short lifetime at the turn off can be simultaneously achieved under reverse bias operation.

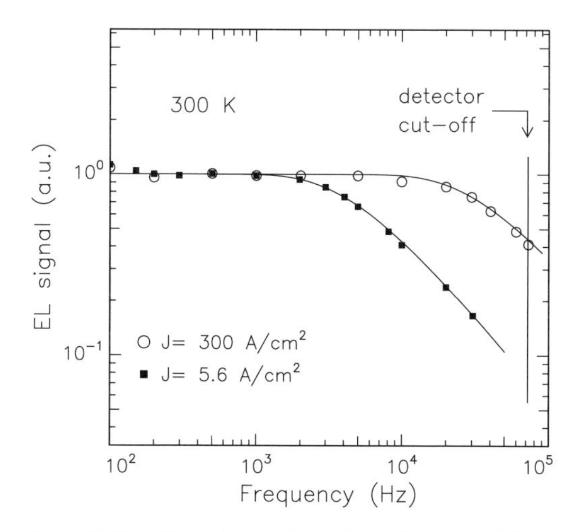

Fig. 5.30. Peak intensity of the 1.54 μm EL during modulation as a function of the frequency of operation for two different current densities passing through the device. Data refer to reverse bias operation. After [41]

Let us now analyze in detail the forward bias operation. As we have seen in Fig. 5.27c under forward bias $\tau_{\rm on}$ and $\tau_{\rm off}$ are both detector limited. This demonstrates that the lifetime both during pumping and during decay is < 12 μs. Indeed, since the carrier density is high in this case (both during pumping and at the turn-off of the device), the Auger process should reduce the lifetime to 200 ns (with 1×10^{19} carriers/cm^3). In contrast, the correspondence between forward bias EL and PL suggests that, the e-h excitation process being the same, also the cross-section should be the same, i.e. $3{\times}10^{-15}\,\text{cm}^2$. The product $\sigma\tau$ yields $6{\times}10^{-22}\,\text{cm}^2\text{s}$ in agreement with the estimate achieved in Fig. 5.26. Therefore in this case, in spite of the high excitation cross-section, de-excitation processes severely decrease the efficiency. Another effect is also responsible for the reduced efficiency under forward bias operation, namely the reduced number of excitable luminescent sites. In fact, from the data in Fig. 5.26 we know that at high current densities the saturation value (EL_{max}) obtained under forward bias is half that obtained under reverse bias. Since $\text{EL}_{\text{max}} \propto N_{\text{Er}}/\tau_{\text{rad}}$, this implies that the number of excitable sites under forward bias is half of that under reverse bias. Though excitation is not restricted to the depletion region but can be achieved in the whole Er-doped region, this result is not surprising since only those Er ions introducing a proper level in the Si band gap can be excited in this case. DLTS and SRP measurements [18, 29] clearly indicated that only 1% of the total amount of Er ions introduce a level at 0.15 eV while most of them introduce much shallower levels, responsible for the donor behavior but possibly unsuitable to achieve Er excitation at high T.

The device performance can be summarized as follows.

Under reverse bias:

- Er ions are excited by impact excitation of hot carriers within the thin depletion layer with an excitation cross-section of $\sim 1 \times 10^{-16}\,\mathrm{cm}^2$;
- most of the Er ions in the depletion region can be pumped; however, a dark zone ($\sim$ 10 nm thick at the edges of the depletion region) exists where Er cannot be excited because carriers are not energetic enough;
- during pumping, within the depletion layer, the Er decay lifetime is quite high ($\sim$ 100 µs) and weakly dependent on T, demonstrating that nonradiative processes are inhibited;
- at the turn-off the EL signal decays in less than 200 ns since the excited Er ions are in this case embedded within the heavily doped ($\sim 1 \times 10^{19}/\mathrm{cm}^3$) neutral regions of the diodes and fast nonradiative Auger processes set in: this phenomenon allows us to modulate the diode at frequencies higher than 10 MHz;
- an internal quantum efficiency of $\sim 10^{-3}$ can be achieved at RT: the major limitation towards the achievement of higher efficiencies is the small areal density of Er incorporated within the depletion layer.

Under forward bias:

- Er ions are excited by e-h pair recombination at an Er-related level within the band gap: the effective excitation cross-section is higher than $3 \times 10^{-15}\,\mathrm{cm}^2$ at RT;
- only a small fraction ($\sim$ 1%) of the Er ions can be pumped;
- the Er decay lifetime is quite low ($<$ 200 ns) due to fast nonradiative processes;
- a quantum efficiency of $\sim 10^{-5}$ is obtained at RT: the major limitations towards the achievement of higher efficiencies are the small number of excitable sites and the strong nonradiative decay processes.

A final remark on electroluminescence is needed. Though reverse bias EL is the most promising route, the presence of a dark zone in the depletion region where Er cannot be pumped [47] poses severe limits towards further improvements and towards the achievement of population inversion. In addition, the fact that current density and carrier energy cannot be controlled independently is a clear disadvantage. A novel device design has hence been made [61] consisting of a n-p-n transistor structure in which Er is embedded in the collector–base depletion region. In this structure electrons injected from the emitter are accelerated by the high electric field in the reverse biased collector–base junction and impact excite Er (which is located deep in the depletion region in order to leave enough space for the carrier to be accelerated). Here current density is controlled by the emitter–base voltage and carrier energy by collector–base voltage. Moreover, no dark region is present and hence population inversion is in principle achievable. This structure is probably the most promising for efficient emission in Er doped crystalline Si.

5.5 Erbium Doped Si Nanocrystals

The nonradiative de-excitation processes may be reduced by widening the Si band gap. In this case the thermally activated back-transfer will become less efficient since the energy mismatch ΔE for the process (see Fig. 5.20) will become larger. Widening of the band gap also produces a reduction in the free carrier concentration thus limiting the Auger process. A way to achieve this goal is to incorporate erbium within Si nanocrystals (NC) [62, 63, 64, 65, 66, 67, 68, 69, 70, 71, 72, 73]. This is well illustrated in Fig. 5.31 showing the room temperature PL spectra of a SiO_2 sample implanted with Er (continuous line) and a sample containing also Si nanocrystals (dash-dotted line).

It should be noted that the PL spectrum for the Er doped Si nanocrystals shows the typical features of the Er luminescence. Moreover, a strong enhancement of the 1.54 μm signal with respect to the Er doped SiO_2 is observed. In fact the PL signal is more than a factor of 10 higher with respect to the SiO_2 sample not containing Si nanocrystals. In addition, another PL peak at 0.98 μm, due to the radiative de-excitation of Er from the $^4I_{11/2}$ to the $^4I_{15/2}$ level, is observed in the sample containing Si nanocrystals while it is absent in the Er doped SiO_2 (at this Er concentration). All of these

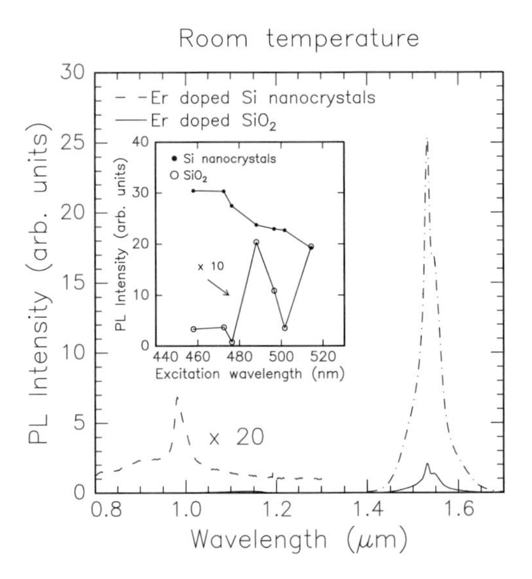

Fig. 5.31. Room temperature photoluminescence spectra for SiO_2 (*continuous line*) and Si nanocrystals (*dash-dotted line*) samples implanted with 300 keV Er to a dose of $3 \times 10^{15}\ cm^{-2}$. Spectra were taken with a laser pump power of 200 mW. In the insert the photoluminescence signal at 1.54 μm for the two samples is reported as a function of the excitation wavelength. Data are shown for both Er doped Si nanocrystals (•) and Er doped SiO_2 (○). The data for Er doped SiO_2 are multiplied by a factor of 10. After [72]

observations indicate that the presence of Si nanocrystals produces a strong enhancement of the Er luminescence.

In order to identify which is the mechanism responsible for the increase in the luminescence yield, we can observe the PL peak intensity at 1.54 μm at different pump wavelengths. This wavelength scan is shown for Er implanted Si nanocrystals (•) as well as for Er implanted SiO2 film (○) in the insert of Fig. 5.31. For the SiO_2 sample two peaks at 488 nm and 514.5 nm are observed corresponding to the direct photon absorption from the ground state of Er ($^4I_{15/2}$) to the $^4F_{7/2}$ and $2H_{11/2}$ levels, respectively. In contrast, for the Er doped Si nanocrystals no resonance is observed and the PL signal monotonically and slightly decreases with increasing the excitation wavelength (due to the different absorption coefficient of Si nanocrystals at the different wavelengths). These data demonstrate that the strong luminescence observed in the sample containing Si nanocrystals comes from Er ions that are pumped through an electron–hole mediated process in the Si nanocrystals in a fashion similar to that shown for single crystal Si.

Figure 5.32 shows that a competition between rare-earth and nanocrystal luminescence is present. In the absence of Er a well-known signal at around 0.85 μm is observed coming from the Si nanocrystals (dashed line) as a result of confined exciton recombination. As soon as Er is introduced, also at doses as small as $2 \times 10^{12}/cm^2$, the signal from Si nanocrystals at 0.85 μm is seen to decrease, while a new peak at around 1.54 μm, coming from Er appears. With increasing Er dose this phenomenon becomes particularly evident with a quenching of the visible nanocrystal luminescence and a simultaneous enhancement in the Er-related luminescence. This effect is a demonstration

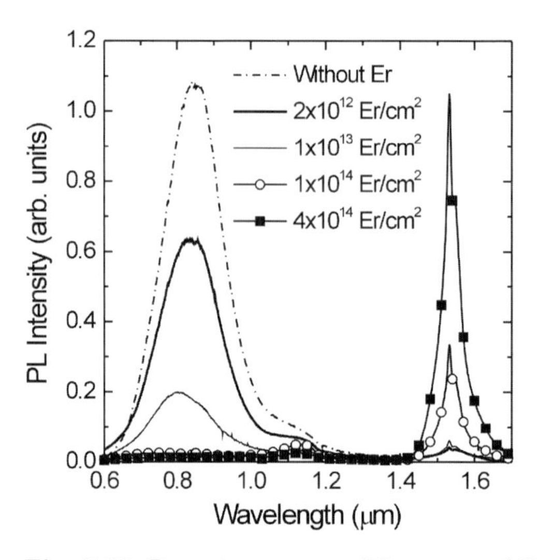

Fig. 5.32. Room temperature PL spectra of Er-implanted Si nanocrystals at different Er doses. The pump power of the laser beam was 50 mW. After [65]

of an energy transfer from the excitons confined in the nanocrystals to the erbium ions.

The observed behavior suggests that the two processes are in competition: in the absence of Er the exciton recombines radiatively; in the presence of Er the energy transfer to the Er $4f$ shell takes over. If this picture is correct one might expect that, as the Er concentration increases, not only the nanocrystal PL signal decreases, but also its decay time. Lifetime measurements of the 0.85 μm nanocrystal signal in the absence of Er and with increasing Er concentration are reported in Fig. 5.33. In the absence of Er the lifetime is around 130 μs. Surprisingly, with increasing Er content, in spite of the decrease of the luminescence at 0.85 μm (see Fig. 5.32), no change in lifetime is observed (see Fig. 5.33).

This result demonstrates that the energy transfer has characteristics very different from what are expected. In fact, it is now generally agreed that the coupling between Er ions and nanocrystals is so strong that when an Er ion is close by a nanocrystal this latter becomes dark, in the sense that as soon as it is excited it immediately gives the energy to the Er. Therefore the decrease in the intrinsic nanocrystal luminescence (at 0.85 μm) with increasing Er content reflects an increasing number of dark nanocrystals. However, those nanocrystals that are still bright do not interact with any Er ion and hence their properties are unchanged, in agreement with the lifetime data of Fig. 5.33.

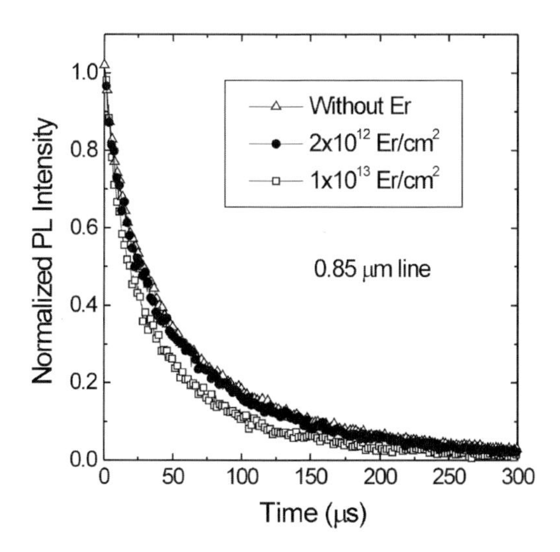

Fig. 5.33. Time decay curves of the nanocrystal-related luminescence at 0.85 μm in the absence of Er and for different Er contents. Data were taken by pumping to steady state at 50 mW and mechanically switching off the laser beam at $t = 0$. After [65]

A further important issue concerns whether Er is inside the nanocrystals or within the oxide matrix outside them. High resolution PL being sensitive to the Er local environment give an answer to this question (Fig. 5.34).

In Er doped crystalline Si several sharp peaks are present with the main emission occurring at 1538 nm. In contrast the high resolution spectrum of Er doped SiO_2 consists of only two lines at 1535 and 1546 nm. These peaks are much wider than those observed in Er doped crystalline Si. The high resolution PL spectrum of Er doped Si NC is identical to that of Er implanted SiO_2 indicating that the environment of the emitting Er ions is the same in the two samples. Since photoluminescence excitation measurements performed on Er doped Si NC demonstrated that Er is excited through an electron–hole mediated process (see insert in Fig. 5.31), as it occurs in Er doped crystalline Si, these data demonstrate that the emitting Er ions in Si NC are pumped by the electron–hole pairs generated within the NC but are located in SiO_2 or at the Si-NC/SiO_2 interface. The fact that Er is located outside the NC is not surprising since, due to the low solid solubility of Er in Si, thermal annealing tends to segregate it outside the NC and within the SiO_2 matrix.

The questions are now what is the process causing excitation at a distance and what is the observed luminescence enhancement due to. Though not yet definitively proven, excitation has been suggested to be due to a dipole–dipole interaction between Er and an excited nanocrystal with the energy transferred from the latter to the former. Since the energy is transferred by the NC at an energy of ~ 1.5 eV (probably to the $^4I_{9/2}$ Er level) a subsequent

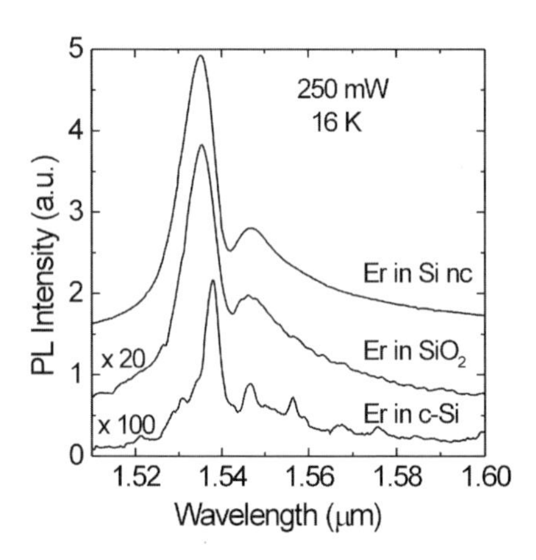

Fig. 5.34. High resolution (~ 1 nm) photoluminescence spectra measured at 16 K by pumping with a laser pump power of 250 mW three different samples: Er+O implanted crystalline silicon; Er implanted SiO_2 and Er implanted Si nanocrystals. After [69]

fast nonradiative decay brings the Er to the first excited state ($^4I_{13/2}$ at 0.8 eV). A back-transfer of the energy to the NC is then impossible.

As far as the luminescence enhancement is concerned this may be due to either a decrease of nonradiative processes or an enhancement in excitation efficiency, or both. Let us first analyze the nonradiative processes. Figure 5.35 shows the luminescence time decay at 1.54 µm of Er at 17 K and at 300 K.

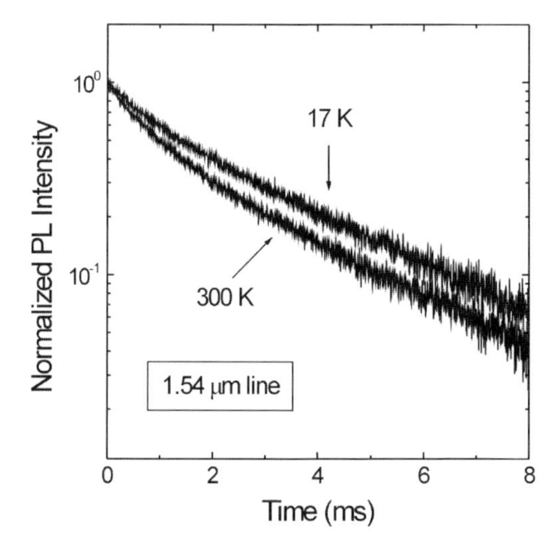

Fig. 5.35. Time decay of the PL intensity at 1.54 µm in Er-doped Si nanocrystals. Data are shown for both room temperature and 17 K. After [65]

The decay time is almost independent of temperature. Indeed, the small T dependence of the lifetime demonstrates that thermally activated nonradiative processes are weak within the nanocrystals. Temperature-activated energy back-transfer is not observed as a result of the enlargement of the gap of Si nanocrystals, and also Auger processes are weak due to the absence of free carriers. Hence nonradiative processes are very weak in contrast with crystalline silicon and in a fashion similar to insulating hosts.

This fact has strong implications for the PL temperature dependence. Figure 5.36 reports the PL signal at 1.54 µm as a function of the reciprocal temperature for Er doped SiO_2 (Δ) and for the Er doped Si nanocrystals ($\bullet$). The PL signal decreases by only a factor of ~ 3 in both nanocrystals and oxide samples by increasing T from 17 K to RT. This is very different from the behavior of Er-doped Si.

The excitation properties of Er ions can be obtained by studying the risetime of the 1.54 µm luminescence as a function of pump power in a way similar to that illustrated in the previous section. Figure 5.37 shows the risetime of the 1.54 µm PL intensity for different pump powers.

τ_{on} decreases with increasing pump power. In fact, $1/\tau_{on}$ vs the photon flux (Fig. 5.38) shows a linear behavior with a slope given by an effective excitation cross-section $\sigma \sim 1.1 \times 10^{-16}$ cm^2 and an intercept given by the

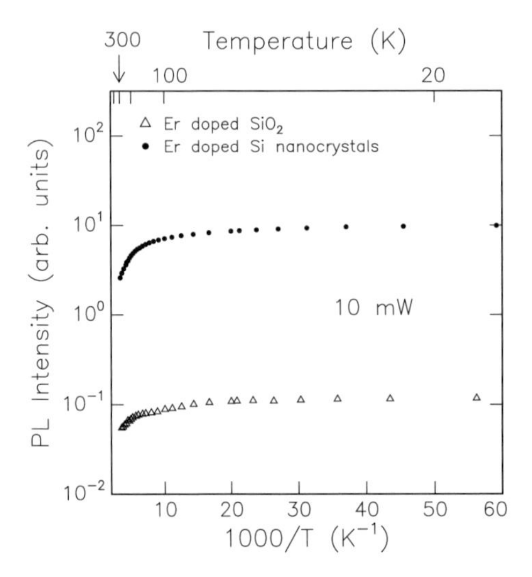

Fig. 5.36. Temperature dependence of the photoluminescence signal at 1.54 μm for Er doped SiO_2 (Δ) and Er doped Si nanocrystals (•). Data were taken at a pump power of 10 mW. After [71]

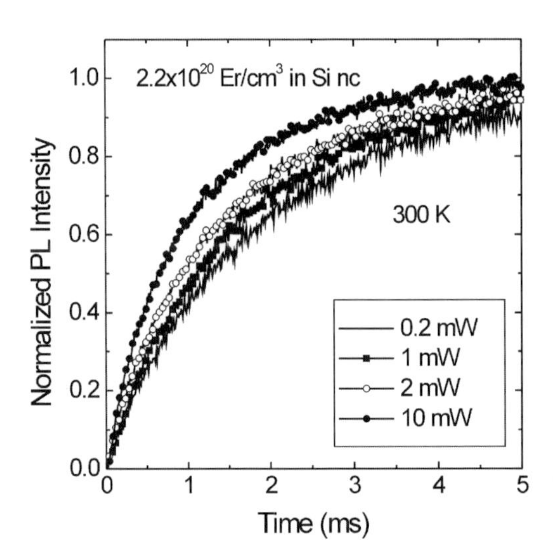

Fig. 5.37. Time resolved PL intensity at 1.54 μm when switching on the pumping laser at $t = 0$ for a sample containing Si NC and 2.2×10^{20} Er/cm^3. Data were taken at room temperature and at different pump powers and are normalized to the maximum intensity. After [72]

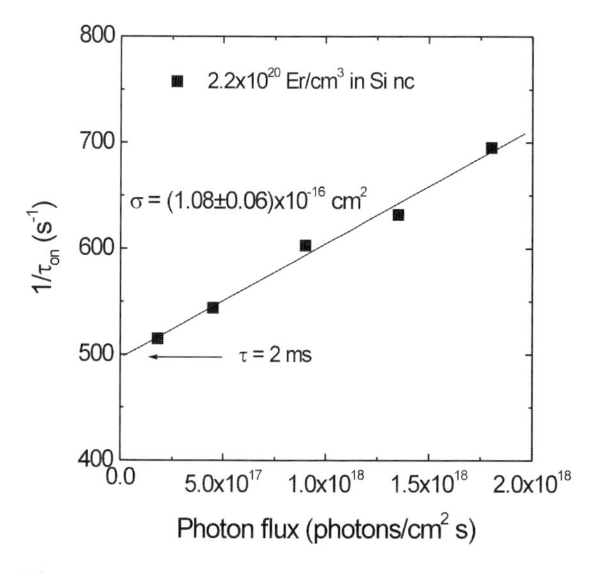

Fig. 5.38. Reciprocal of the τ_{on} time as obtained from the data in Fig. 5.10 as a function of the pump laser photon flux. The slope gives the effective cross-section for Er excitation. After [72]

Er lifetime (2 ms). It is now interesting to compare this number with the excitation cross-section of the Si NC themselves. For these Si NC an effective excitation cross-section of $1.8\times10^{-16}\text{cm}^2$ has been determined. The excitation cross-section of Si NC and that of Er in the presence of Si NC are almost identical. This demonstrates that energy transfer from NC to Er is extremely efficient.

These data demonstrate that Er takes advantage of the presence of the NC in the excitation process but then it behaves as in an insulating host as far as decay processes are concerned. For Er in insulating hosts it is quite well known that by increasing the concentration a phenomenon called concentration quenching sets in. In this process an excited Er ion de-excites nonradiatively giving its energy to another Er ion close by. Hence, the excitation travels along the sample being transferred resonantly from one Er ion to another. Since the travel of the energy along the sample makes it easier to reach quenching centers in which the energy is lost forever the decay time of the luminescence should decrease. The decay times of the 1.54 μm luminescence as a function of Er concentration are shown in Fig. 5.39. It is quite clear that the decay time is almost constant up to a concentration of 2.2×10^{20} Er/cm^3 and then rapidly decreases. The values of the decay time, obtained as a fit to the data of Fig. 5.39, are reported in Fig. 5.40.

These data suggest that the optimum Er content to take advantage of the sensitizer effect of the NC without being affected by quenching phenomena is $\sim 2\times10^{20}$ Er/cm^3. This result is extremely important for applications.

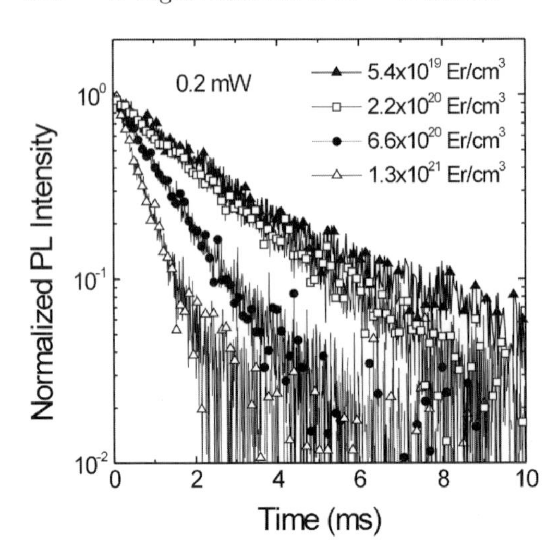

Fig. 5.39. Room temperature measurements of the time-decay of the PL signals at 1.54 μm for samples containing Si NC and different Er concentrations. The pump power of the excitation laser was 0.2 mW. After [72]

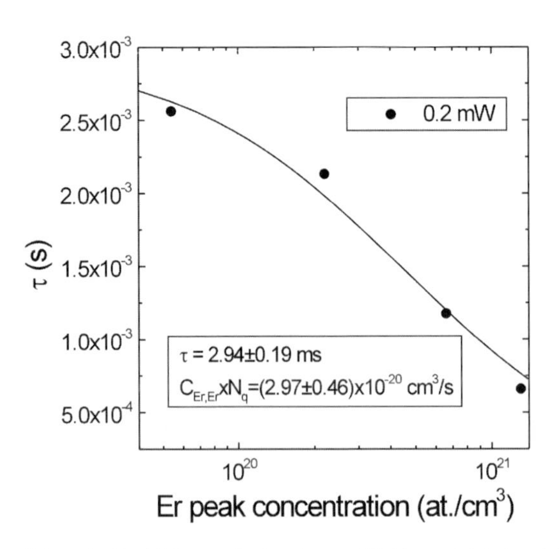

Fig. 5.40. Values of the decay lifetime (obtained from a fit to the data reported in Fig. 5.13) as a function of the Er concentration. The *continuous line* is a fit to the data according to the concentration quenching model. After [72]

Another interesting issue is the behavior of Si NC and Er within an optical microcavity. As an example in Fig. 5.41 a cross-sectional TEM image of one microcavity is shown.

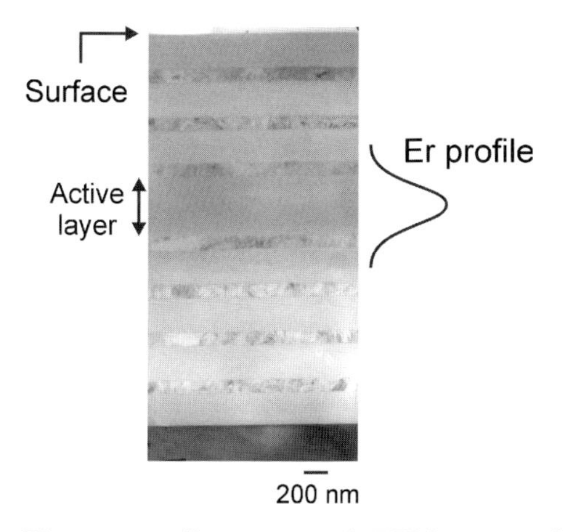

Fig. 5.41. Cross-sectional TEM image of a typical Si/SiO_2 microcavity with Si nanocrystals and Er in the active region

The cavities consist of two distributed Si/SiO_2 Bragg reflectors and a central region containing both Si NC and Er. They have been grown by sequential plasma enhanced chemical vapor deposition on top of a Si substrate. The thicknesses of the layers and of the active region match $\lambda_0/4n$ and $\lambda_0/2n$, respectively (λ_0 being the intended value of the resonance and n the refractive index of the medium).

The room temperature PL spectra of the Er and Si NC within the microcavity compared with that of Er-doped Si NC outside the microcavity are reported in Fig. 5.42. The spectrum is much sharper ($\Delta\lambda = 5.5$ nm in the microcavity to be compared with $\Delta\lambda \sim 40$ nm in Er-doped Si NC). Moreover its intensity is enhanced by over an order of magnitude. Since the room temperature luminescence intensity of Er-doped Si NC is already $\sim$ two orders of magnitude above that of Er-doped SiO_2, the coupling of Si NC and a microcavity gives a net enhancement of more than three orders of magnitude. It should be noted that the emission in this case is also extremely directional along the vertical axis of the microcavity.

In summary, Er in Si NC benefits from the advantages of both silicon (efficient excitation) and SiO_2 (weak nonradiative processes), while it avoids their disadvantages (low excitation efficiency in SiO_2 and strong nonradiative processes in silicon). This system holds strong promises not only for light emitting devices.

Indeed MOS light emitting devices operating at room temperature have been made with this system [74]. The main application for Er in Si NC will probably be waveguide amplifiers. In waveguide amplifiers one needs to amplify a signal at 1.5 μm by passing it through a waveguide in which Er population inversion has been achieved. Er in Si NC can present several important advantages. First of all, the presence of Si NC modifies the refrac-

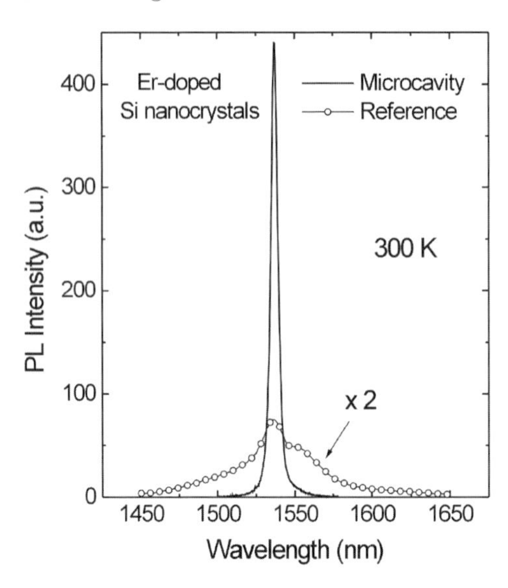

Fig. 5.42. Room temperature PL spectra for an Er-doped Si nanocrystal microcavity (*continuous line*). The spectrum of Er-doped Si nanocrystals outside the cavity is also shown for comparison (*circles*). After [73]

tive index of SiO_2 clearly defining a waveguide. Secondly, the sensitizer action of the nanocrystals excites Er very efficiently and thus population inversion is reached at lower pump power with respect to insulating hosts. Thirdly, inversion can be obtained by pumping Er with white light in contrast to insulating hosts for which, as a consequence of direct photon absorption, a laser matching the Er absorption band is needed. Finally the possibility of electri-

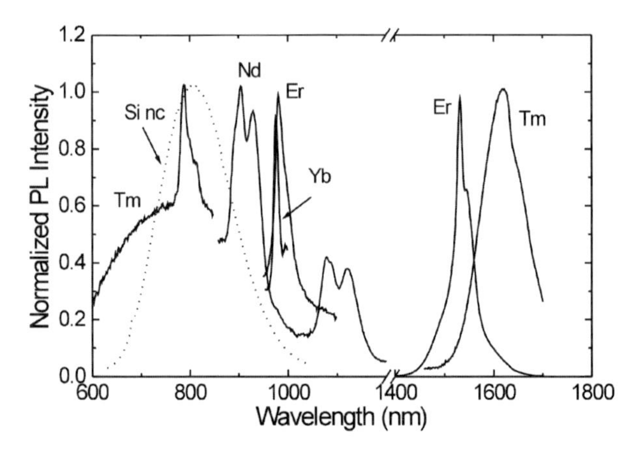

Fig. 5.43. Room temperature PL spectra of Si NC containing Er, Tm, Yb or Nd. After [65]

cal pumping might allow the future fabrication of electrically driven optical amplifiers.

As a final remark for this chapter, it should be noted that the processes described for Er and Si nanocrystals are also true for several other rare earths. In Fig. 5.43 the luminescence for Er, Tm, Nd and Yb in Si nanocrystals is reported. In all cases a preferential transfer is observed from the NC to the rare earth with a subsequent radiative de-excitation within the $4f$ shell and a decay typical of the particular rare earth. Emission in the wavelength range 0.9–1.7 μm can be achieved. These systems can be used in all those cases in which different wavelengths are needed and hold strong promise towards future optoelectronics applications.

References

1. G.S. Pomrenke, P.B. Klein, D.W. Langer: Mater. Res. Soc. Proc. **301** (1993)
2. S. Coffa, A. Polman, R.N. Schwartz: Mater. Res. Soc. Proc. **422** (1996)
3. J.M. Zavada, T. Gregorkiewicz, A.J. Steckl: Materials Science and Engineering **B81**, (2001)
4. J. Linnros, F. Priolo, L.T. Canham: J. Lumin. **80** (1999)
5. A. Polman, S. Coffa, J. Cunningham, R. Soref: Mater. Res. Soc. Symp. Proc. **486** (1998)
6. S. Coffa, L. Tsybeskov: MRS Bullettin **23**, no. 4 (1998)
7. A.J. Steckl, J. Zavada: MRS Bullettin **24**, no. 9 (1999)
8. H. Ennen, J. Schneider, G. Pomrenke, A. Axmann: Appl. Phys. Lett. **43**, 943 (1983)
9. H. Ennen, G. Pomrenke, A. Axmann, K. Eisele, W. Haydl, J. Schneider: Appl. Phys. Lett. **46**, 381 (1985)
10. F.Y.G. Ren, J. Michel, Q. Sun-Paduano, B. Zheng, H. Kitagawa, D.C. Jacobson, J.M. Poate, L.C. Kimerling: Mater. Res. Soc. Symp. Proc. **422**, 87 (1996)
11. D.J. Eaglesham, J. Michel, E.A. Fitzgerald, D.C. Jacobson, J.M. Poate, J.L. Benton, A. Polman, Y.-H. Xie, L.C. Kimerling: Appl. Phys. Lett. **58**, 2797 (1991)
12. J. Michel, J.L. Benton, R.F. Ferrante, D.C. Jacobson, D.J. Eaglesham, E.A. Fitzgerald, Y.-H. Xie, J.M. Poate, L.C. Kimerling: J. Appl. Phys. **70**, 2667 (1991)
13. J.L. Benton, J. Michel, L.C. Kimerling, D.C. Jacobson, Y.-H. Xie, D.J. Eaglesham, E.A. Fitzgerald, J.M. Poate: J. Appl. Phys. **70**, 2767 (1991)
14. D.L. Adler, D.C. Jacobson, D.J. Eaglesham, M.A. Marcus, J.L. Benton, J.M. Poate. P.H. Citrin: Appl. Phys. Lett. **61**, 2181 (1992)
15. P.N. Favennec, H. l'Haridon, D. Moutonnet, M. Salvi, M. Gauneau: Jpn. J. Appl. Phys. **29**, L524 (1990)
16. Y.-H. Xie, E.A. Fitzgerald, Y.J. Mii: J. Appl. Phys. **70**, 3233 (1991)
17. A. Polman, J.S. Custe, E. Snoeks, G.N. van den Hoven: Appl. Phys. Lett. **62**, 507 (1993)
18. J.S. Custer, A. Polman, M.H. van Pinxteren: J. Appl. Phys. **75**, 2809 (1994)

19. S. Coffa, F. Priolo, G. Franzó, V. Bellani, A. Carnera, C. Spinella: Phys. Rev. B **48**, 11782 (1993)
20. M. Matsuoka, S. Tohno: J. Appl. Phys. **78**, 2751 (1995)
21. M. Morse, B. Zheng, J. Palm, X. Duan, L.C. Kimerling: Mat. Res. Soc. Symp. Proc. **422**, 41 (1996)
22. W.-X. Ni, K.B. Joelsson, C.-X. Du, I.A. Buyanova, G. Pozina, W.M. Chen, G.V. Hansson, B. Monemar, J. Cardenas, B.G. Svensson: Appl. Phys. Lett. **70**, 3383 (1997)
23. F. Priolo, S. Coffa, G. Franzó, C. Spinella, A. Carnera, V. Bellani: J. Appl. Phys. **74**, 4936 (1993)
24. S. Coffa, G. Franzó, F. Priolo, A. Polman, R. Serna: Phys. Rev. B **49**, 1631 (1994)
25. J. Michel, F.Y.G. Ren, B. Zheng, D.C. Jacobson, J.M. Poate. L.C. Kimerling: Mat. Sci. Forum **143–147**, 707 (1994)
26. M. Needles, M. Schlüter, M. Lannoo: Phys. Rev. B. **47**, 15533 (1993)
27. I.N. Yassievich, L.C. Kimerling: Semicond. Sci. Technol. **8**, 718 (1993)
28. S. Libertino, S. Coffa, G. Franzó, F. Priolo: J. Appl. Phys. **78**, 3867 (1995)
29. F. Priolo, G. Franzó, S. Coffa, A. Polman, S. Libertino, R. Barklie, D. Carey: J. Appl. Phys. **78**, 3874 (1995)
30. L. Palmetshofer, Yu. Suprun-Belevich, M. Stepikhova: Nucl. Instrum. Meth. B **127–128**, 479 (1997)
31. H. Przybylinska, G. Hendorfer, M. Bruckner, L. Palmetshofer, W. Jantsch: Appl. Phys. Lett. **66**, 490 (1995)
32. H. Przybylinska, W. Jantsch, Yu. Suprun-Belevitch, M. Stepikhova, L. Palmetshofer, G. Hendorfe, A. Kozanecki, R.J. Wilson, B.J. Sealy: Phys. Rev. B **54**, 2532 (1996)
33. J.D. Carey, J.F. Donegan, R.C. Barklie, F. Priolo, G. Franzó, S. Coffa: Appl. Phys. Lett. **69**, 3854 (1996)
34. A. Terrasi, G. Franzó, S. Coffa, F. Priolo, D. D'Acapito, S. Mobilio: Appl. Phys. Lett. **70**, 1712 (1997)
35. U. Wahl, A. Vantomme, J. De Wachter, R. Moons, G. Langouche, J.G. Marques, J.G. Correia and the ISOLDE Collaboration: Phys. Rev. Lett. **79**, 2069 (1997)
36. P.G. Kik, M.J.A. de Dood, K. Kikoin, A. Polman: Appl. Phys. Lett. **70**, 1721 (1997)
37. G. Franzó, F. Priolo, S. Coffa, A. Polman, A. Carnera: Appl. Phys. Lett. **64**, 2235 (1994); Nucl. Instrum. Meth. B **96**, 374 (1995)
38. B. Zheng, J. Michel, F.Y.G. Ren, L.C. Kimerling, D.C. Jacobson, J.M. Poate: Appl. Phys. Lett. **64**, 2842 (1994)
39. S. Coffa, G. Franzó, F. Priolo: Appl. Phys. Lett. **69**, 2077 (1996)
40. J. Stimmer, A. Reittinger, J.F. Nútzel, G. Abstreiter, H. Holzbrecher, Ch. Buchal: Appl. Phys. Lett. **68**, 3290 (1996)
41. G. Franzó, S. Coffa, F. Priolo, C. Spinella: J. Appl. Phys. **81**, 2784 (1997)
42. C.X. Du, W.X. Ni, K.B. Joelsson, G.V. Hansson: Appl. Phys. Lett. **71**, 1023 (1997)
43. M. Matsuoka, S. Tohno: Appl. Phys. Lett. **71**, 96 (1997)
44. N.A. Sobolev, A.M. Emel'yanov, K.F. Shtel'makh: Appl. Phys. Lett. **71**, 1930 (1997)
45. J. Palm, F. Gan, B. Zheng, J. Michel, L.C. Kimerling: Phys. Rev. B **54**, 17603 (1996)

46. F. Priolo, G. Franzó, S. Coffa, A. Carnera: Phys. Rev. B **57**, 4443 (1998)
47. S. Coffa, G. Franzó, F. Priolo, A. Pacelli, A. Lacaita: Appl. Phys. Lett. **73**, 93 (1998)
48. A. Polman: J. Appl. Phys. **82**, 1 (1997)
49. S. Lombardo, S.U. Campisano, G.N. van den Hoven, A. Polman: J. Appl. Phys. **77**, 6504 (1995)
50. G.N. van den Hoven, J.H. Shin, A. Polman, S. Lombardo, S.U. Campisano: J. Appl. Phys. **78**, 2642 (1995)
51. M.S. Bresler, O.B. Gusev, V. Kh. Kudoyarova, A.N. Kuznetsov, P.E. Pak, E.I. Terukov, I.N. Yassievich, B.P. Zakharchenya, W. Fuhs, A. Sturm: Appl. Phys. Lett. **67**, 3599 (1995)
52. J.H. Shin, R. Serna, G.N. van den Hoven, A. Polman, W.G.J.H.M. van Sark, A.M. Vredenberg: Appl. Phys. Lett. **68**, 46 (1996)
53. O.B. Gusev, A.N. Kuznetsov, E.I. Terukov, M.S. Bresler, V.Kh. Kudoyarova, I.N. Yassievich, B.P. Zakharchenya, W. Fuhs: Appl. Phys. Lett. **70**, 240 (1997)
54. R. Serna, M. Lohmeier, P.M. Zagwijn, E. Vlieg, A. Polman: Appl. Phys. Lett. **66**, 1385 (1995)
55. A. Cavallini, B. Fraboni, S. Pizzini: Appl. Phys. Lett. **72**, 468 (1998)
56. E.F. Kennedy, L. Csepregi, J.W. Mayer, T.W. Sigmund: J. Appl. Phys. **48**, 4241 (1977)
57. J.M. Langer, Le Wan Hong: J. Phys. C **17**, L923 (1984)
58. A. Suchocki, J.M. Langer: Phys. Rev. B **39**, 7905 (1989)
59. J.M. Langer: J. Lumin. **40/41**, 589 (1988)
60. K. Takahei, A. Taguchi, H. Nakagome, K. Uwai, P.S. Whitney: J. Appl. Phys. **66**, 4941 (1989)
61. C.X. Du, F. Duteil, G.V. Hansson, W.X. Ni: Appl. Phys. Lett. **78**, 1697 (2001)
62. A.J. Kenyon, P.F. Trwoga, M. Federighi, C.W. Pitt: J. Phys. Condens. Matter **6**, L319 (1994)
63. M. Fujii, M. Yoshida, Y. Kanzawa, S. Hayashi, K. Yamamoto: Appl. Phys. Lett. **71**, 1198 (1997)
64. M. Fujii, M. Yoshida, S. Hayashi, K. Yamamoto: J. Appl. Phys. **84**, 4525 (1998)
65. G. Franzó, V. Vinciguerra, F. Priolo: Appl. Phys. A **69**, 3 (1999)
66. C.E. Chryssou, A.J. Kenyon, T.S. Iwayama, C.W. Pitt, D.E. Hole: Appl. Phys. Lett. **75**, 2011 (1999)
67. A.J. Kenyon, C.E. Chryssou, C.W. Pitt: Appl. Phys. Lett. **76**, 688 (2000)
68. H. Shin Jung, S. Se-young, K. Sangsig, S.G. Bishop: Appl. Phys. Lett. **76**, 1999 (2000)
69. G. Franzó, D. Pacifici, V. Vinciguerra, F. Priolo, F. Iacona: Appl. Phys. Lett. **76**, 2167 (2000)
70. P.G. Kik, M.L. Brongersma, A. Polman: Appl. Phys. Lett. **76**, 2325 (2000)
71. G. Franzó, V. Vinciguerra, F. Priolo: Phil. Mag B **80**, 719 (2000)
72. F. Priolo, G. Franzó, D. Pacifici, V. Vinciguerra, F. Iacona, A. Irrera: J. Appl. Phys. **89**, 264 (2001)
73. F. Iacona, G. Franzó, E.C. Moreira, F. Priolo: J. Appl. Phys. **89**, 8354 (2001)
74. F. Iacona, D. Pacifici, A. Irrera, M. Miritello, G. Franzó, F. Priolo, D. Sanfilippo, G. Di Stefano, P.G. Fallica: Appl. Phys. Lett. **81**, 3242 (2002)

6 Silicon-Based Photonic Crystals*

The optical analogoue of a normal crystal is a *photonic crystal* (PC), where photons are constrained to propagate on energy bands and within a dielectric lattice. Photonic bands are the result of the multiple scattering of photons in the dielectric lattice. Photons in a photonic crystal also exhibit new properties which cannot be found in free space. For example, the coupling of the electronic excitations and photons is drastically changed and enhanced or inhibited spontaneous recombination, or threshold-less lasing can be achieved. Also photon propagation is changed and subliminal velocity, omnidirectional reflectors, extremely sharp bends in propagation, short nonlinear effective interaction lengths, etc., can all be observed. If one could exploit photonic crystals at the integration level of what has been achieved with today's ultralarge scale integrated circuits, then many limitations of today's microelectronics can be overcome. In fact photonics will replace electronics in signal processing and, eventually, in computing. Moreover photons do many things better than electrons: they do not dissipate heat, are insensitive to cross-talk (in the linear regime), show very low propagation losses and many can be sent over the same physical channel by using different wavelengths. All these factors motivate the growing research work on photonic crystals. Photonic crystals fabricated with silicon have many advantages with respect to those fabricated with compound semiconductors: they can be built by using conventional microelectronics technology and are inherently CMOS compatible, show better optical properties due to the high refractive index difference between silicon and silica or air, and the starting materials are cheaper. However compound semiconductors have some advantages with respect to silicon. In particular they have better light emission properties and are easy to pattern with current microelectronics technology. The patterning process is a delicate step because it can lead to high surface recombination that can have a detrimental effect on device efficiency. The deep reactive ion etching (RIE) technique used to produce structures with a suitable depth, leaves a non-negligible wall roughness that enhances losses by scattering. As we will discuss, self-assembling methods or macroporous silicon formation are free from these drawbacks.

* This chapter has been written with the help of Paolo Bettotti

One of the possible uses of silicon-based PC is to help defeat the low emission efficiency of silicon. Some steps towards this goal have already been made, e.g. the enhanced emission efficiency of microcavities or the high power efficiency of textured LED (Sect. 1.2.2).

In this chapter we will review the properties of photonic crystals based on silicon. We will deal both with fabrication techniques and with the PC properties. The first part is a short introduction to the new physics involved in photonic crystals. Then we will discuss various kinds of photonic crystals by separating them by their dimensionality: 1D, 2D and 3D photonic crystals. Books, conference proceedings and review articles have been published on photonic crystals [1]. In view of the large number of results published in the literature these last few years, a selection based on the author preferences has been necessary.

6.1 Engineering Photonic States: Basics

This field has been pioneered by Yablonovitch [2] and John [3] who suggested that light propagation can be controlled in suitable dielectric media. Following these reports, a great number of results has produced photonic structures in the world of real applications, thanks to the possibility of precise control of the dispersion relations of photons inside solid media [4]. As has already happened in the 1980s with the electronic properties of multilayered semiconductor system (band gap engineering), a new technology is now available, where the photon propagation in matter can be engineered at will. The term *engineering photonic state* is used to describe this technology. Acting only with dielectric properties of the matter, it is possible to tailor the states of the light inside a structure, similarly to the modification of the electronic states in the crystalline lattice (Fig. 6.1). For electrons in a solid, an effective mass smaller than that of electrons in vacuum and the energy band structure are due to the presence of a periodic arrangement of ions in the crystalline

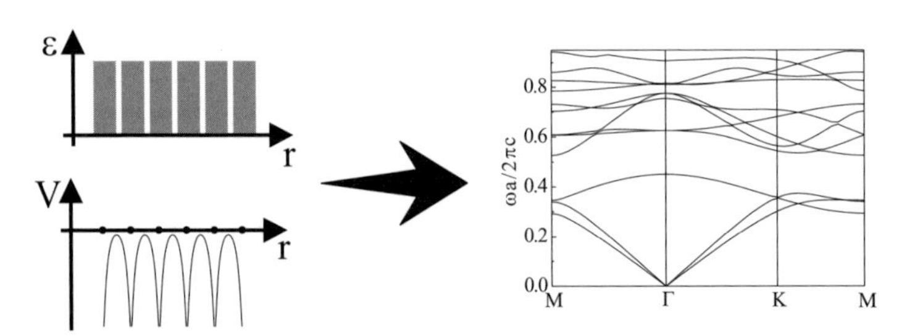

Fig. 6.1. The modulation of the dielectric constant $\varepsilon(r)$ has the same effect, on photons, as the potential modulation V in a solid has on electrons. In fact the resulting banddiagram on the right is almost the same for electrons and for photons

lattice. For photons, a suitable potential field can be created by varying the dielectric constant. A band structure for photons in the material is created and the material itself is called a *photonic crystal*. Thus a photonic crystal is a material where some kind of periodic arrangement of dielectrics in space causes a variation of the photon propagation. Particular attention has been devoted to the study of periodic structures because they were the first to be investigated and demonstrate the concept of "molding the flow of light" in complex dielectric structures [5]. Then a great deal of research has been devoted to the study of defects (both point and line defects) in PC due to their potential applications in optical integrated circuits. However, nowadays also the concept of light propagation in disordered dielectrics [6] or in dielectric quasicrystals [7, 8] is emerging because of the interesting properties of light propagation in these systems: photon localization [9], random laser action [10], optical Bloch oscillations [11], etc.

Getting back to periodic systems, these are classified on the basis of the number of spatial dimensions where the photon propagation is influenced: the dispersion law for photons is modified in one, two or three dimensions by a lattice in which the dielectric constant is varied along one, two or three directions [12, 13, 14] (see Fig. 6.2).

A rigorous treatment of PC is presented in many books [5]. Here it is useful just to recall the basic properties of these materials. A photonic lattice works on photons as the atomic lattice works on electrons. This analogy is quite good as the theoretical framework for both systems is similar. For photons, one starts from the Maxwell equations for the electromagnetic field and writes down an eigenvalue equation where the magnetic field acts as the *wavefunction* and the eigenvalues are the energy of (photon) modes. The eigenvalues then define the photonic band structure of the system. For some PC, also band gaps appear in the photonic band structure, which are named photonic band gaps. For photon frequencies corresponding to the photonic band gap, no photons can propagate in the PC. If the photonic band gap is opened in all directions, a complete photonic band gap material is formed. Two main differences exist between the formal treatment of electrons and photons:

- the wavefunction in a PC is a vectorial field reflecting the polarization properties of light;
- the Maxwell equations are scalable, thus the band structures for a lattice geometry is the same once it has been scaled by the lattice dimensions. This can be easily understood by looking at the master equation:

$$\nabla \times \left(\frac{1}{\epsilon(\mathbf{r})} \nabla \times \mathbf{H}(\mathbf{r}) \right) = \left(\frac{\omega}{c} \right) \mathbf{H}(\mathbf{r}), \tag{6.1}$$

where $\epsilon(\mathbf{r})$ is the periodic dielectric function, ω the eigenvalue and $\mathbf{H}$ the magnetic field. If the length is scaled by s: $\epsilon(\mathbf{r}) \rightarrow \epsilon(\mathbf{r}/\mathbf{s})$, then it is easy

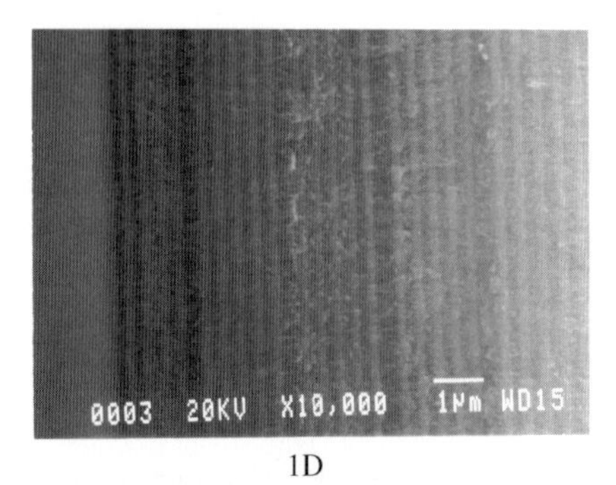

1D

2D

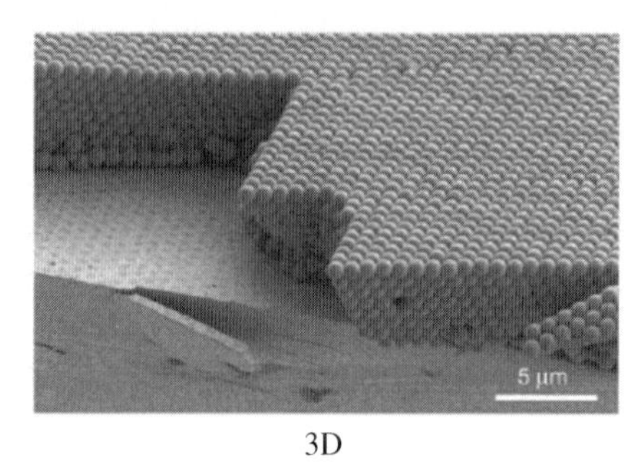

3D

Fig. 6.2. The three main types of PC are classified using the number of directions in which they modify photon dispersion. 1D: coupled microcavities fabricated in porous silicon. The thick defect layers are evident. After [13]. 2D macroporous silicon sample formed on p-type doped substrate. After [14]. 3D a very good quality opal sample. After [12]

to demonstrate that the eigenvalue scales as $\omega \to \omega/s$. Similarly if the dielectric function is scaled: $\epsilon(\mathbf{r}) \to \epsilon(\mathbf{r})/\mathbf{s^2}$ then $\omega \to s\omega$.

6.2 One-Dimensional Photonic Crystals

A periodic stack of layers of different dielectric materials realizes a one-dimensional PC. These are often called dielectric thin film multilayers and their properties were studied many years before the concept of photonic crystals [15]. Optical filters with different one-dimensional dielectric function distribution were suggested. Among them the periodic stacking of $\lambda/4$-thick films are called distributed Bragg reflectors (DBR) because they perform as mirrors where the reflection is caused by the multiple interference of the light reflected at each dielectric interface of the whole stacking [16]. DBR are narrow band mirrors whose reflectivity is very high in a well-defined spectral range (named the stop-band). The stop-band spectral position is varied by changing the thickness and/or the refractive index of each layer. Indeed the interference depends on the optical path, i.e. the product $n \cdot d$, where n is the refractive index and d the thickness of the layer [17, 18, 19] (see Fig. 6.3).

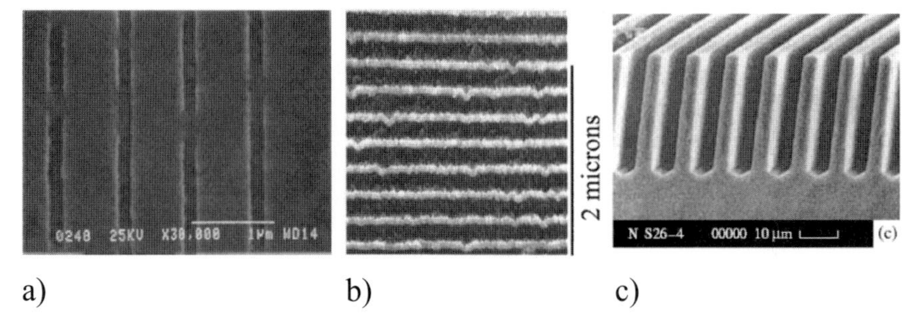

Fig. 6.3. (**a**) Example of a DBR produced by low pressure LPCVD. DBR are composed of poly-Si and SiO_2 layers about 220 and 660 nm thick, respectively. After [17]. (**b**) DBR made by PS on p^+-oxidized PS. Overall thickness of 2.4 μm. After [18]. (**c**) DBR produced by vertical etching of silicon. After [19]

The stop-band, which in the classical picture is a consequence of the constructive interference of multiple reflected light beams, can also be explained by the PC concept. By using quantum field concepts, it is simple to explain the photonic band gap formation. A standing wave in a multilayer structure can have two different shapes: one with wave maxima and the other with wave minima in the layers of high dielectric constant. Thus from the electromagnetic variational theorem [5]:

$$E_{\mathrm{f}}(\mathbf{H}) = \left(\frac{1}{\mathbf{2(H,H)}} \int \mathrm{d}\mathbf{r} \frac{\mathbf{1}}{\epsilon} \left| \frac{\omega}{\mathbf{c}} \mathbf{D} \right|^2 \right) \tag{6.2}$$

it is clear that the modes whose wavefunctions have maxima in the high ϵ regions have lower energies while the modes whose wavefunctions have maxima in the region of low ϵ have higher energies. The two regions are separated by an energy gap. It is customary to call the high energy modes the air band and the low energy modes the dielectric band.

The analogy between stop-bands and photonic band gaps shows that the PC can also be explained classically by interference of light which is reflected, refracted or diffracted by the periodic variations in the dielectric functions. Light with a frequency within a photonic band gap cannot propagate in the structure because it is reflected back. These arguments are also valid for PC of different dimensions. In these PC, the photon modes have different dispersion depending on the photon propagation direction, hence the concept of a Brillouin zone typical of crystalline lattice is extended to the PC. Photonic band gaps could appear only for some wavevector directions.

One-dimensional PC have been produced in Si by different technologies widely used in microelectronics. Deposition by using chemical vapor deposition (CVD) [17] or molecular beam epitaxy (MBE) [20] or sputtering and evaporation [21] of various materials yield particularly interesting results. Materials used were Si, SiO_2, Si_3N_4, $Si_xN_yO_z$, YF_3, etc.

Another approach is silicon etching. Etching of vertical trenches produces Si/air multilayers with photonic confinement direction parallel to the plane. Here the process requires a careful lithography step and a very oriented dry-etching step. These structures have maximum refractive index contrast ($\Delta n \approx 3.5$) and the possibility to obtain PC with a large number of periods [19].

A much simpler technique to obtain dielectric multilayers in silicon is to exploit the current dependence of porous silicon (PS) formation (see Sect. 3.1.10). PC formed in this way show optical properties comparable to those of PC produced by other methods. With PS, the maximum refractive index contrast is ≈ 0.7 in order to avoid cracks during the drying process of the sample [22]. This is not a true limitation as 1D PC with up to 300 layers can be produced. Increasing the number of layers allows compensating the low refractive index mismatch (see more later).

It is useful to note that for 1D PC the band gap is formed as soon as $n_{\mathrm{H}} - n_{\mathrm{L}} \neq 0$, where n_{H} and n_{L} refer to the high and low refractive indexes. This is a condition satisfied only in the 1D structure; for 2D and 3D PC the condition on the refractive index contrast to open a band-gap is much more stringent [5].

DBR are fabricated by stacking layers of thickness $\lambda/4n$ that give a stop-band centered at λ. The mirror reflectivity R is:

$$R = 1 - 4\frac{n_{\mathrm{ext}}}{n_{\mathrm{c}}}\left(\frac{n_{\mathrm{L}}}{n_{\mathrm{H}}}\right)^{2N} \tag{6.3}$$

where n_{ext} and n_{c} are the refractive indexes of the media surrounding the DBR; and $2N$ is the number of periods. The width $\Delta\omega$ of the stop-band is a function of frequency, ω, and refractive indexes [23]:

$$\Delta\omega = \omega\frac{2}{\pi}\arcsin\left(\frac{n_{\mathrm{H}} - n_{\mathrm{L}}}{n_{\mathrm{H}} + n_{\mathrm{L}}}\right) . \tag{6.4}$$

For light incident on the DBR in a direction different from the normal, the stop-band becomes narrower and shifts to the blue due to the variation of the phase with the angle of incidence [23]. From a different point of view: for $k_{\parallel} \neq 0$, photonic bands curve upwards and when all $k_{\parallel}$ are taken into account no band gap exists. A classical explanation of this phenomenon can be found in [23].

In 1D PC the band gap can be experimentally accessed by reflectance (transmittance) measurements (see Figs. 6.4 and 6.5).

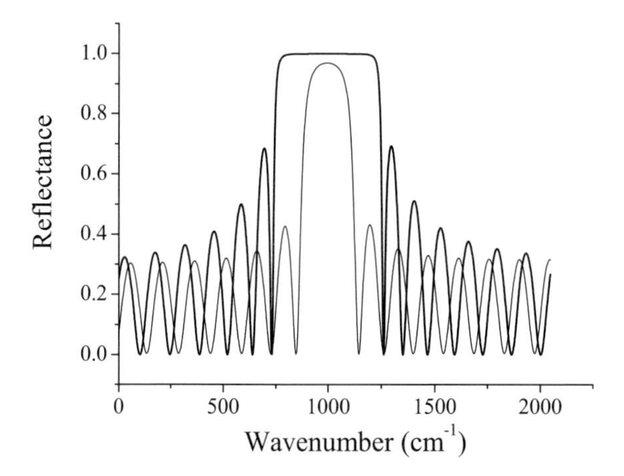

Fig. 6.4. Simulated reflectance spectra of two Bragg reflectors. Both mirrors have 12 periods. The *thick line* corresponds to a $\Delta n = 0.80$; the *thinner* has $\Delta n = 0.36$

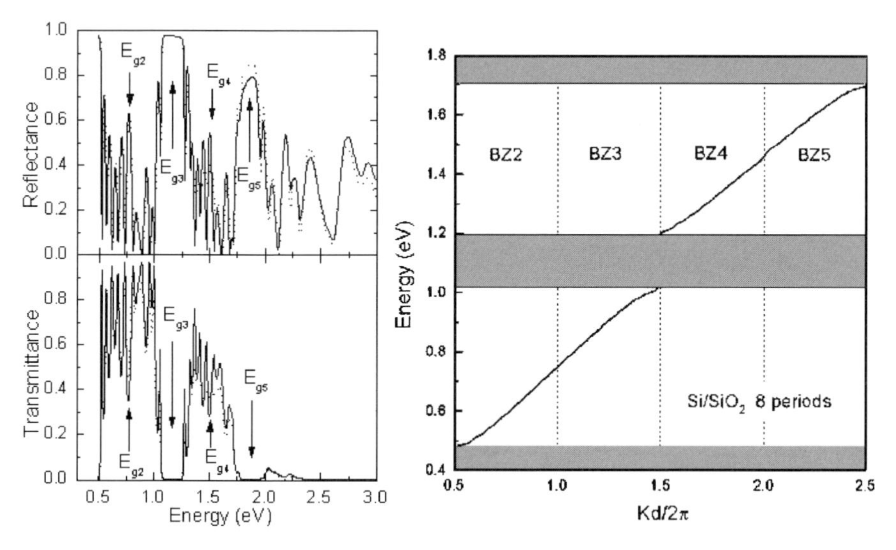

Fig. 6.5. *Left*: Reflection (R) and transmission (T) of three-period Si/SiO$_2$ DBR (*dotted lines*). The best-fit curves within the full multilayer model are also displayed (*solid lines*). *Arrows* mark the photonic band gaps clearly visible in both R and T spectra. *Right*: $\mathbf{k} - \omega$ dispersion of an eight-period Si/SiO$_2$ DBR measured by interferometry. After [24]

The photonic band dispersion of 1D Bragg reflectors can be directly measured by means of white-light interferometry [24]. An original setup based on a Mach–Zehnder interferometer allowed the measurement of the frequency-dependent relative phase shift introduced by the sample placed in one arm of the interferometer. This yields a direct measure of the $\mathbf{k}(\omega)$ dispersion in a broad spectral range from 0.5 to 1.5 eV. The photonic band dispersion of several 1D photonic crystals with different number of periods has been measured up to the fifth Brillouin zone, showing very good agreement with theoretical calculations [24]. It is also important to note that the theoretical approach used to describe the properties of PC are based on the assumption that the PC are perfectly periodic, i.e. they extend to infinity. In reality they have a finite extension. This produces some deviations from the perfectly periodic PC (see Sect. 6.2.4 and [17]). Nevertheless the main properties derived in the case of infinite crystals apply also to finite structures.

As shown in Fig. 6.6, at the band edge the photon mode density of a DBR grows enormously which allows one to study light localization [25], anomalous dispersion phenomena, like superluminal pulse propagation and abnormal

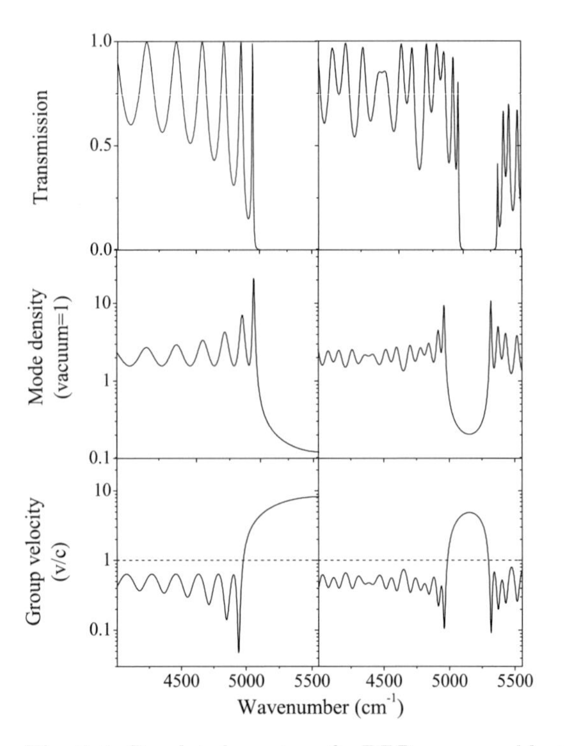

Fig. 6.6. Simulated spectra of a DBR composed by 25 periods, for $\lambda = 1.8$ μm (*left column*), and for a Fibonacci structure (a multilayer with layers thickness following Fibonacci series) of 12th order. *First row* shows transmission spectra, the *middle row*, mode density and at the *bottom* group velocity

dispersion of refractive index and phase velocity [26, 27]. These are exciting problems still open to investigation and to applications such as ultrafast optical switching using the Kerr effect at the band edge [28].

6.2.1 Optical Microcavities

The analogy between electronics and photonics can be exploited further by using the concept of *doping* the PC. Acceptor or donor impurities in semiconductors introduce localized impurity states in the band gap of the semiconductor. The same can be realized in a PC introducing a rupture in the periodicity of the dielectric function either by adding or by removing a constituent: a defect state (a photon mode) appears in the photonic band gap. Adding some dielectric material at the defect site has the same effect as adding an attractive potential on a quantum well, with the result of pulling a state from the *air band* (the photonic conduction band) into the photonic band gap, and vice versa removing some dielectric material has the effect of pushing up a state from the *dielectric band* (the photonic valence band) into the photonic band gap, like adding a repulsive potential to the state. For 1D PC the simplest defect is the insertion of a layer with a thickness different from those used to form the multilayered structure. This defect is spatially localized, the photon mode has an allowed frequency in the band gap through which a photon can propagate. In optics these structures are called Fabry–Perot filters, the defect state is the Fabry–Perot resonance; while in laser-physics they are called optical microcavities (MC).

In Fig. 6.7 simulated transmittance spectra of different porous silicon MC, whose defect thicknesses are varied between 1.6 and 2.4 $\lambda/4$, are plotted together. Usually, MC are symmetric structures about the defect, i.e. two equal DBR are separated by a central thicker layer. As soon as the thickness of the central layer is varied from those typical of the DBR, a *defect state* appears inside the photonic band gap [29]. The most common configuration is showed in Fig. 6.7 with the thicker line. The DBR have the optical thickness of the layers equal to $\lambda/4$, and the defect (central) layer has an optical thickness of $\lambda/2$. In this case the photon mode (resonance) is centered in the middle of the photonic band gap. Figure 6.8 shows the transmittance around the resonance of a MC made of PS: the resonance is at 1.6045 μm and is wide about 0.5 nm with a cavity quality factor $Q = \lambda/\Delta\lambda \simeq 3150$ [13].

An MC profoundly affects the light emission properties of an active medium embedded in it by enhancing or inhibiting its luminescence [30]. This is a consequence of the modified density of photon states (DOS) inside the cavity which affects the spontaneous emission (SE) rate of the active medium. Using Fermi's golden rule:

$$\gamma_{\mathrm{sp}} = \frac{1}{\tau_{\mathrm{sp}}} = \frac{2\pi}{\hbar}|M_{fi}|^2\rho(\omega) \tag{6.5}$$

where γ_{sp} is the spontaneous emission rate (the inverse of the lifetime τ_{sp}),

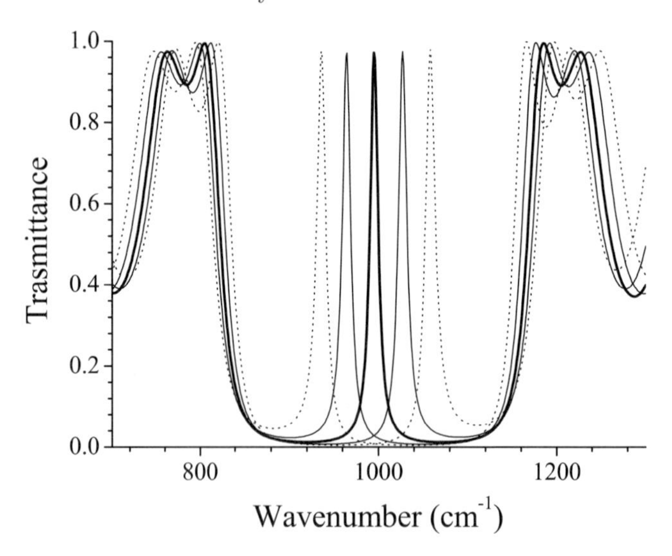

Fig. 6.7. Transmission spectra of a simulated cavity made on porous silicon. The refractive indexes are 1.83 and 2.19 for the high and low porosity layer, respectively. The different resonant peaks correspond to different thicknesses of the defect layer: from 2.4 (*left peak*) to 1.6 (*right peak*) $\lambda/4$ thickness, $\lambda = 1.55$ μm. The porous silicon was assumed to be nonabsorbing

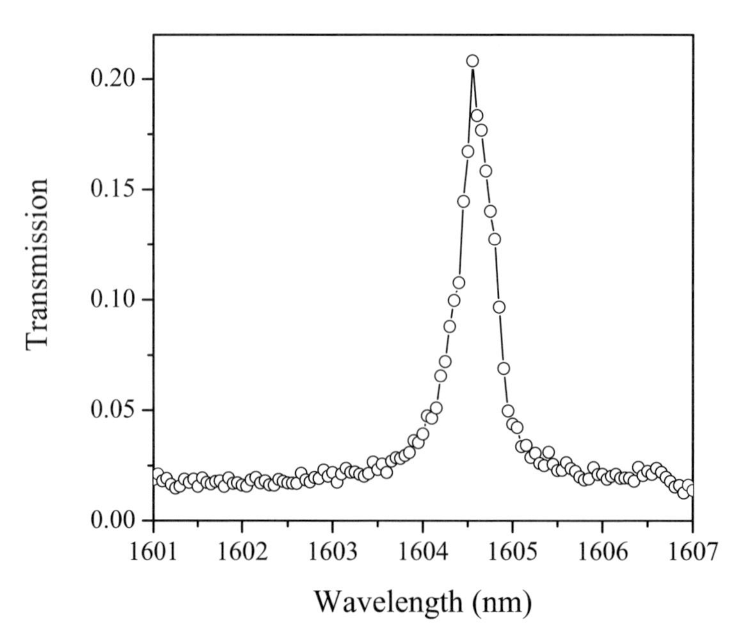

Fig. 6.8. Transmission spectra of a microcavity made on porous silicon. Spectra are recorded using a focused beam with a diameter of 35 μm.

M_{fi} is the matrix element coupling the states involved in the transition and ρ is the photon DOS. From the equation it follows that a reduced DOS increases the lifetime or, which is the same, reduces the SE, whereas where the DOS is large as for transitions resonant with a defect mode, the SE will be enhanced. Within the photonic band gap the DOS is reduced down to zero if the refractive index contrast and the number of layers are high enough [31]. Many theoretical and experimental proofs confirm these simply arguments [11]. In Fig. 6.9 a comparison between the optical and luminescence properties of a bare film of PS and a PS active layer embedded into an all-PS MC is shown [32]. The effect of the cavity on the PS luminescence is (i) a dramatic narrowing of the emission band, (ii) an enhanced emission intensity at the resonance of orders of magnitude, and (iii) a directional emission in a narrow cone around the normal. Even better results than those reported in the figure have been recently published [33].

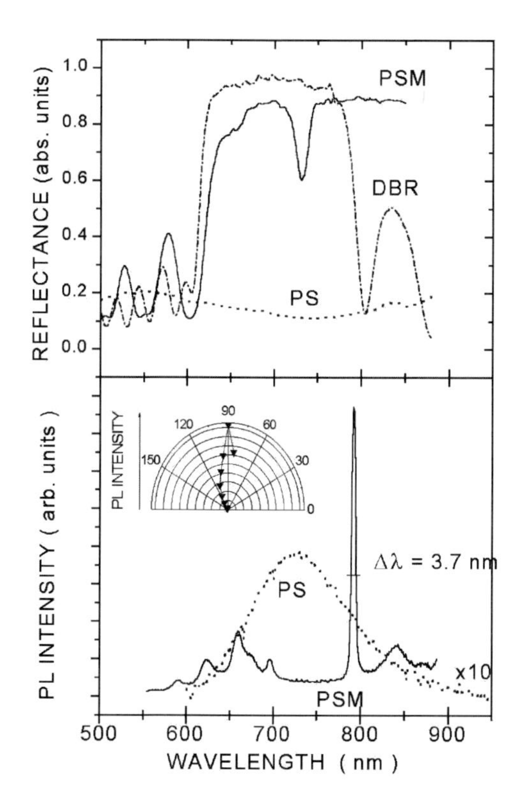

Fig. 6.9. Reflectance (*top*) and luminescence (*bottom*) of a reference PS (*dots*), a distributed Bragg reflector (*DBR, dashes*) and porous silicon MC (*line*) samples at room temperature. The *inset* shows the emission pattern. All have been formed on *p*-type 0.02 Ωcm Si wafers. The two MC in the *top* and *bottom panels* have different parameters. After [32]

In general different parameters contribute to determine the optical behavior of an MC:

- the central layer optical thickness, which determines the resonance of the cavity;
- the polarization of the active center. If the active centers are assimilable to emitting dipoles, when these dipoles are oriented parallel to the mirrors they feel the cavity structure heavily both for enhancement and suppression of SE;
- the reflectivity values of the cavity mirror, which determine the size of the effects on SE;
- the position of the active centers with respect to the resonant optical mode. If they are placed within a thin layer, the position of this layer with respect to the node (antinode) of the standing optical wave of the resonant mode of the cavity, reduces (enlarges) the SE.
- the wavelength of the MC defect, λ, scales with the incident angle as follows:

$$\lambda = 2nd\left[\left(1 - \frac{\sin\theta}{n}\right)^2\right]^{\frac{1}{2}} . \qquad (6.6)$$

An important property of an MC is the high directionality of the emitted radiation [34, 35]. This is a consequence of the two flat Bragg mirrors that recirculate photons propagating on the normal. The optical field strength is greatly increased at the resonance wavelength which is used to enhance the efficiency in extracting light emitted from the active layer [36, 37].

Not only the SE, but also the nonlinear optical (NLO) properties of the active material can be greatly enhanced in a 1D PC with a defect. This can happen by two different mechanisms. The first relies on the increased photon DOS at the band edge which increases the effective coherent length which, in turn can be viewed as a better phase matching condition. The second is due to the increased field strength at the resonance wavelength which increases the nonlinearity [38, 39, 40].

Possible use of microcavities ranges from a vertical cavity surface emitting layer (VCSEL) [41, 42, 43], in which most of the photons produced by SE go into the laser mode [35]; to sharp pass-band optical filters; to enhanced LED [44]. A review of other devices that can benefit from being embedded in a microcavity is found in [45], even though the review is mainly related to compound semiconductors.

6.2.2 Coupling Phenomena

An MC allows the study of exciton–photon coupling or cavity–polariton states. Two extreme regimes are usually considered: the strong and the weak coupling regimes. Strong and weak refer to the lifetimes of the photons and excitons in the cavity. When the photon lifetime is longer than the exciton lifetime the coupling can be effective and a new coupled state is formed, the cavity polariton, and the coupling regime is named strong. The polariton state is formed because of the impossibility of coupling the exciton state with the photon continuum outside the cavity. Once the exciton is created it can couple only with quantized photon states inside the cavity, till a photon leakage occurs. The opposite is true for the weak regime. Cavity polaritons are easier to observe than electronic polaritons, because polariton spectra are not affected by the scattering process, so the dispersion curve can be determined directly from reflectivity or luminescence spectra [46]. The coupling strength is determined by the vacuum Rabi splitting, which for an atom in a cavity reads

$$\Omega_{\mathrm{i}} = \frac{P_{12}}{\hbar}\left(\frac{\hbar\omega}{2\epsilon_0 V_0}\right)^{1/2} \tag{6.7}$$

where P_{12} is the atomic dipole moment and V_0 is the optical cavity volume. When Ω_{i} is greater than both the cavity and exciton linewidth, the system is in the strong coupling limit. If not, it is in the weak coupling limit, which

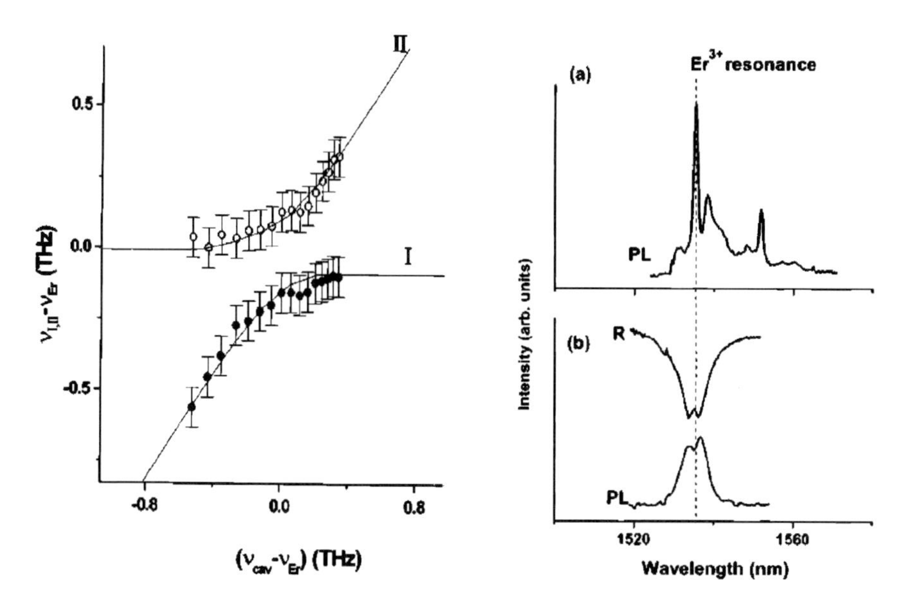

Fig. 6.10. (*left*) Frequencies of the two resonances in the reflectivity spectra, relative to the Er resonance, as a function of the microcavity-Er detuning. These polaritonic dispersion bands show a clear anticrossing behavior. (*right*) (**a**) Photoluminescence of a thin Er_2O_3 layer grown on top of a distributed Bragg reflector; (**b**) photoluminescence and reflectivity spectra of an Er_2O_3 layer embedded in a strongly confined Si/SiO_2 microcavity ($Q \simeq 500$). After [47]

is the case shown in Fig. 6.9. In the strong coupling limit, cavity polariton dispersion is observed in reflectance spectra. The exciton and photon polariton branches show an anticrossing behavior with an energy separation of Ω_{i}. An example is reported in Fig. 6.10, where the optical properties of a silicon based microcavity, where Er ions have been placed in the cavity, are shown [47].

In the weak coupling regime, it is possible to study the effect of multiple cavities in analogy with the electronic case where multiple quantum wells can be grown. This happens because the optical mode of each cavity resonantly couples with the mode of another cavity: splitting occurs. In Fig. 6.11 an example of ten coupled microcavities formed by PS are shown. Using coupled MC it was possible to determine the oscillator strength of spontaneous emission from the silicon nanocrystals in PS, without applying time resolved measurements [48].

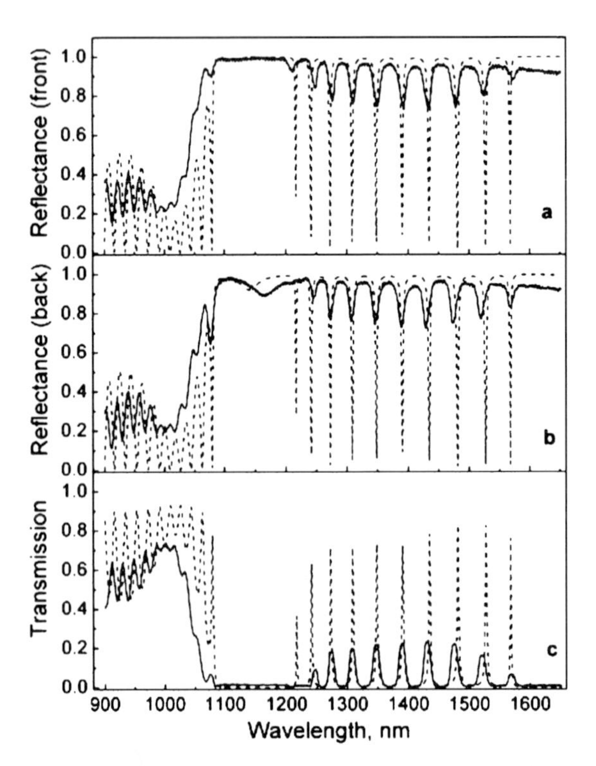

Fig. 6.11. Reflectance spectra (*solid lines*) from the front side (**a**) and backside (**b**) of a compensated free-standing ten coupled MC sample with 1.5 periods of coupling DBR centered at 1350 nm. The *dashed lines* are the results of numerical simulations considering no-drift in the layer optical thicknesses. (**c**) Transmission spectrum (*solid line*) and simulation (*dashed line*) of the same sample. After [13]

6.2.3 Omnidirectional Reflector

It was already mentioned that a PC modifies the photon dispersion in a number of directions equal to the number on which a modulated refractive index profile exists. However it is possible to tailor photonic dispersion properties of a simple multilayer structure in order to obtain a perfect mirror whatever the incident light direction [49, 50]. This can be achieved by designing a structure in which an overlapping gap for both polarization modes (TE and TM) exists over the light line (Fig. 6.12). The light line is defined as $\omega = ck_{\|}$. The light line defines the cone of light above which optical modes are not confined in a photonic structure but can couple to the continuum.

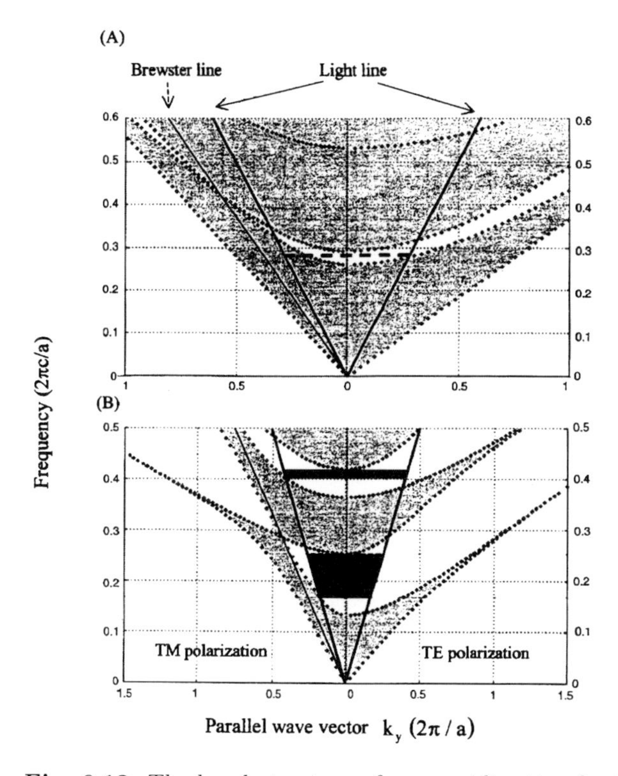

Fig. 6.12. The band structure of an omnidirectional reflector shows a forbidden gap extended along all the incident angles. (**A**) band structure of a normal DBR, (**B**) band structure of an omnireflector. After [49]

The photonic band gap is found in reflection spectra when $k_y = 0$, whereas for $k_y \neq 0$ the bands curve upward. When $k_y \to \infty$ modes are completely localized in the high index layer and cannot couple between different layers. The difference between DBR and the omnireflector is due to the k dependence of the photonic band gap: for a DBR the photonic gap does not extend through both the light lines, as it does for the omnireflector.

6.2.4 Effects of Finite Size of PC on Optical Properties

The calculated reflection spectrum of an infinite DBR shows a reflectance equal to 1, whereas in real structures $R \to 1$ because mirrors always have a finite transmission.[1] Figure 6.13 shows the transmittance of a Si/SiO_2 DBR versus the number of periods.

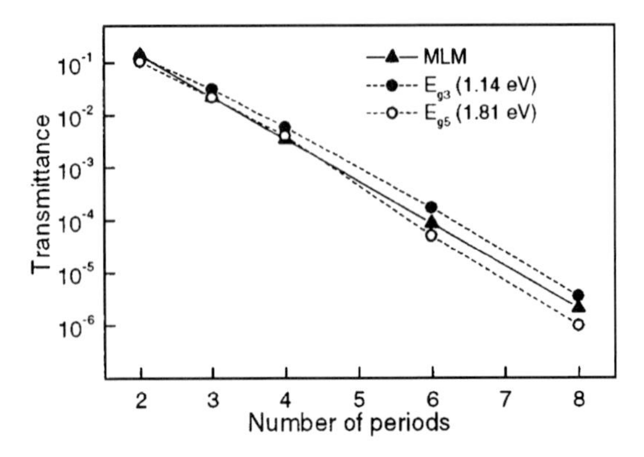

Fig. 6.13. DBR transmittance as a function of period number. After [17]

In the same way the width of the resonance in an ideal MC is described by a δ function, while a real cavity mode has a finite width given by:

$$\Delta = \frac{c(1-R)}{n_c L_{\text{eff}}} \tag{6.8}$$

where Δ is the FWHM of the resonance, and L_{eff} is an effective width of the cavity layer that takes into account the optical mode penetration into the mirrors which causes an effective cavity length longer than the layer thickness [46, 51]. These effects are also observed with photonic crystals of more than one dimension.

6.3 Two-Dimensional Photonic Crystals

6.3.1 Why More Than One-Dimensional Confinement?

PC based on 1D structures have an intrinsic limit: in the plane losses are very high because the system is homogeneous in the (x, y)-plane; in addition

[1] This can be easily deduced from 6.3, where R is a function of the number of periods.

the integration within an optical chip is not easy. On the other hand, 2D and 3D PC can be easily integrated within an optical chip and show very interesting properties that cannot be achieved in the 1D case. The various properties described for 1D PC extend on more directions for 2D and 3D PC and are enhanced by better light confinement: photonic DOS are much more strongly modified, SE can be completely inhibited in 3D PC, and various kind of defects can be inserted, etc.

6.3.2 Basics

In two dimensions PC are formed by modulating the dielectric constant periodically in the (x, y) plane, so that only along the z-direction is the material homogeneous. Different geometrical lattices can be produced: the most common ones are composed of air holes in a solid matrix, but also the opposite structure works fine, made by solid pillars in air [2] [52, 53]. Air can be replaced by another dielectric material [54] at the expense of a smaller dielectric index contrast. Generally it is desired to have the maximum possible refractive index contrast to open the widest photonic band gap.

In 2D PC photon dispersion is strongly modified in the (x, y) plane with the appearance of photonic band gaps. As for the 1D, as soon as the refractive index contrast is different from zero, a photonic band gap appears. In general, the photonic band gap opens only for one light polarization (TE or TM) and for a limited solid angle. If one desires a photonic band gap whatever the incident light direction (e.g. a complete gap for one polarization), then all the band gaps opened at different points of the Brillouin zone (BZ) have to overlap. This can be satisfied when the BZ is different from the spherical shape shown in the free photon dispersion case ($h\nu \propto ck$). The preferred lattice geometries to open a complete gap are the triangular and the honeycomb ones [55, 56]. The possibility of opening partial gaps for different polarizations allows the fabrication of polarization sensitive devices. The partial or complete character of the opened photonic band gap is defined, for a given lattice symmetry, by the relative volume of the different dielectric materials that compose the PC (see more below).

A few rules of thumb can be drawn in order to optimize the material for maximum photonic band gap size and polarization (see also Fig. 6.14).

1. Connected high-ϵ lattice sites (e.g. veins) facilitate the opening of photonic band gaps for TE modes, whereas isolated high-ϵ structures (e.g. pillars) for TM modes.

[2] Alternatively two-dimensional modulation of ϵ in the plane can be achieved using lattices of veins instead of single column elements. This structure is difficult to fabricate and does not have any real advantage if compared with the other two mentioned.

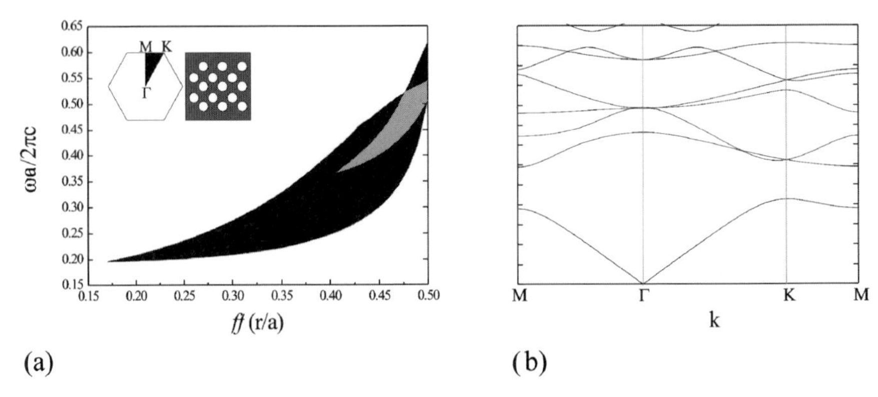

Fig. 6.14. (**a**) Photonic band gap map versus the filling factor ff $= r/a$. The *dark gray* is the region where the TE band gap appears, the *black* indicates the region where band gaps open for TM modes, and the *light gray* refers to the region where both TE and TM photonic band gaps exist. The *inset* shows the triangular lattice of holes in real space and in the reciprocal space with the main directions of the 2D Brillouin zone. (**b**) Band dispersion of a triangular lattice of holes ($\epsilon = 11.3$) for a filling factor of 0.3

2. A square geometry shows a complete photonic band gap neither for holes nor for pillars, whereas triangular and honeycomb lattices can have complete photonic band gaps.
3. If we define the *filling fraction* (ff) as the ratio between the pore radius (r) and the lattice pitch (a), then for a lattice of holes in a high-ϵ material the widest gap opens for ff ≤ 0.5, whereas for a lattice of pillars the best ff is at about 0.2 (more data can be found in [57]).
4. As for 1D, the frequency at mid-gap is defined by the lattice dimensions and, in particular, by the period. Moreover secondary gaps appear at multiples of this wavenumber.
5. Adding a small quantity of a third component enlarges the gap [58].

In 2D PC disorder effects play a fundamental role on the optical properties of the final structures. High control is needed on the processing of the PC in order to maintain the required periodicity and structures. Nevertheless it is interesting to note that the opening of the fundamental photonic band gap is strictly related only to the periodicity of neighboring scatterer elements [59], while long-range disorder has no major influence. On the contrary, the secondary gaps are very sensitive to disorder in the lattice and disappear as the randomness increases. These arguments are supported by the fact that complete photonic band gaps were demonstrated in 12-fold quasicrystal structures [60, 61] even in materials with low dielectric constant. This observation indicates that the high symmetry of quasicrystals helps the overlap of the photonic band gap existing at different k-values. One of their main advantages is the isotropic behavior of the gap with respect to the incident photon direction; in turn this allows the appearance of photonic crystal properties even for structures of very low volume and exploitation of the complete photonic band gap in extremely sharply bent waveguide. New devices become possible like, e.g. polarization insensitive, direction independent WDM (wavelength division multiplexing) filters and multiplexers.

Band calculations on PC are performed with similar methods as for electronic band calculations (see Chap. 2) [5]. To calculate the band dispersion of a simple infinite perfect lattice plane wave expansion methods (PWM) can be used. In the PWM the dielectric lattice and the related Bloch waves are approximated by a truncated Fourier series. Advantages of this method are its flexibility and simplicity that allow handling whatever geometry of the unit cell. For finite sized PC the PWM method can have convergence problems. In this case, real-space methods have to be used. These are based on a discretization of real space, in the time or in the frequency domains, which allows us to solve the Maxwell equations. 2D systems have a symmetry with respect to specular reflection with respect to the plane of the system which enables us to decouple the TE and TM fields. This allows obtaining two different eigenvalue equations that can be solved independently.

PC with defects are difficult to simulate because the lattice symmetry is broken and no easy definition of the boundary conditions exists. For localized defects, such as a microcavity, a supercell method can be employed: defects are repeated periodically in the space where an effective lattice is formed with the unit cell containing the defect. If line defects are considered, either enormous supercells have to be computed or other approaches have to be used. One of the most frequently used is the finite difference time domain (FDTD) discretizing method to solve Maxwell's equations. Another method relies upon expansion into a localized basis set, like Wannier wavefunctions for the electronic case. Other information about band calculations can be found in [62] and references therein.

Various methods have been used to produce 2D PC [63, 64]. The most common utilizes a lithographic step to define the geometry of the lattice, followed by an etch step to form the vertical features that produce the modulation of ϵ in the plane. Depending on the material, either dry or wet etching have been used. Dry etching processes on III-V materials are well exploited and optimal results were obtained using the reactive ion etching RIE technique [65, 66, 67] or the chemical assisted ion beam etching CAIBE method [68].

6.3.3 Macroporous Silicon

As already introduced in Chap. 3, 2D PC can be produced by using the anisotropic etching of Si in HF solution. Electrochemistry of this element allows, after a suitable optimization of the etching conditions (in particular the solvent, HF concentration and current density), the variation of pore dimensions in a continuous way from nanopores (3–5 nm) to macropores (several μm) [69]. The complete process of obtaining a suitable template in silicon for PC is summarized in Fig. 6.15.

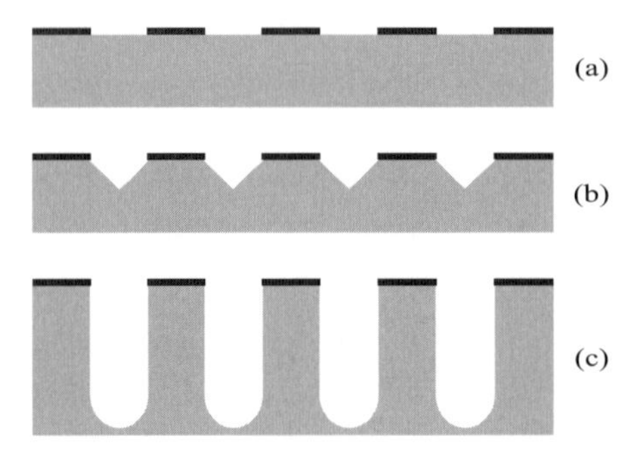

Fig. 6.15. Process steps needed in order to fabricate 2D PC on silicon by electrochemical etching: (**a**) Lattice definition; (**b**) alkaline etch pit formation; (**c**) electrochemical etching

The lithographic step defines the PC lattice that is transferred into Si by the etching. As the frequency of the photonic band gap is inversely related to the lattice pitch, the minimum PC period is determined by the Si absorption edge, $\simeq 1.1$ μm. For instance to open the first photonic band gap at 1.55 μm on a triangular lattice a pitch of about 600 nm is needed.

The second step is alkaline etching, normally performed in KOH at moderate temperature for a few seconds, to form well-defined etch pits, shaped as inverted pyramids with sharp tips. As mentioned in Chap. 3, Si etching starts at the etch pits due to the enhanced electric field present there. Some effects related with etch tip shapes have been reported [70].

Then the HF electrochemical etch step produces vertical pores with smooth walls and high aspect ratio (see Fig. 3.8). If compared with luminescent nanoporous silicon, etching conditions are modified lowering both HF concentration and current density. To obtain high aspect ratio stable macropores it is important to distinguish between *p*- and *n*-type doped silicon. The first macroporous PC were formed on *n*-type Si [71]. Etching of *n*-type doped Si substrate is very anisotropic when assisted by back illumination. Vertical pores aligned along the [100] direction without any significant lateral etching are produced [72] (see Fig. 6.16). Up to now aspect ratios of up to 400 were reported in macroporous silicon.

On *p*-type substrates, it is more difficult to obtain structures with high aspect ratio due to the non-negligible dissolution of the pore walls during the etching. Within a simplified model this can be connected with a high density of holes that reduces the space charge region which, in turn, limits the pore radius growth (see Sect. 3.1.7). Lateral pore growth can be inhibited by using *ad hoc* electrolytes, where water is substituted with aprotic organic molecules

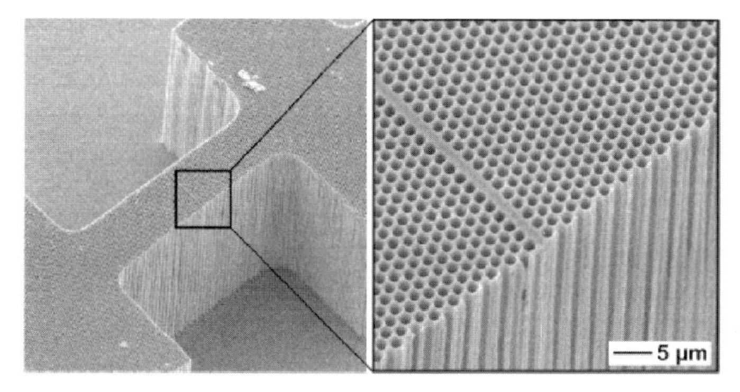

Fig. 6.16. Photonic crystals produced on n-type silicon. After [72]

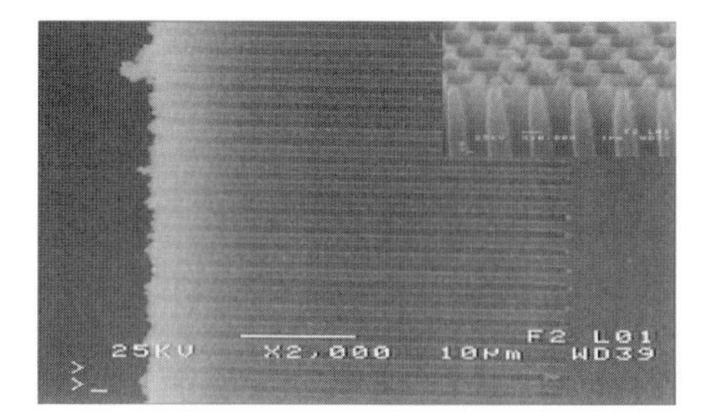

Fig. 6.17. Photonic crystals produced on p-type silicon. After [14]

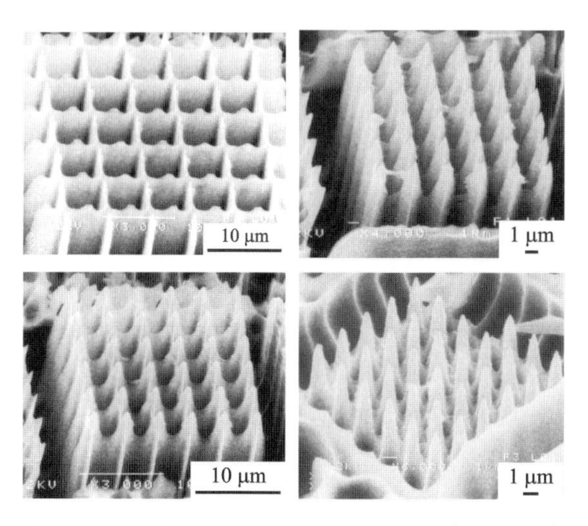

Fig. 6.18. 2D PC formed on a p-type (20 Ω cm) Si substrate by macroporous silicon etching. Square symmetry with different lattice constants (from *top left* to *bottom right*: 6, 4, 3, 2 µm). The initial filling factor was 0.25. After [14]

like DMSO and DMF [14, 73, 74]. In this way structures with extremely high aspect ratio are possible (Fig. 6.17).

An advantage of p-type with respect to n-type Si is the simplified experimental set-up to produce PC, as illumination is not required. Moreover, this substrate is more flexible due to the possibility of using controlled lateral dissolution to achieve different final structures from the same lithographic mask. It is possible to change between a lattice of pores to a lattice of pillars, simply varying HF concentration or etching time [14] as shown in Fig. 6.18.

This is because lateral dissolution proceeds faster on the center of the wall between two pores and leaves unattacked the vertex among four adjacent pores.

6.3.4 Optical Characterization

To measure photonic band structures, transmission or reflection are usually performed. Care has to be taken to avoid straight light in the detector. Transmission or reflection of light incident at a given symmetry direction can give information about the photonic band gap which is then compared with theory to achieve a full photonic band diagram [75] (Fig. 6.19).

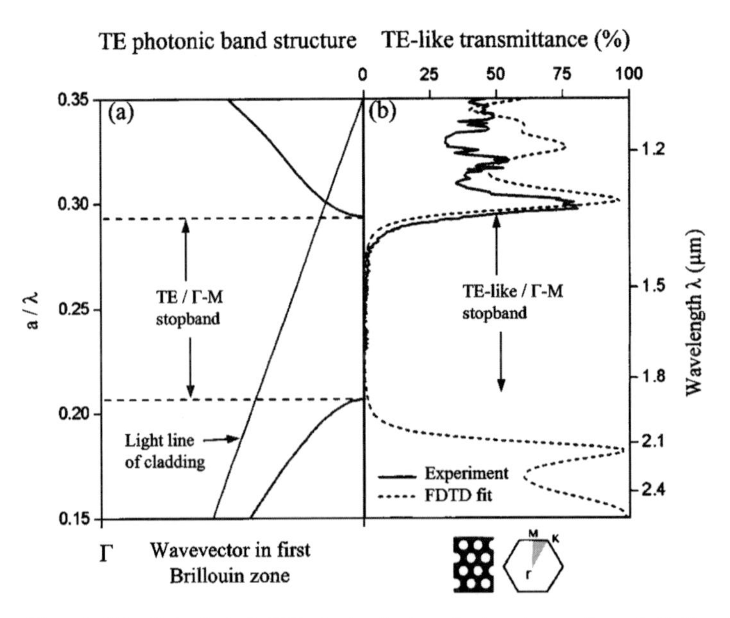

Fig. 6.19. Calculated TE photonic band structure of an infinite silicon PC slab in a symmetric oxidized SOI configuration. Measured (*solid line*) and calculated (*dashed line*) TE-like transmission of a six-row 2D triangular lattice of air holes PC. After [75]

A more direct method relies on resonant excitation of photonic modes by coupling of the $k_{\|}$ component of the incident light with a photonic mode in variable angle reflectance measurements. The excitation of a PC photonic mode produces a dispersive contribution in the reflectance spectrum. Recording the wavelength and knowing the incident angle the value of k is recovered by $k_{\|} = \omega/c \sin(\theta)$. All the points $(k_{\|}, \omega)$ describe the dispersion band [76, 77]. An example of the data and the comparison with band-structure calculations are shown in Fig. 6.20. Not all the bands can be investigated in such experiment because of the broken mirror symmetry of the PC lattice at the surface of the sample. In fact a detailed analysis of mode symmetry reveals that the only bands that can be probed by an external source are the ones with the same symmetry of the electromagnetic incident field [76].

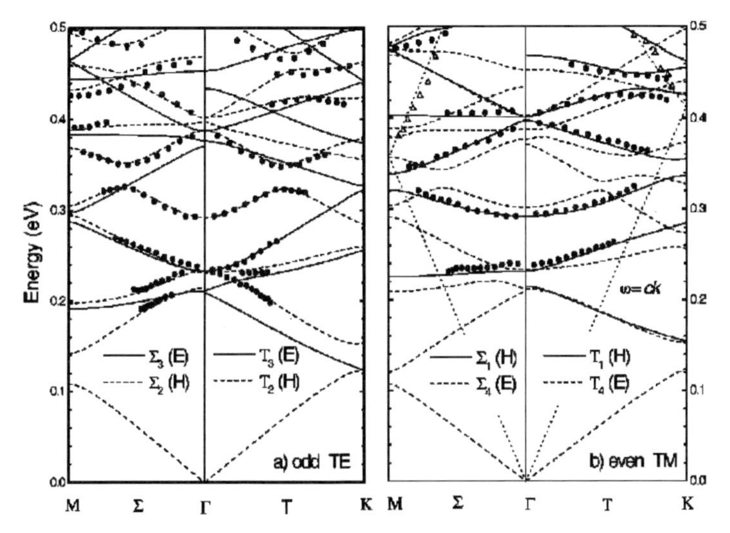

Fig. 6.20. Simulated and experimental measurements of dispersion bands for a 2D macroporous sample, with triangular lattice geometry. Experimental data were taken by variable angle reflectance. After [76]

Many different experiments have to be used to analyze the complex phenomena related with photonic crystals. For example, large dispersive effects in PC can be analyzed by using phase-sensitive ultrashort pulse interferometry. This gives information about group velocity and its dispersion [78, 79]: in PC, the speed of photons can be dramatically slowed down hundreds of times.

6.3.5 Waveguides

Introduction of defects in two-dimensional PC allows for applications. A linear defect is formed when the lattice symmetry is broken along one direction.

For a 2D PC this means that a line of pores (pillars) is removed and filled with dielectric (air) forming the line defect. As a result a defect state forms in the photonic band gap which behaves as a channel where photons of the defect energy are guided. In fact the photonic lattice surrounding the defect confines laterally the photons, as their wavelength falls inside the PC band gap. Photons are forced to move only along the channel formed by the line defect. One of the most interesting uses of this kind of defect is in waveguide applications where arbitrary shaped waveguide can be realized, e.g. with sharp bends. Three-dimensional light confinement is achieved by using a planar waveguide to confine light in the vertical direction and a line defect formed in a PC embedded in the core of the planar waveguide. Typical systems are those where SOI substrates are used: the silicon epilayer is processed to form the PC and the layered SiO_2/Si structure forms the planar waveguide [75]. An example is given in Fig. 6.21 [80].

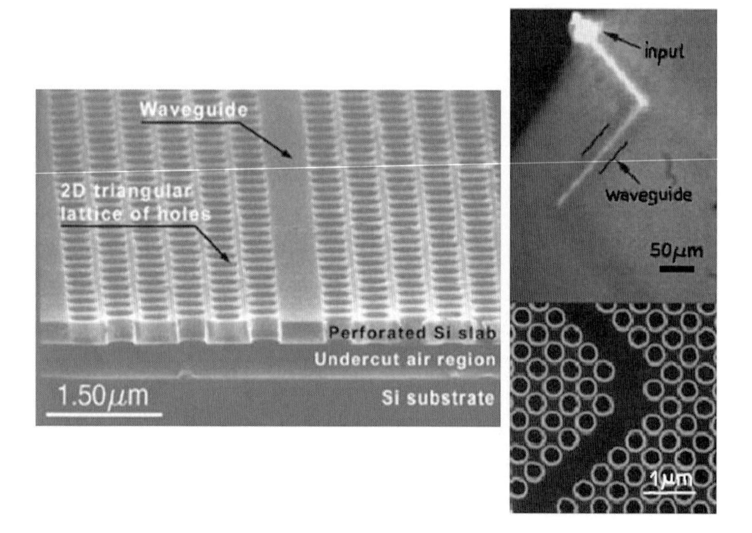

Fig. 6.21. *Left*: SEM image of a waveguide of triangular symmetry fabricated on a SOI substrate. *Right*: Light guided in the waveguide around a 90° bend (*top view*). It can be seen that the light is confined to the waveguide. *Inset* in the lower-right corner shows the SEM micrograph of the corner design in the square lattice. After [80]

In the usual waveguides, light confinement is due to total internal reflection, whereas in PC waveguide light confinement is due to something similar to a diffraction effect [81]. This allows changing abruptly the propagation direction without significant losses. However, reflection and interference effects can lower the transmission across the bend and the defects have to be accurately designed [82] to achieve almost 100% transmission [83]. In a SOI sample, propagation through 120° bent waveguide was demonstrated; the estimated losses were less than 35 dB [84].

Optimal guiding properties do not match the condition of widest photonic band gap due to out of plane losses induced by diffraction at every dielectric/air interface in the PC [56]. Many parameters should be optimized in order to reduce these losses. A PC with a high filling factor shows big losses because refraction at interfaces produces a diverging beam. This problem can be reduced either by using a low contrast index structure [85], or by using a small ff. In structures with medium to low refractive index contrast, losses are proportional to $(\Delta\epsilon)^2 = (n_{\mathrm{core}}^2 - n_{\mathrm{clad}}^2)^2$, where n_{core} and n_{clad} are respectively the refractive index of the core and the cladding of the planar waveguide. If the refractive index contrast is very high (e.g. at the Si/air interface), lossless Bloch modes inside the PC band gap can be excited. In this case, the losses are only due to processing imperfections (wall roughness, finite size of the sample) [86]. Dielectrics with low-ϵ are in principle more suitable in PC devices with high packing density of defects (cavities, filters), where photons have to propagate for small distances before being scattered by the defect. But low index materials require very deep holes (pillars), to behave as a PC. High-ϵ dielectrics are better materials for photonic circuits with long interconnection distances. Minimal losses of PC waveguides realized by 2D PC on SOI substrates were 4 dB/mm at 1.55 μm [87].

An interesting property of quasicrystalline PC is the independence of the band gap position on the photon propagation direction.[3] This allows for the fabrication of waveguides that have not to be engineered in order to minimize losses across bends [88].

6.3.6 Microcavities

2D PC offers the possibility to build small-volume, high-Q MC, easily integrable in optoelectronic devices with small dimensions ($\simeq$ 20 μm). When coupled with an active medium, the microscopic sizes of the cavity allow for ultrasmall lasers with very low power consumption and threshold-less operation [89, 90]. In 2D, PC microcavities are composed by localized point-defects, obtained either by removing holes (pillars) or by modifying the dimension of one or more holes (pillars) in the PC lattice (Fig. 6.22).

The frequency and quality factor (Q) of the cavity are defined by the dimensions of the defect. As in 1D PC the thickness of the central layer of a MC defines the cavity resonance. An example of a MC realized on a PC waveguide by using macroporous silicon [91] is shown in Fig. 6.23.

When an emitter is placed in the 2D microcavity the SE can be enhanced by a factor:

[3] Due to the lack of a real Brillouin zone for aperiodic structures. In fact, *in general*, every states is allowed in a quasicrystal, but a small number of them defines the principal points of the reciprocal lattice, and, thus, defines the properties of the crystal.

Fig. 6.22. Examples of a two-dimensional microcavity, obtained by enlarging a single hole (*left*) or removing holes (*right*) from the periodic 2D lattice

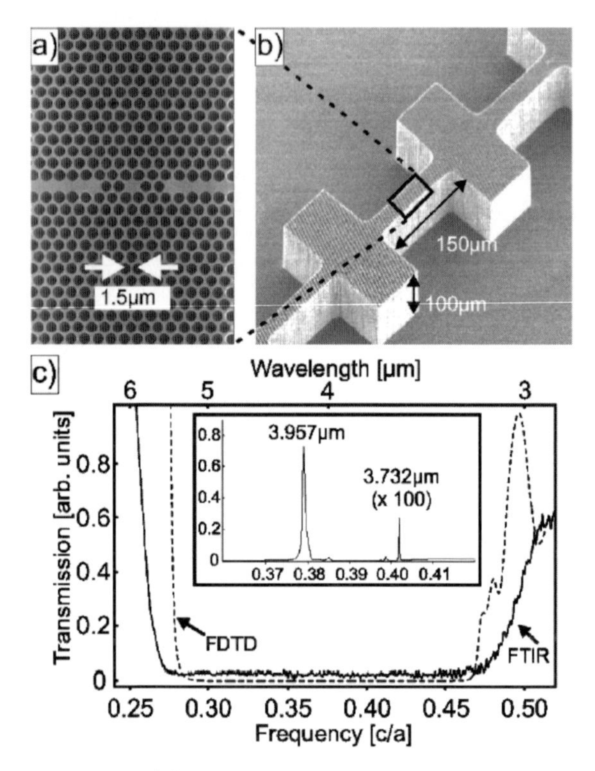

Fig. 6.23. (**a**) Top view of a zoom into the region of the crystal containing the microresonator. (**b**) Overview of the photonic crystal substrate. (**c**) The *solid curve* shows the band gap spectrum of the structure in (**b**) measured by FTIR. The *dashed curve* displays the outcome of numerical simulations (FDTD) for a crystal without defects. When defects are introduced two resonances appear, as shown in the *insert*. The horizontal axis is in reduced units of c/a where c is the speed of light and a is the lattice constant. The corresponding values of the wavelength in our experiment are indicated on the upper horizontal axis. After [91]

$$\eta = \frac{Q}{4\pi V}\left(\frac{\lambda}{n}\right)^3 \tag{6.9}$$

where V is the mode volume of the cavity. Q is related to the number of oscillations of the optical mode before it escapes from the cavity. Q scales with the refractive index contrast. Free standing 2D PC microcavities have larger Q than systems based on the SOI system. Also the vertical mode confinement, defined by hole depth, influences the Q of the cavity. It is also interesting to note that the Q-factor is determined both by the vertical $Q_{\perp}$, which is limited by the losses due to scattering at the semiconductor/hole interface, and by the $Q_{\parallel}$, the in plane Q-factor, which grows monotonically with the number of PC periods surrounding the point defect.

The symmetry and size of the defect determine the number and the type of the sustained optical modes. Simply removing an element from the lattice produces a cavity with many degenerate modes. To have a monomode cavity, either one has to introduce an asymmetry to lift the degeneracy of the modes, e.g. by shifting the positions of the holes around the defect or reducing their dimensions; or one has to select intrinsically nondegenerate modes, such as the monopole for a triangular lattice.

PC based MC are also used to study phenomena related to quantum electrodynamics where an excited atom is trapped and coupled with the cavity mode [92].

6.3.7 Coupled Defects

More advanced optoelectronic devices can be fabricated by using coupled defects, mainly waveguides and microcavities. For example for WDM application, couplers can be realized by designing two nearby parallel PC waveguides. As in normal waveguides, the number of wavelength channels (N_λ) which can be coupled in or out of a waveguide grows linearly with the ratio L/L_B, where L is the coupler length and L_B is the beating length. $L_\mathrm{B} = 2\pi/(\beta_\mathrm{odd} - \beta_\mathrm{even})$, where β is the waveguide mode propagation constant. In PC waveguide, L_B can be very short, resulting in compact devices [93].

Another WDM compact component that can be realized with PC is a channel add/drop filter. It is based on a single microcavity coupled with a waveguide [94] (Fig. 6.24).

A WG optical mode is coupled with the localized defect states, where it is emitted out of the plane. The microcavity size and the separation between defect and waveguide determine the frequencies that can couple in/out (Fig. 6.25a). An accurate design is needed to couple only the desired modes with good efficiency.

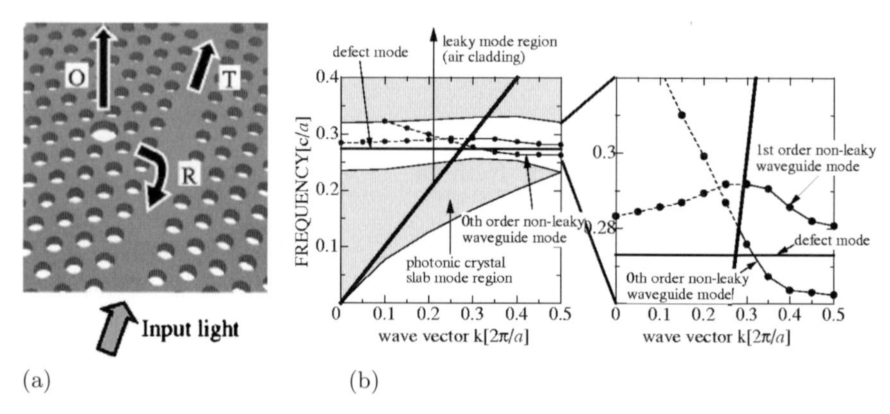

Fig. 6.24. (**a**) Scheme of a channel drop filter; (**b**) band diagram of the structure with magnification of the zone inside the gap. After [94]

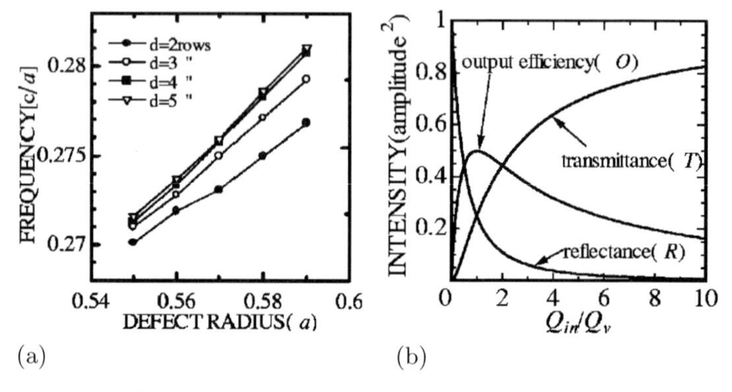

Fig. 6.25. (**a**) Scaling of resonant frequencies versus coupling strength; (**b**) intensity redistribution compare quality factors. After [94]

6.4 Three-Dimensional Photonic Crystals

3D PC were built in the early 1990s in the microwave region. Then much progress has been made, and nowadays they are widely available. 3D PC are extremely difficult to fabricate and most people think that they are even unusable because of the impossibility to design well-controlled defect schemes.

The most successful approach is based on the self-assembly technique which exploits the slow sedimentation of spheres (often made by SiO_2 or polystyrene) in a close packed structure, named *opal.* The 3D PC properties are maintained over a small length scale or domains because of disorder. Other approaches, which are based on layer-by-layer growth to form 3D PC, have limited extension in one dimension (e.g. a woodpile). Almost all techniques have large difficulties in inserting point or line defects.

The first 3D PC was proposed by E. Yablonovitch (from which the name *yablonovite*, which is shown in Fig. 6.26) and has a fcc symmetry [95, 96]. To grow the 3D PC focused ion beam (FIB) technology was used [96]. From a triangular lattice of holes patterned on the Si surface deep holes are produced by FIB at angles of 35.26° from the surface normal and spaced by 120° on the azimuth. With this method a yablonovite structure with a photonic band gap centered at 1.55 μm and a gap to midgap ratio of about 19% was produced [96].

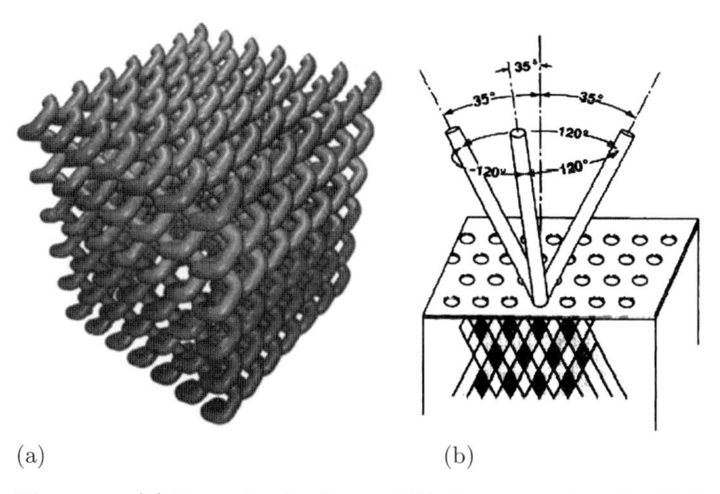

Fig. 6.26. (**a**) Example of a diamond-like lattice produced with the glancing angle technique. After [95]. (**b**) Scheme of a drilling method to obtain the yablonovite structure. After [97]

Another family of 3D lattices is built using self-organizing materials, like opals. Sedimentation by gravity [98, 99] allows fabricating the best quality opals, but takes weeks for every sample. Faster methods are based on centrifugation of the suspension [100]. Opals can easily be infiltrated with active optical materials to study the modification induced by the 3D PC of their emission properties [101]. The main problem of opals as 3D PC is the poor control of their uniformity: defects, often in the form of stacking faults or polycrystallinity, limit the maximum domain size in which the PC lattice can be considered perfect. Direct opals, formed after sedimentation, do not show a complete photonic band gap due to the low refractive index difference and air fraction in the lattice. To overcome this limitation, inverse opal, formed by infiltration of the opal structure with a high refractive index material (like Si) and subsequent dissolution of the spheres, have been developed [102]. Figure 6.27 reports reflectance spectra of samples produced starting with spheres of silica and subsequently inverted using bare silicon [103].

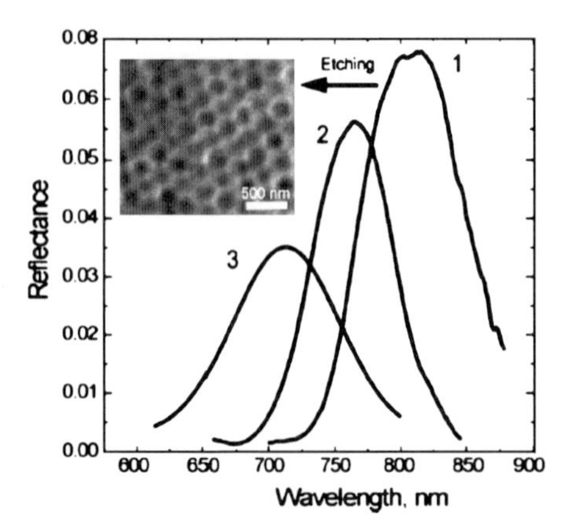

Fig. 6.27. Reflectance spectra of samples with different degree of inversion: (1) direct opal; (2) partially inverted opal; (3) Inverted opal. After [103]

Very similar to the last fabrication method is the *replica technique*. A positive crystal is created that is then inverted by filling the voids of the direct structure with a dielectric material. The removal of the template leaves the 3D PC [104, 105].

The easiest way to obtain a 3D PC is by using the *superposition* of a 1D and a 2D lattice . If compared with the previous technique, this 3D lattice is much more easy to build, it can be *doped* with defect states, but it does not have a real 3D gap, due to the lack of a true three-dimensional periodicity. Examples of structures based on this idea are etched Bragg reflectors [106] and multilayers grown on a patterned substrate, so that each layer maintains the initial patterning [107].

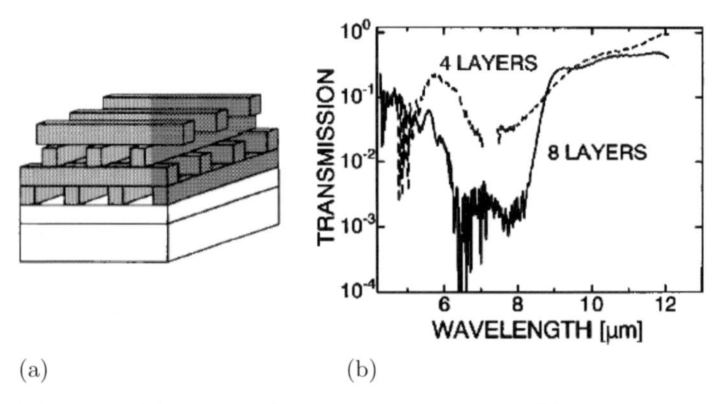

Fig. 6.28. (**a**) Scheme of the woodpile structure; (**b**) transmission spectra of woodpile samples for different layer numbers. After [108]

Other methods are based on the element by element construction of the lattice. The *glancing angle technique* [95] allows for the growth of very interesting structures with fcc symmetry; an example is show in Fig. 6.26. The crystal is formed by 3D spirals on a square lattice. This PC shows a complete photonic band gap which can have $\Delta\omega/\omega \simeq 24\%$. A *woodpile structure* is obtained by repeated wafer fusions to obtain a stacking of 2D lattices made of parallel wires [108, 109]. Each layer is rotated by $\pi/2$, as shown in Fig. 6.28. The growing technique is very complicated because of the need of wafer bonding, precise wafer allegement and a very large number of processing steps, but the obtained structures are extremely ordered and defects (localized or extended) can be easily inserted (e.g. by removing a single linear stripe from one layer, a waveguide is formed, or by enlarging the size of a rod it is possible to tailor the guiding properties).

References

1. C.M. Soukoulis: *Photonic Band Gaps and Localization* (Plenum, New York, 1993); J. Joannopoulos, R. Meade, J. Winn: *Photonic Crystals* (Princeton Press, Princeton 1995); C.M. Soukoulis: *Photonic Band Gap Materials* (Kluwer Academic, Dordrecht 1996); J. Rarity, C. Weisbuch: *Microcavities and Photonic Bandgaps, Physics and Applications* (Kluwer Academic, Dordrecht 1996)
2. E. Yablonovitch: Phys. Rev. Lett. **58**, 2059 (1987)
3. S. John: Phys. Rev. Lett. **58**(23), 2486 (1987)
4. T.F. Krauss, R.M. De La Rue: Progress in Quantum Electronics **23**, 51 (1999)
5. K. Sakoda: *Optical Properties of Photonic Crystals* (Springer Verlag, Berlin 2001) (For a rigorous theoretical treatment.)
6. P. Sheng: *Introduction to Wave Scattering, Localization, and Mesoscopic Phenomena* (Academic Press, New York 1995)
7. M. Hase, H. Miyazaki, M. Egashira, N. Shinya, K.M. Kojima, S.-I. Uchida: Phys. Rev. B **66**, 214205 (2002)
8. L. Dal Negro, C. Oton, Z. Gaburro, L. Pavesi, P. Johnson, A. Langendijk, R. Righini, M. Colocci, D.S. Wiersma: Phys. Rev. Lett. **90**, 055501 (2003)
9. D.S. Wiersma, P. Bartolini, A. Lagendijk, R. Righini: Nature **390**, 671 (1997)
10. D.S. Wiersma: Nature **406**, 132 (2000)
11. C.J. Oton, L. Dal Negro, P. Bettotti, L. Pancheri, Z. Gaburro, L. Pavesi, 'Photon states in one dimensional photonic crystals based on porous silicon multilayers'. In: *Radiation–Matter Interaction in Confined Systems*, ed. by L.C. Andreani, G. Benedek, E. Molinari (SIF, Bologna 2002), p. 302
12. Yu. A. Vlasov, X.-Z. Bo, J.C. Sturm, D.J. Norris: Nature **414**, 289 (2001)
13. M. Goulinhyan, C. Oton, Z. Gaburro, L. Pavesi: Appl. Phys. Lett. **82**, 1550 (2003); J. Appl. Phys. **93**, 9724 (2003)
14. P. Bettotti, Z. Gaburro, L. Dal Negro, L. Pavesi: J. Appl. Phys. **92**, 6966 (2002)
15. H.A. MacLeod: *Thin-Film Optical Filters* (Adam Hilger, London 1969)
16. H. Yokoyama, K. Ujihara: *Spontaneous Emission and Laser Oscillations in Microcavities* (CRC Press 1995)

17. M. Patrini, M. Galli, M. Belotti, L.C. Andreani, G. Guizzetti, G. Pucker, A. Lui, P. Bellutti, L. Pavesi: J. Appl. Phys. **92**(4), 1816 (2002)
18. S. Chan, P.M. Fauchet: Proc. SPIE **3630**, 144 (1999)
19. V.A. Tolmachev, L.S. Granitsyna, E.N. Vlasova, B.Z. Volchek, A.V. Nashchekin, A.D. Remenyuk, E.V. Astrova: Semiconductors **36**, 932 (2002)
20. G.W. Pickrell, H.C. Lin, K.L. Chang, K.C. Hsieh, K.Y. Cheng: Appl. Phys. Lett. **78**, 1044 (2001)
21. D. Sotta, E. Hadji, N. Magnea, E. Delamadeleine, P. Besson, P. Renard, H. Moriceau: J. Appl. Phys. **92**, 2207 (2002)
22. O. Bisi, S. Ossicini, L. Pavesi: Surf. Sci. Rep. **38**, 5 (2000)
23. J. Lekner: J. Opt. A **2**, 349 (2000)
24. M. Galli, D. Bajoni, F. Marabelli, L.C. Andreani, L. Pavesi, G. Pucker: Phys. Rev. B (2004) to be published
25. Yu. A. Vlasov, M.A. Kaliteevski, V.V. Nikolaev: Phys. Rev. B **60**, 1555 (1999)
26. G. D'Aguanno, M. Centini, M. Scalora, C. Sibilia, M.J. Bloemer, C.M. Bowden, J.W. Haus, M. Bertolotti: Phys. Rev. E **63**, 036610–1 (2001)
27. D.-Y. Jeong, Y.H. Ye, Q.M. Zhang: J. Appl. Phys. **92**, 4194 (2002)
28. A. Haché, M. Bourgeois: Appl. Phys. Lett. **77**(25), 4089 (2000)
29. R.P. Stanley, R. Houdré, U. Oesterle, M. Ilegems, C. Weisbuch: Phys. Rev. A **48**, 2246 (1993)
30. V. Pellegrini, A. Tredicucci, C. Mazzoleni, L. Pavesi: Phys. Rev. B **52**, R14328 (1995)
31. C.H.R. Ooi, T.C.A. Yeung, T.K. Lim, C.H. Kam: Phys. Rev. E **62**, 7405 (2000)
32. L. Pavesi, V. Mulloni: J. Lumin. **80**, 43 (1999)
33. P.J. Reece, G. Lerondel, W.H. Zheng, M. Gal: Appl. Phys. Lett. **81**, 4895 (2002)
34. G. Bijörk, S. Machida, Y. Yamamoto, K. Igeta: Phys. Rev. A **44**(1), 669 (1991)
35. S.M. Dutra, P.L. Knight: Phya. Rev. A **53**, 3587 (1996)
36. E.F. Schubert, N.E.J. Hunt, A.M. Vredenberg, T.D. Harris, J.M. Poate, D.C. Jacobson, Y.H. Wong, G.J. Zydzik: Appl. Phys. Lett. **63**, 2603 (1993)
37. A.M. Vredenberg, N.E.J. Hunt, E.F. Schubert, D.C. Jacobson, J.M. Poate, G.J. Zydzik: Phys. Rev. Lett. **71**, 517 (1993)
38. T.V. Dolgova, A.I. Maĭdikovskiĭ, M.G. Martem'yanov, G. Marovsky, G. Mattei, D. Schuhmacher, V.A. Yakovlev, A.A. Fedyanin, O.A. Aktsipetrov: JETP Lett. **73**, 6 (2001)
39. G. D'Aguanno, M. Centini, C. Sibilia, M. Bertolotti, M. Scalora, M.J. Bloemer, C.M. Bowden: Opt. Lett. **24**, 1663 (1999)
40. Y. Dumeige, P. Vidakovic, S. Sauvage, I. Sagnes, J.A. Levensona, C. Sibilia, M. Centini, G. D'Aguanno, M. Scalora: Appl. Phys. Lett. **78**, 3021 (2001)
41. M. Bayer, T.L. Reinecke, F. Weidner, A. Larionov, A. McDonald, A. Forchel: Phys. Rev. Lett. **86**, 3168 (2001)
42. B. Ohnesorge, M. Bayer, A. Forchel, J.P. Reithmaier, N.A. Gippius, S.G. Tikhodeev: Phys. Rev. B **56**, R4367 (1997)
43. D.G. Deppe, D.L. Huffaker, T.-H. Oh, H. Deng, Q. Deng: IEEE J. of Selec. Topics In Quantum Electr. **3**, 893 (1997)
44. E.F. Schubert, N.E. Hunt, M. Micovic, R.J. Malik, D.L. Sivco, A.Y. Cho, G.J. Zydzik: Science **265**, 943 (1994)

45. M.S. Ünlü, S. Strite: J. Appl. Phys. **78**, 607 (1995)
46. M.S. Skolnick, T.A. Fisher, D.M. Whittaker: Semicond. Sci. Technol. **13**, 645 (1998)
47. M. Lipson, L.C. Kimerling: Appl. Phys. Lett. **77**, 1150 (2000)
48. L. Pavesi, G. Panzarini, C. Andreani: Phys. Rev. B **58**, 15794 (1998)
49. Y. Fink, J.N. Winn, S. Fan, C. Chen, J. Michel, J.D. Joannopoulos, E.L. Thomas: Science **282**(27), 1679 (1998)
50. D.N. Chigrin, A.V. Lavrinenko, D.A. Yarotsky, S.V. Gaponenko: Appl. Phys. A **68**, 25 (1999)
51. V. Savona, L.C. Andreani, P. Schwendimann, A. Quattropani: Solid State Commun. **93**, 733 (1995)
52. T. Tada, V.V. Poborchii, T. Kanayama: Jpn. J. Appl. Phys. **38**, 7253 (1999)
53. V. Grigaliunas, V. Kopustinskas, S. Meskinis, M. Margelevicius, I. Mikulskas, R. Tomasiunas: Opt. Mater. **17**(1–2), 15 (2001)
54. O. Hanaizumi, Y. Othera, T. Sato, S. Kawakami: Appl. Phys. Lett. **74**, 777 (1999)
55. M. Plihal, A.A. Maradudin: Phys. Rev. B **44**, 8565 (1991)
56. H. Benisty, C. Weisbuch, D. Labilloy, M. Rattier: Appl. Surf. Sci. **164**, 205 (2000)
57. R.D. Meade, O. Alerhand, J.D. Joannopoulos: *Handbook of Photonic Band Gap Materials* (JAMteX I.T.R. 1993)
58. X. Zhang, Z.-Q. Zhang, L.-M. Li: Phys. Rev. B **61**(3), 1892 (2000)
59. C. Jin, X. Meng, B. Cheng, Z. Li, D. Zhang: Phys. Rev. B **63**, 195107 (2001)
60. M.E. Zoorob, M.D.B. Charlton, G.J. Parker, J.J. Baumberg, M.C. Netti: Mat. Sci. and Eng. B **74**, 168 (2000)
61. X. Zhang, Z.-Q. Zhang, C.T. Chan: Phys. Rev. B **63**, 081105 (2002)
62. K. Bush: C.R. Physique **3**, 53 (2002)
63. H. Masuda, M. Ohya, H. Asoh, K. Nishio: Jpn, J. Appl. Phys. Part 2 **40**, L1217 (2001)
64. H.W.P. Koops, O.E. Hoinkis, M.E.W. Honsberg, R. Schimdt, R. Blum, G. Böttger, A. Kuligk, C. Liguda, M. Eich: Microlectr. Engin. **57–58**, 995 (2001); I.B. Divliansky, A. Shishido, I.C. Khoo, T.S. Mayer, D. Pena, S. Nishimura, C.D. Keating, T. Mallok: Appl. Phys. Lett. **79**, 3392 (2001); T. Kondo, S. Matsuo, S. Juodkazis, H. Misawa: Appl. Phys. Lett. **79**, 725 (2001); C.C. Cheng, A. Scherer, R.C. Tyan, Y. Fainman, G. Witzgall, E. Yablonovitch: J. Vac. Sci. Technol. B **15**, 2764 (1997); S.Y. Chou, C. Keimel, J. Gu: Nature **417**, 835 (2002); Some results are also available with the solgel technology as shown In: S. Shimada, K. Miyazawa, M. Kuwabara: Jpn. J. Appl. Phys. Part 2 **41**, L291 (2002)
65. C. Reese, B. Gayral, G.B. Gerardot, A. Imamoglu, P.M. Petroff, E. Hu: J. Vac. Sci. Technol. B **19**, 2749 (2001)
66. T.F. Krauss, R.M. De La Rue: Progress in Quantum Electronics **23**, 51 (1999)
67. T. Zijstra, E. van der Drift, M.J.A. de Dood, E. Snoeks, A. Polman: J. Vac. Sci. Technol. B **17**, 2734 (1999)
68. M. Loncar, D. Nedeljkovic, T.P. Pearsall: Appl. Phys. Lett. **77**, 1937 (2000)
69. X.G. Zhang: *Electrochemistry of Silicon and its Oxide* (Kluwer Academic-Plenum Publisher, New York 2001)
70. H. Ohji, P.J. French, S. Izuo, K. Tsutsumi: Sensors and Actuators **85**, 390 (2000)

71. V. Lehmann, H. Föll: J. Electrochem. Soc. **137**, 653 (1990)
72. A. Birner, R.B. Wehrspohn, U.M. Gösele, K. Busch: Adv. Mater. **13**, 377 (2001)
73. S. Lust, C. Levy-Clement: Phys. Stat. Sol (a) **182**, 17 (2000)
74. S. Lust, C. Levy-Clement: J. Electrochem. Soc. **149**, C338 (2002)
75. M. Zelsmann, E. Picard, T. Charvolin, E. Hadji, B. Dalzotto, M.E. Nier, C. Seassal, P. Rojo-Romeo, X. Letartre: Appl. Phys. Lett. **81**, 2340 (2002)
76. M. Galli, M. Agio, L.C. Andreani, M. Belotti, G. Guizzetti, F. Marabelli, M. Patrini, P. Bettotti, L. Dal Negro, Z. Gaburro, L. Pavesi, A. Lui, P. Bellutti: Phys. Rev. B **65**, 113111 (2002).
77. H.T. Miyazaki, H. Miyazaki, K. Ohtaka, T. Sato: J. Appl. Phys. **87**, 7152 (2000)
78. A. Imhof, W.L. Vos, R. Sprik, A. Lagendijk: Phys. Rev. Lett. **83**, 2942 (1999)
79. M. Notomi, K. Yamada, A. Shinya, J. Takahashi, C. Takahashi, I. Yokohama: Phys. Rev. Lett. **87**, 253902 (2001)
80. M. Loncar, D. Nedeljkovic, J. Vuckovic, A. Scherer, T.P. Pearsall: Appl. Phys. Lett. **77**, 1937 (2000)
81. H. Benisty, C. Weisbuch, D. Labilloy, M. Rattier, C.J.M. Smith, T.F. Krauss, R.M. De La Rue, R. Houdré, U. Oesterle, C. Jouanin, D. Cassagne: IEEE J. Lightwave Technol. **17**(11), 1 (1999)
82. A. Talneaua, L. Le Gouezigou, N. Bouadma, M. Kafesaki, C.M. Soukoulis, M. Agio: Appl. Phys. Lett. **80**, 547 (2002)
83. A. Mekis, J.C. Chen, I. Kurland, S. Fan, P.R. Villeneuve, J.D. Joannopoulos: Phys. Rev. Lett. **77**, 3787 (1996)
84. M. Tokushima, H. Kosaka, A. Tomita, H. Yamada: Appl. Phys. Lett. **76**, 952 (2000)
85. J. Arentoft, M. Kristensen, T. Sondergaard, A. Boltasseva: Proc. 27th Eur. Conf. on Opt. Comm. (ECOC'01 - Amsterdam) 592 (2001)
86. W. Bogaerts, P. Bienstman, D. Taillaert, R. Baets, D. De Zutter: IEEE Photon. Technol. Lett. **13**, 565 (2001)
87. J. Arentoft, T. Sondergaard, M. Kristensen, A. Boltasseva, M. Thorhange, L. Frandsen: Electron. Lett. **38**, 274 (2002)
88. C. Jin, B. Cheng, B. Man, Z. Li, D. Zhang, S. Ban, B. Sun: Appl. Phys. Lett. **75**, 1848 (1999)
89. M. Lončar, T. Yoshie, A. Scherer, P. Gogna, Y. Qiu: Appl. Phys. Lett. **81**, 2680 (2002)
90. H.-Y. Ryu, H.-G. Park, Y-H. Lee: IEEE J. of Selec. Topics in Quantum Electron. **8**, 891 (2002)
91. P. Kramper, A. Birner, M. Agio, C.M. Soukoulis, F. Müller, U. Gosele, J. Mlynek, V. Sandoghdar: Phys. Rev. B **64**, 233102 (2001)
92. J. Vučović, M. Lončar, H. Mabuchi, A. Scherer: Phys. Rev. E **65**, 016008 (2002)
93. S. Boscolo, M. Midrio, C.G. Someda: IEEE J. Quantum Electron. **38**, 47 (2002)
94. A. Chutinan, M. Mochizuki, M. Imada, S. Noda: Appl. Phys. Lett. **79**, 2690 (2001)
95. O. Toader, S. John: Phys. Rev. E **66**, 016610 (2002)
96. A. Chelnokov, K. Wang, S. Rowson, P. Garoche, J.-M. Lourtioz: Appl. Phys. Lett. **77**, 2943 (2000)

97. E. Yablonovitch, T.J. Gmitter, K.M. Leung: Phys. Rev. Lett. **67**, 2295 (1991)
98. R. Mayoral, J. Requena, J.S. Moya, C. Lopez, A. Cintas, H. Miguez: Adv. Mater. **9**, 257 (1997)
99. M. Dongbin, L. Hongguang, C. Bingying, L. Zhaolin, Z. Daozhong, D. Peng: Phys. Rev. B **58**, 35 (1998)
100. N.P. Johnson, D.W. McComb, A. Richel, B.M. Treble, R.M. De La Rue: Synthetic Metals **116**, 469 (2001)
101. S.V. Gaponenko, A.M. Kapitonov, V.N. Bogomolov, A.V. Prokofiev, A. Eychmüller, A.L. Rogach: JETP Lett. **68**, 142 (1998)
102. A. Blanco, E. Chomski, S. Grabtchak, M. Ibisate, S. John, S.W. Leonard, C. Lopez, F. Meseguer, H. Miguez, J.P. Mondia, G.A. Ozin, O. Toader, H.M. van Driel: Nature **405**, 437 (2000)
103. V.G. Golubev, J.L. Hutchison, V.A. Kosobukin, D.A. Kurdyukov, A.V. Medvedev, A.B. Pevtsov, J. Sloan, L.M. Sorokin: J. Non-Cryst. Solids **299–302**, 1062 (2002)
104. M. Meier, A. Dodabalapur, J.A. Rogers, R.E. Slusher, A. Mekis, A. Timko, C.A. Murray, R. Ruel, O. Nalamasu: J. Appl. Phys. **86**(7), 3502 (1999)
105. S.G. Romanov, T. Maka, C.M.S. Torres: Appl. Phys. Lett. **79**, 731 (2001)
106. E. Pavarini, L.C. Andreani: Phys. Rev. E **66**, 036602 (2002)
107. T. Kawashima, K. Miura, T. Sato, S. Kawakami: Appl. Phys. Lett. **77**, 2613 (2000)
108. S. Noda, N. Yamamoto, M. Imada, H. Kobayashi, M. Okano: J. Lightwave Technol. **17**, 1948 (1999)
109. S. Noda, N. Yamamoto, H. Kobayashi, M. Okano, K. Tomoda: Appl. Phys. Lett. **75**, 905 (1999)

7 Conclusions and Future Outlook

Initiated by technical investigations in the 1940s and started as an industry in the late 1960s, CMOS technology has distinguished itself by the rapid improvements in its products [1]. In a four-decade history, most of the figures of merit of the Si industry have been marked by exponential growth, principally as a result of the ability to exponentially decrease the minimum feature sizes of integrated circuits (ICs). As of today, the development of devices and processors appears to stand and possibly even overshoot Moore's law [2, 3].

On the other hand, it is becoming more apparent that development of silicon electronic IC will be limited by the rate at which data can be transmitted between devices and chips [4]. The use of copper and low k materials will allow scaling of the intermediate wiring levels, and minimize the impact on wiring delay. RC delay, however, is dominated by global (i.e. extended over the whole chip size) interconnect and the benefit of materials changes alone is insufficient to meet overall performance requirements. Repeaters can be incorporated to mitigate the delay in global wiring but consume power and chip area. Inductive effects will also become increasingly important as frequency of operation increases, and additional metal patterns or ground planes may be required for inductive shielding. As supply voltage is scaled or reduced, cross-talk has become an issue for all clock and signal wiring levels. In the long term, disruptive new design or technology solutions will be needed to overcome the performance limitations of traditional interconnect.

Optical interconnects are considered a primary option for replacing the conductor/dielectric system for global interconnects [5]. It is meaningful to assume that the optical solution will meet speed requirements because the signal travels at the speed of light. Moreover, optical signals have other advantages, such as cross-talk immunity. Optical interconnect technology has many variants. The simplest approach is to have emitters off-chip and only waveguides and detectors in top layers on-chip [6]. Such a simple solution–although not ideal for several reasons, such as reliability, cost or power consumption–motivates the development of all-Si passive photonic and optoelectronic devices. Progressively more complex options culminate in monolithic emitters, waveguides, and detectors.

7.1 Silicon Photonics

Silicon microphotonics, or photonics for short, is the optical equivalent of microlectronics for integrated circuits [6]. Integration of various optical functionalities with and within integrated circuit electronics is the overall goal of this technology. Silicon microphotonics aims to use silicon-based materials that are process compatible with standard integrated circuit fabrication methods [7].

Various applications can be found for silicon photonics: circuits for optical network access, driving circuits for displays, and, most important, optical interconnects within chips. The rationale and the challenges which have to be faced to realize optical interconnects have been reviewed in [5].

7.1.1 Silicon Photonic Components

It was predicted in the early 1990s that silicon based optoelectronics will be a reality before the end of the century [8]. Indeed, all the basic components have already been demonstrated [9], but for a silicon laser. The first essential component in silicon microphotonics is the medium through which light propagates: the waveguide. To realize low-loss optical waveguides, various approaches have been followed [10]: low dielectric mismatch structures (e.g. doped silica [11], silicon nitride [12] or silicon oxynitride on oxide [13], or differently doped silicon [14]) or high dielectric mismatch structures (e.g. silicon on oxide [6]). Low loss silica waveguides are characterized by large dimensions, typically 50 μm of thickness, due to the low refractive index mismatch ($\Delta = 0.1$–0.75%). Silica waveguide have a large mode spatial extent and thus are interesting for coupling with optical fiber but not to be integrated into/within electronic circuits because of the significant difference in sizes (see Fig. 7.1). The large waveguide size also prevents the integration of a large number of optical components in a single chip. Similar problems exist for silicon on silicon waveguides where the index difference is obtained by varying the doping density [14]. Silicon on silicon waveguides are very effective for realizing free-carrier injection active devices as well as fast thermo-optic switches thanks to the high thermal conductivity of silicon. A major problem with these waveguides is the large free carrier absorption which causes optical losses of some dB/cm for single-mode waveguides at 1.55 μm. Silicon nitride based waveguides [12] and silicon oxynitride waveguides [13] show losses at 633 nm lower than 0.5 dB/cm and bending radii of less than 200 μm. Both these waveguides are extremely flexible with respect to the wavelength of the signal light: both visible and IR.

At the other extreme, silicon on insulator (SOI) or polysilicon-based waveguides allow for a large refractive index mismatch and, hence, for small size waveguides in the submicrometer range. This allows a large number of optical components to be integrated within a small area. Optical losses as low as 0.1 dB/cm at 1.55 μm have been reported for channel waveguides in SOI

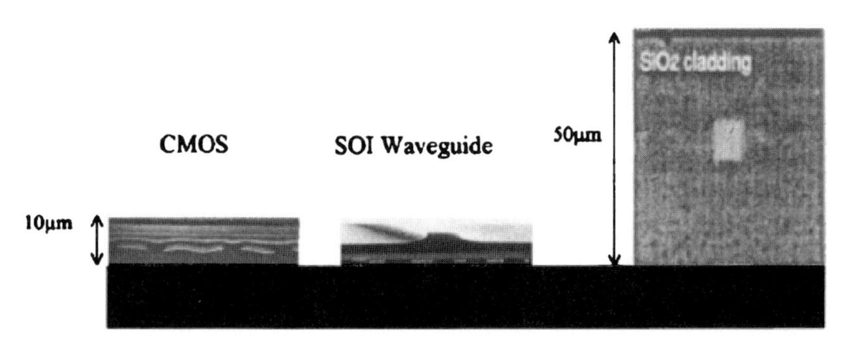

Fig. 7.1. Comparison of the cross-sections of a typical CMOS circuit, of a SOI waveguide and of a typical silica waveguide. After [15]

(optical mode cross-section 0.2×4 μm^2) [16]. Ideal for on-chip transmission, SOI waveguides have coupling problems with silica optical fiber due to both the large size difference and the different optical impedance of the two systems. Various techniques have been proposed to solve these problems, among which adiabatic tapers, V-grooves and grating couplers [17, 18]. Large single-mode stripe-loaded waveguides on SOI can be achieved provided that the stripe and the slab are both made of silicon [15]. This SOI system provides low loss waveguides (≤ 0.2 dB/cm) with single mode operation with large rib structures (optical mode cross-section 4.5×4 μm^2) and low birefringence ($\leq 10^{-3}$). Appropriate geometry with the use of asymmetric waveguide allows bend radii as short as 0.1 mm [19]. A number of photonic components in SOI have been demonstrated [15] and commercialized [19]: directional couplers, dense WDM arrayed waveguide grating, Mach–Zehnder filters, star couplers, etc.

The optical signal is converted into an electrical signal by using silicon based photodetectors. Detectors for silicon photonics are based on three different approaches [20]: silicon photoreceivers for $\lambda \leq 1.1$ μm, hybrid systems and heterostructure based systems. High speed (up to 8 Gb/s) monolithically integrated silicon photoreceivers at 850 nm have been fabricated by using 130-nm CMOS technology on a SOI wafer [21]. Other recent results confirm the ability of silicon integrated photoreceivers to detect signals with a high responsivity of 0.46 A/W at 3.3 V for 845 nm light and 2.5 Gb/s data rate [22]. The heterostructure approach is mainly based on the heterogrowth of Ge rich SiGe alloys: Ge-on-Si photodetectors have been reported with a responsivity of 0.89 A/W at 1.3 μm and 50 ps response time [23, 24]. 1% quantum efficiency at 1.55 μm in a MSM detector based on a Si/SiGe superlattice shows that promising developments are possible [25]. Similarly waveguide photodetectors with Ge/Si self-assembled islands show responsivities of 0.25 mW at 1.55 μm with zero bias [26].

Almost all the other photonics components have been demonstrated in silicon microphotonics. Optical modulator, optical routers, and optical switching systems have all been integrated into silicon waveguides by using various effects such as the plasma dispersion effect [27]. Discussion of a series of photonics components realized within SOI waveguide is reported in [15] which includes the plasma dispersion effect based active grating, evanescent waveguide coupled silicon-germanium based photodetectors, and Bragg cavity resonant photodetectors.

7.1.2 Silicon Photonic Integrated Circuits

Based on the technologies reported in the previous section, various demonstrators of fully integrated silicon photonic circuits have been reported. Here we discuss some examples.

Hybrid integration of active components and silica-based planar lightwave circuits provide a full scheme for photonic component integration within a chip [28]. Passive components are realized by using silica waveguides while active components are hybridized within the silica (see Fig. 7.2).

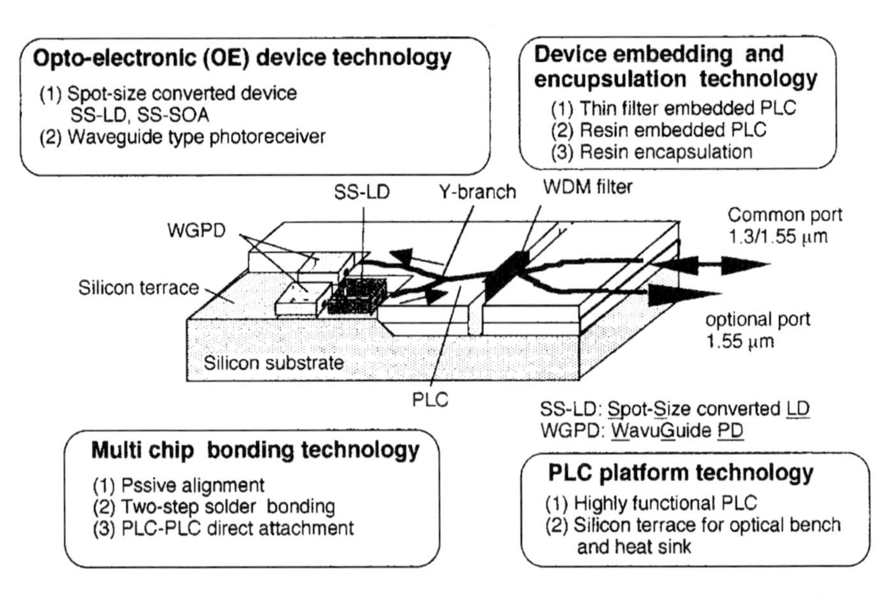

Fig. 7.2. Example of the various devices that can be integrated on a silica based lightwave circuit. SS-LD stands for laser diode, WGPD stands for photodetectors. After [28]

Active components (laser diodes, semiconductor optical amplifiers and photodiodes) are flip-chip bonded on silicon terraces where also the optical waveguides are formed. By using this approach various photonic components have been integrated such as multiwavelength light sources, optical wavelength selectors, wavelength converters, all optical time-division multiplexers, etc. [28]. Foreseen applications are WDM transceiver modules for fiber-to-the-home.

A fully integrated optical system based on silicon oxynitride waveguides, photodetectors and CMOS transimpedance amplifiers has been realized [13]. Coupling of visible radiation to a silicon photodetector can be achieved by using mirrors at the end of the waveguide. These are obtained by etching the end of the waveguide with an angle so that the light is reflected at almost 90 degrees into the underlying photodetector. A schematic of the approach is shown in Fig. 7.3.

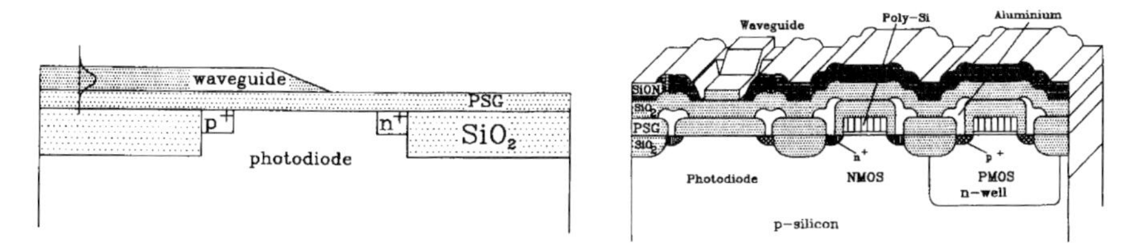

Fig. 7.3. *Left* Principle of the method of coupling of light from the waveguide into the photodiode with a curved mirror. *Right* Cross-section of the device indicating all the different parts of the CMOS circuits, in particular, the waveguide and the photodiode and a test CMOS. After [13]

Commercial systems for the access network telecom market has been realized by using SOI waveguides and the silicon optical bench approach to interface the waveguides with both III-V laser sources and III-V photodetectors. Lasers and photodetectors are lodged into etched holes in silicon and bump soldered in place. The system operates at 1.55 μm with a typical bit rate of 155 Mb/s [19]. A further advantage of the use of large optical mode waveguide is the ease of interfacing to a single-mode optical fiber. In the approach described here these are located in V-grooves etched into silicon.

A fully integrated system working at 1.55 μm has been demonstrated based on silicon waveguides with extremely small optical mode (cross-section 0.5×0.2 μm^2) which allows extremely small turn radii (1 μm) [6]. In this way a large number of optical components can be integrated on a small surface ($\simeq 10\,000$ components per cm^2). Detectors are integrated within silicon by using Ge hetero-growth on silicon itself. Responsivity of 250 mA/W at 1.55 μm and response times shorter than 0.8 ns have been achieved [23]. A scheme for optical clock distribution within integrated circuits based on this approach is shown in Fig. 7.4.

A realistic bidirectional optical bus architecture for clock distribution on a Cray T-90 supercomputer board based on polyimide waveguides (loss of 0.21 dB/cm at 850 nm), GaAs VCSEL and silicon MSM photodetectors has been tested [29]. By using a 45° TIR mirror, coupling efficiencies as high

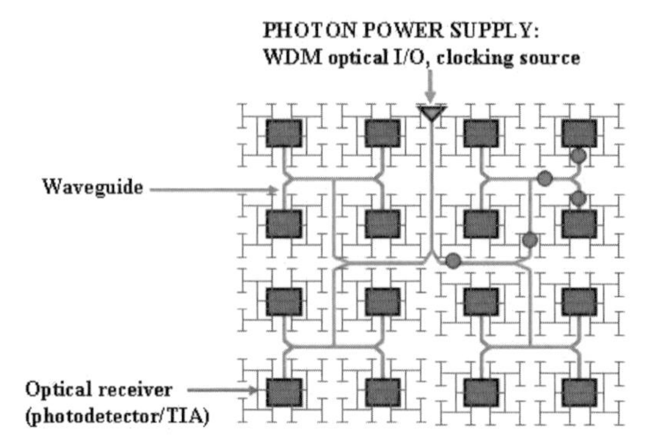

Fig. 7.4. Scheme for an integrated optical circuit to distribute the clock signal on a chip. After [6]

as 100% among the sources or the detectors and the waveguides have been demonstrated. Examples of the connection scheme are shown in Fig. 7.5.

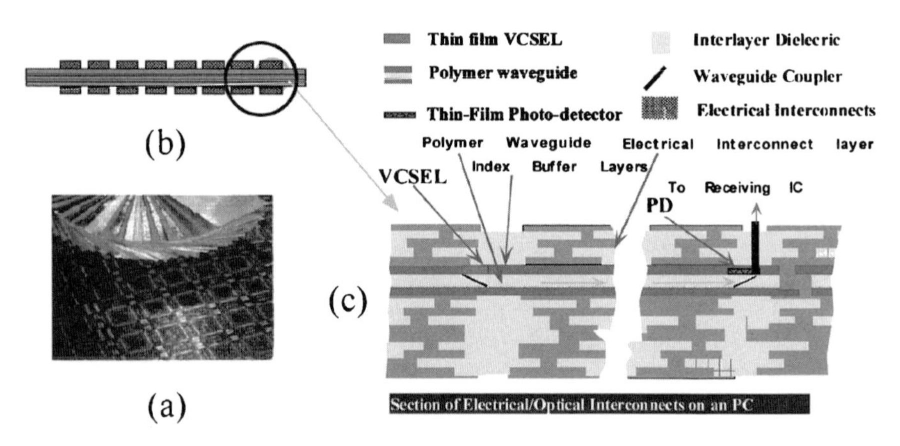

Fig. 7.5. The optical interconnect scheme proposed in [29]. (**a**) 52 integration layers of a supercomputer board; (**b**) schematic of the side view of the vertical integration layers; (**c**) details of the schematic shown in (**b**). After [29]

7.2 Towards the First Silicon Laser

To achieve monolithically integrated silicon microphotonics, the main limitation is the lack of any practical Si-based light sources: either efficient light emitting diodes (LED) or Si lasers. To overcome this, one has to use either an external "photon" source as in Fig. 7.4 or the hybrid integration of III-V laser on silicon. On the other hand in this book we have discussed the progress towards an efficient silicon based light source. Let us discuss now the state-of-the-art of the research towards a silicon laser. In order to be compatible with the various schemes discussed in the previous section the laser should be either in the visible or at 1.55 µm. In Fig. 7.6, the various approaches currently under research are summarized [30].

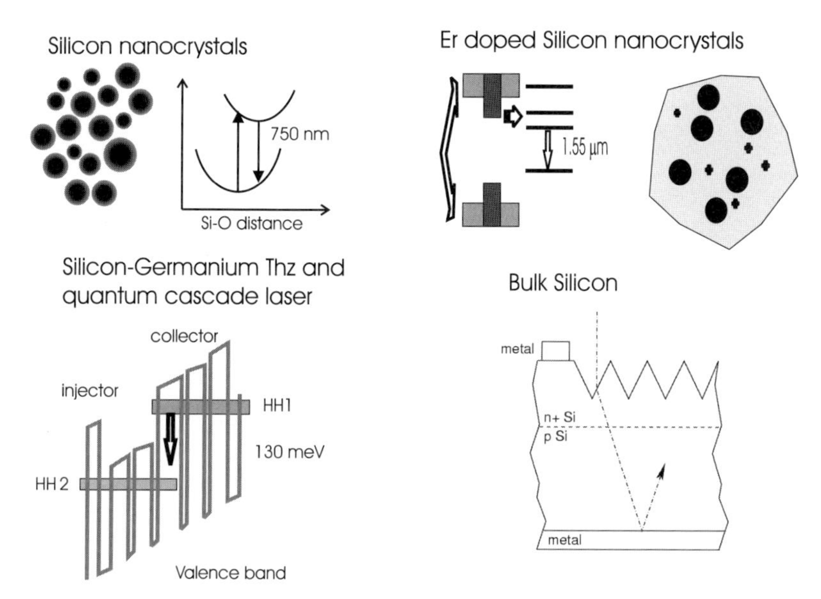

Fig. 7.6. The various approaches proposed to achieve a silicon laser

7.2.1 Bulk Silicon

Silicon is an indirect band gap material, thus the probability for a radiative transition is very low. In order to defeat this low radiative probability, one can decrease to zero the nonradiative recombination probability: thus the only recombination path available for injected carriers is radiative recombination (see Sect. 1.2.2). In addition if one succeeds in getting all the generated photons out of the semiconductor by using surface texturing or the photonic crystal approach, one can achieve extremely high LED plug-in efficiency. This route was indeed followed and LED with power efficiency approaching 1% were demonstrated [31]. In addition a fully integrated opto-coupler device (LED coupled to a photodetector) was also demonstrated on the basis of this technology [32]. The main drawbacks of this approach for an integrated laser or light emitting diode are:

- the need of both high purity (low doping concentration) and of surface texturing which renders the device processing incompatible with standard CMOS processing;

- the strong and fast free carrier absorption typical of bulk Si that can prevent reaching the condition for population inversion [33];
- suitable integration of the active bulk Si into an optical cavity to achieve the required optical feedback to sustain the laser action can be a problem;
- the modulation speed of the device which can be limited by the long lifetime of the excited carriers (ms).

A somewhat different approach exploits the strain produced by localized dislocation loops to form energy barriers for carrier diffusion (see Sect. 1.2.2). A remarkable feature of this device is the high injection efficiency into the confined regions [34]. The main problem of this approach for a silicon laser is that it does not remove the two main problems of silicon which prevent population inversion, i.e. Auger recombination [35, 36] and free carrier absorption [33]. Still the power efficiency of the reported diodes are lower than those required in real devices.

Finally a problem is also related to the wavelength of emission of these bulk silicon LED which is resonant with the silicon band gap: it is very difficult to control the region where the light is channelled in silicon if one uses these LED as a source for optical interconnects. Light will propagate through the wafer and will be absorbed in unwanted places.

7.2.2 Silicon Nanocrystals

Another way to increase the emission efficiency of silicon is to turn it into a low-dimensional material and, hence, to exploit quantum confinement effects to increase the radiative probability of carriers (see Sect. 1.3.5).

Porous silicon was among the first systems studied (see Chap. 3). It has however a drawback in the high reactivity of its sponge-like structure which causes the rapid ageing of LED and uncontrollable variations of the LED performances with time. No optical gain was reported in bulk PS. Silicon nanocrystals can be obtained by scrapping or ultrasonically dispersing porous silicon [37]. Evidence of amplification in these materials has been presented [38].

An alternative way is to produce silicon nanocrystals (Si-nc) in a silica matrix to exploit the quality and stability of the SiO_2/Si interface and the improved emission properties of low-dimensional silicon (see Sect. 4.3). In these systems optical gain has been observed (Sect. 4.3.5) [39]. Almost all the authors agree on the fact that the gain is due to localized state recombination either in the form of silicon dimers or in the form of Si=O bonds formed at the interface between the Si-nc and the oxide or within the oxide matrix [38, 39, 40, 42, 43, 44, 45]. The suggested scheme to explain population inversion and hence gain is a four-level model where a large lattice relaxation of the localized center gives rise to the four levels (see Fig. 4.47) [40, 41].

Optical gain in Si-nc has been revealed either as a superlinear increase of the luminescence intensity [43], or as the measurement of amplified spontaneous emission in a waveguide geometry (see Fig. 4.44) [39, 40, 42, 44, 45], or as probe amplification in transmission experiments under high pumping excitation [39], or as collimated and speckled patterned emissions from Si-nc [38]. Some concerns have been raised about the methods used to measure gain [46].

The Si-nc system is very promising to achieve a visible silicon based laser. The wavelength of emission is particularly attractive as it overlaps that of the more common VCSEL laser based on III-V semiconductors and, hence, such a laser can easily be integrated within the microphotonics circuits which are already under development based on visible VCSEL. Other key ingredients for a silicon nanocrystal laser have already been demonstrated. Vertical optical micro-cavities based on a Fabry–Perot structure with mirrors constituted by distributed Bragg reflectors (DBR) have already been fabricated (see Fig. 6.8). The presence of a thick SiO_2 layer needed to form the DBR can be a problem for electrical injection when current has to flow through the DBR. Lateral injection schemes can avoid these problems. On the other hand the electrical injection into the Si-nc is a delicate task by itself. Bipolar injection is extremely difficult to achieve. Despite some claims, most of the reported Si-nc LED are impact ionization devices: electron–hole pairs are generated by impact ionization by the energetic free carriers injected through the electrode. By exploiting impact ionization Si-nc LED have been demonstrated with electroluminescence spectra overlapping luminescence spectra, onset-voltage as low as 5 V and efficiencies in excess of 0.1% (see Sect. 4.3.6). Some unconfirmed claims of near-laser action of Si-nc LED have appeared in the literature [47, 48].

7.2.3 Er Coupled Silicon Nanocrystals

Several breakthroughs have recently been achieved in the field of Er doping of crystalline Si that allowed the fabrication of LED operating at room temperature (see Sect. 5). In addition, Er coupled Si-nc benefits from the advantages of both silicon (efficient excitation) and SiO_2 (weak nonradiative processes, i.e. negligible temperature quenching of the luminescence), while it avoids their disadvantages (low excitation efficiency in SiO_2 and strong nonradiative processes in bulk Si) (Sect. 5.5). Indeed MOS light emitting devices operating at room temperature have been made with this system, where a quantum efficiency larger than 1% was demonstrated [49, 50]. Even higher efficiencies (10%) are reported for Er in silicon-rich oxide films; however in this system reliability is still an issue [50]. When Er coupled silicon nanocrystals are inserted into a waveguide structure, evidence of signal enhancement at 1.55 μm has been reported [51]. What makes this finding interesting is the possibility of pumping the amplifier with a high brightness LED and to significantly reduce the cavity length in an optical amplifier or Er doped silicon laser compared to that usually employed in silica doped fiber systems. To summarize the very interesting properties of the Er^{3+} coupled Si-nc system, Table 7.1 compares the main cross-sections of Er^{3+} in silica, silicon and in Si-nc.

Table 7.1. Summary of the various cross-sections related to Er^{3+} in various materials

	Er in SiO_2 (cm^2)	Er in Si (cm^2)	Er in Si-nc (cm^2)	Refs. for Er in Si-nc
Effective excitation cross-section of luminescence at a pumping energy of 488 nm	$1-8\times 10^{-21}$	3×10^{-15}	$1.1-0.7\times 10^{-16}$	[52, 53]
Effective excitation cross-section of electroluminescence		4×10^{-14}	1×10^{-14} by impact ionization	[49]
Emission cross-section at 1.535 μm	6×10^{-21}		2×10^{-19}	[51]
Absorption cross-section at 1.535 μm	4×10^{-21}	2×10^{-20}	8×10^{-20}	[54]

From Table 7.1, it emerges that the system Er^{3+} coupled to Si-nc is very promising for laser applications. The technology to produce the material is very compatible with CMOS processing. Microcavities with excellent luminescence properties have also been demonstrated (see Fig. 6.10), and these allow designing both edge emitting or vertical emitting laser structures. The issue related to electrical pumping of the active material, which was believed to be a major short-coming of this approach, is no longer a problem as extremely high efficiency LED have been demonstrated. The major issue still open is to engineer the waveguide losses in order to be able to measure net optical gain and not only signal enhancement in pump and probe experiments.

7.2.4 Si/Ge Quantum Cascade Structures

One route to avoid the fundamental limitations to lasing in silicon, i.e. its indirect band gap, is to avoid using interband transitions. Indeed if one exploits only intraband transitions, e.g. intravalence band transitions, no fundamental problems exist to impede lasing in silicon. This is indeed the approach of the quantum cascade (QC) Si/Ge system, where one is trying to use the concept that was already successful in III-V semiconductors, which is advancing as a viable option for mid-IR emission, covering today a large wavelength range, 3–24 μm (see Sect. 1.3.4).

The SiGe system has many advantages with respect to III-V semiconductors, but also a fundamental limit posed by the high lattice mismatch between Si and Ge. This forces one to use simple cascade structures and prevents the increase at will of the number of QW cascades, i.e. of the gain. Hence, even though these devices show interesting EL properties for the prospect of the development of a Si-based laser, highly evolved cascade structures to engineer the strain have to be realized. As the gain per single element is low due to the nature of the intraband transition, a large number of cascading structures will be needed to accumulate a macroscopic gain. In addition, all these have to be integrated within a waveguide cavity. The possible solutions may come from the use of Si-on-insulator (SOI) substrates or thick, relaxed SiGe graded buffers. No stimulated emission in SiGe QC structures has been reported to date. In addition the emission wavelength is different from those commonly used for optical interconnects. Despite some authors proposing to use quantum cascade lasers for free-air optical interconnects, such a Si/Ge quantum cascade lasers will be of little use for silicon photonics if all other compatible elements, such as detectors and waveguides, are not developed in an integrated approach.

7.3 Conclusion

It turns out that, about once a decade, a new technology comes along that completely reshapes the information landscape. Just before 1980, that key enabling technology was the microprocessor, and its arrival set off a decade-long processing revolution symbolized by the personal computer. Then just as the 1980s were closing, another new enabling technology came along to displace the centrality of the microprocessor – cheap lasers. Much as the microprocessor slipped into our lives hidden in PCs a decade earlier, lasers slipped into the lives of ordinary citizens hidden in everyday appliances – compact disc music players, CD-ROMs, and long-distance optical fiber phone lines. The consequence was a shift in emphasis from processing to access.

The question is about the driving technology for the current decade. It is becoming clear that integrated logic (computing elements, sensors, actuators) is pervading everyday objects. The interface with the outside world has here a primary role. The world in general is analog, so you need to provide a set of components that sense the inputs from the world in analog form and act on an analog world. Much of this input is either intrinsically based on optical signals (an example is image acquisition), or has characters which make strategic the handling with optical signals (on example is gas recognition using infrared absorption spectrum) [55]. If we want silicon to play a key role in this new scenario we have to increase the brightness of silicon itself.

As we have detailed in this book, the state of light emission in silicon is extremely alive. After ten years since the report of bright room temperature luminescence from porous silicon, observation of optical gain and new high-efficiency LED structures, which cover such different emission wavelength from the UV to the NIR, demonstrate that this field is moving ahead very fast. Many breakthroughs should be expected in the near future which will keep silicon at the center of the technological development as it uses to be now.

References

1. G.E. Moore: *Progress in Digital Integrated Electronics*, IEEE Intle Electron Device Meeting Tech. Digest (IEEE Press, 1975) p. 11
2. J.D. Meindl, Qiang Chen, J.A. Davis: Science **293**, 2044 (2001)
3. J.D. Meindl: Computing in Science and Engineering (January/February 2003) p. 21
4. T.N. Theis: IBM J. Res. Develop. **44**, 379 (2000)
5. D.A. Miller: Proc. of IEEE **88**, 728 (2000)
6. L.C. Kimerling: Appl. Surf. Sci. **159–160**, 8 (2000)
7. O. Bisi, S.U. Campisano, L. Pavesi, F. Priolo: *Silicon Based Microphotonics: from Basics to Applications* (IOS press, Amsterdam 1999)
8. R. Soref: Proc. of IEEE **81**, 1687 (1993)
9. G. Masini, L. Colace, G. Assanto: Mat. Sci. Eng. B **89**, 2 (2002)
10. B.P. Pal: *Progress in Optics XXXII*, ed. by E. Wolf (Elsevier Science Publishers B.V. 1993) p. 1
11. T. Miya: IEEE J. Sel. Top. Quantum Electr. **6**, 38 (2000)
12. D.A.P. Bulla et al.: IMOC'99 Proceedings IEEE (1999) p. 454
13. U. Hilleringmann, K. Goser: IEEE Trans. Electron. Dev. **42**, 841 (1995)
14. G. Cocorullo, F.G.D. Corte, M. Iodice, I. Rendina, P.M. Sarro: IEEE J. Sel. Top. Quantum Electron. **4**, 983 (1998)
15. B. Jalali, S. Yegnanarayanan, T. Yoon, T. Yoshimoto, I. Rendina, F. Coppinger: IEEE J. Sel. Top. Quantum Electron. **4**, 938 (1998)
16. K.K. Lee, D.R. Lim, H.-C. Luan, A. Agarwal, J. Foresi, L.C. Kimerling: Appl. Phys. Lett. **77**, 1617 (2000)
17. M.A. Rosa, N.Q. Ngo, D. Sweatman, S. Dimitrijev, H.B. Harrison: IEEE J. Sel. Top. Quantum Electr. **5**, 1249 (1999)
18. T.W. Ang, G.T. Reed, A. Vonsovici, A.G.R. Evans, P.R. Routley, M.R. Josey: Electron. Lett. **35**, 977 (1999)
19. T. Bestwick: Electronic Components and Technology Conference. 48th IEEE, 25–28 May 1998 (1998) p. 566
20. H. Zimmermann: *Integrated Silicon Optoelectronics* (Springer-Verlag, New York 2000)
21. S.M. Csutak, J.D. Schaub, W.E. Wu, R. Shimer, J.C. Campbell: J. Lightwave Technol. **20**, 1724 (2002)
22. M. Yang, K. Rim, D.L. Rogers, J.D. Schaub, J.J. Wleser, D.M. Kuchta, D.C. Boyd, F. Rodier, P.A. Rabidoux, J.T. Marsh, A.D. Ticknor, Q. Yang, A. Upham, S.C. Ramac: IEEE Elect. Dev. Lett. **23**, 395 (2002)
23. G. Masini, L. Colace, G. Assanto, K. Wada, L.C. Kimerling: Electron. Lett. **35**, 1467 (1999)
24. L. Colace, G. Masini, G. Assanto, H.C. Luan, K. Wada and L.C. Kimerling: Appl. Phys. Lett. **76**, 1231 (2000)
25. S. Winnerl, D. Buca, S. Lenk, Ch. Buchal, S. Mantl, D.X. Xu: Mat. Sci: Eng. B **89**, 73 (2002)
26. M. El Kuedi et al.: J. Appl. Phys. **92**, 1858 (2002)
27. G. Coppola, A. Irace, G. Breglio, M. Iodice, L. Zeni, A. Cutulo, P.M. Sarro: Optics and Laser Eng. **39**, 317 (2003)
28. K. Kato, Y. Tohmori: IEEE J. Sel. Top. Quantum Electron. **6**, 4 (2000)
29. R.T. Chen et al.: Proc. of IEEE **88**, 780 (2000)

30. L. Pavesi, S. Gaponenko, L.D. Negro: *Towards the First Silicon Laser*, NATO Science series vol. 93 (Kluwer, Dordrecht 2003)
31. M.A. Green, J. Zhao, A. Wang, P.J. Reece, M. Gal: Nature **412**, 805 (2001)
32. J. Zhao, M.A. Green, A. Wang: J. Appl. Phys. **92**, 2977 (2002)
33. W.P. Dumke: Phys. Rev. **127**, 1559 (1962)
34. W.L. Ng et al.: Nature **410**, 192 (2001)
35. P. Jonsson, H. Bleichner, M. Isberg, E. Nordlander: J. Appl. Phys. **81**, 2256 (1997)
36. C. Delerue, M. Lannoo, G. Allan, E. Martin, I. Mihalcescu, J.C. Vial, R. Romestain, F. Muller, A. Bsiesy: Phys. Rev. Lett. **75**, 2229 (1995)
37. J.L. Heinrich, C.J. Curtis, G.M. Credo, K.L. Kavanagh, M.J. Sailor: Science **25**, 66 (1998)
38. M.H. Nayfeh, S. Rao, N. Barr: Appl. Phys. Lett. **80**, 121 (2002)
39. L. Pavesi, L. Dal Negro, C. Mazzoleni, G. Franzó, F. Priolo: Nature **408**, 440 (2000)
40. L.D. Negro, M. Cazzanelli, N.Daldosso, Z.Gaburro, L. Pavesi, F. Priolo, D. Pacifici, G. Franzó, F. Iacona: Physica E **16**, 297 (2003)
41. S. Ossicini, C. Arcangeli, O. Bisi, E. Degoli, M. Luppi, R. Magri, L.D. Negro, L. Pavesi: 'Gain theory and models in silicon nanostructures'. In: *Towards the First Silicon Laser*, ed. by L. Pavesi, S. Gaponenko, L.D. Negro, Nato Science Series vol. 93 (Kluwer Academic Publisher, Dordrecht 2003) pp. 261–280
42. L. Khriachtchev, M. Rasanen, S. Novikov, J. Sinkkonen: Appl. Phys. Lett. **79**, 1249 (2001)
43. M.H. Nayfeh et al.: Appl. Phys. Lett. **78**, 1131 (2001)
44. P.M. Fauchet, J. Ruan: 'Optical amplification in nanocrystalline silicon superlattices'. In: *Towards the First Silicon Laser*, ed. by L. Pavesi, S. Gaponenko, L. Dal Negro, Nato Science Series vol. 93 (Kluwer Academic Publisher, Dordrecht 2003) pp. 197–208
45. M. Ivanda, U.V. Densica, C.W. White, W. Kiefer: 'Experimental observation of optical amplification in silicon nanocrystals'. In: *Towards the First Silicon Laser*, ed. by L. Pavesi, S. Gaponenko, L. Dal Negro, Nato Science Series vol. 93 (Kluwer Academic Publisher, Dordrecht 2003) pp. 191–196
46. J. Valenta, I. Pelant, J. Linnros: Appl. Phys. Lett. **81**, 1396 (2002)
47. C.-F. Lin, P.-F. Chung, M.-J. Chen, W.-F. Su: Optics Lett. **27**, 713 (2002)
48. L.T. Heikkila, T. Kuusela, H.P. Hedman: Superl. Microstruct. **26**, 157 (1999)
49. F. Iacona, D. Pacifici, A. Irrera, M. Miritello, G. Franzó, F. Priolo, D. Sanfilippo, G. Di Stefano, P.G. Fallica: Appl. Phys. Lett. **81**, 3242 (2002)
50. M.E. Castagna, S. Coffa, L. Caristia, M. Monaco, A. Mascolo, A. Messina, S. Lorenti: Mat. Res. Soc. Symp. Proc. **770**, I2.1.1. (2003)
51. H.-S. Han, S.-Y. Seo, J.H. Shin, N. Park: Appl. Phys. Lett. **81**, 3720 (2002)
52. F. Priolo, G. Franzó, D. Pacifici, V. Vinciguerra, F. Iacona, A. Irrera: J. Appl. Phys. **89**, 264 (2001)
53. A.J. Kenyon et al.: J. Appl. Phys. **91**, 367 (2002)
54. P.G. Kik, A. Polman: J. Appl. Phys. **91**, 534 (2002)
55. J. Barry: Interview: Alberto Sangiovanni-Vincentelli, 2003. (http://www.edatoolscafe.com/DACafe/TECHNICAL/Papers/Cadence/archive/vol2No2/albertoS-V.html)

Index

Printing: Mercedes-Druck, Berlin
Binding: Stein+Lehmann, Berlin

Springer Tracts in Modern Physics

154 **Applied RHEED**
Reflection High-Energy Electron Diffraction During Crystal Growth
By W. Braun 1999. 150 figs. IX, 222 pages

155 **High-Temperature-Superconductor Thin Films at Microwave Frequencies**
By M. Hein 1999. 134 figs. XIV, 395 pages

156 **Growth Processes and Surface Phase Equilibria in Molecular Beam Epitaxy**
By N.N. Ledentsov 1999. 17 figs. VIII, 84 pages

157 **Deposition of Diamond-Like Superhard Materials**
By W. Kulisch 1999. 60 figs. X, 191 pages

158 **Nonlinear Optics of Random Media**
Fractal Composites and Metal-Dielectric Films
By V.M. Shalaev 2000. 51 figs. XII, 158 pages

159 **Magnetic Dichroism in Core-Level Photoemission**
By K. Starke 2000. 64 figs. X, 136 pages

160 **Physics with Tau Leptons**
By A. Stahl 2000. 236 figs. VIII, 315 pages

161 **Semiclassical Theory of Mesoscopic Quantum Systems**
By K. Richter 2000. 50 figs. IX, 221 pages

162 **Electroweak Precision Tests at LEP**
By W. Hollik and G. Duckeck 2000. 60 figs. VIII, 161 pages

163 **Symmetries in Intermediate and High Energy Physics**
Ed. by A. Faessler, T.S. Kosmas, and G.K. Leontaris 2000. 96 figs. XVI, 316 pages

164 **Pattern Formation in Granular Materials**
By G.H. Ristow 2000. 83 figs. XIII, 161 pages

165 **Path Integral Quantization and Stochastic Quantization**
By M. Masujima 2000. 0 figs. XII, 282 pages

166 **Probing the Quantum Vacuum**
Pertubative Effective Action Approach in Quantum Electrodynamics and its Application
By W. Dittrich and H. Gies 2000. 16 figs. XI, 241 pages

167 **Photoelectric Properties and Applications of Low-Mobility Semiconductors**
By R. Könenkamp 2000. 57 figs. VIII, 100 pages

168 **Deep Inelastic Positron-Proton Scattering in the High-Momentum-Transfer Regime of HERA**
By U.F. Katz 2000. 96 figs. VIII, 237 pages

169 **Semiconductor Cavity Quantum Electrodynamics**
By Y. Yamamoto, T. Tassone, H. Cao 2000. 67 figs. VIII, 154 pages

170 **d–d Excitations in Transition-Metal Oxides**
A Spin-Polarized Electron Energy-Loss Spectroscopy (SPEELS) Study
By B. Fromme 2001. 53 figs. XII, 143 pages

171 **High-T_c Superconductors for Magnet and Energy Technology**
By B. R. Lehndorff 2001. 139 figs. XII, 209 pages

172 **Dissipative Quantum Chaos and Decoherence**
By D. Braun 2001. 22 figs. XI, 132 pages

173 **Quantum Information**
An Introduction to Basic Theoretical Concepts and Experiments
By G. Alber, T. Beth, M. Horodecki, P. Horodecki, R. Horodecki, M. Rötteler, H. Weinfurter, R. Werner, and A. Zeilinger 2001. 60 figs. XI, 216 pages

174 **Superconductor/Semiconductor Junctions**
By Thomas Schäpers 2001. 91 figs. IX, 145 pages

Springer Tracts in Modern Physics

175 **Ion-Induced Electron Emission from Crystalline Solids**
By Hiroshi Kudo 2002. 85 figs. IX, 161 pages

176 **Infrared Spectroscopy of Molecular Clusters**
An Introduction to Intermolecular Forces
By Martina Havenith 2002. 33 figs. VIII, 120 pages

177 **Applied Asymptotic Expansions in Momenta and Masses**
By Vladimir A. Smirnov 2002. 52 figs. IX, 263 pages

178 **Capillary Surfaces**
Shape – Stability – Dynamics, in Particular Under Weightlessness
By Dieter Langbein 2002. 182 figs. XVIII, 364 pages

179 **Anomalous X-ray Scattering for Materials Characterization**
Atomic-Scale Structure Determination
By Yoshio Waseda 2002. 132 figs. XIV, 214 pages

180 **Coverings of Discrete Quasiperiodic Sets**
Theory and Applications to Quasicrystals
Edited by P. Kramer and Z. Papadopolos 2002. 128 figs., XIV, 274 pages

181 **Emulsion Science**
Basic Principles. An Overview
By J. Bibette, F. Leal-Calderon, V. Schmitt, and P. Poulin 2002. 50 figs., IX, 140 pages

182 **Transmission Electron Microscopy of Semiconductor Nanostructures**
An Analysis of Composition and Strain State
By A. Rosenauer 2003. 136 figs., XII, 238 pages

183 **Transverse Patterns in Nonlinear Optical Resonators**
By K. Staliūnas, V. J. Sánchez-Morcillo 2003. 132 figs., XII, 226 pages

184 **Statistical Physics and Economics**
Concepts, Tools and Applications
By M. Schulz 2003. 54 figs., XII, 244 pages

185 **Electronic Defect States in Alkali Halides**
Effects of Interaction with Molecular Ions
By V. Dierolf 2003. 80 figs., XII, 196 pages

186 **Electron-Beam Interactions with Solids**
Application of the Monte Carlo Method to Electron Scattering Problems
By M. Dapor 2003. 27 figs., X, 110 pages

187 **High-Field Transport in Semiconductor Superlattices**
By K. Leo 2003. 164 figs.,XIV, 240 pages

188 **Transverse Pattern Formation in Photorefractive Optics**
By C. Denz, M. Schwab, and C. Weilnau 2003. 143 figs., XVIII, 331 pages

189 **Spatio-Temporal Dynamics and Quantum Fluctuations in Semiconductor Lasers**
By O. Hess, E. Gehrig 2003. 91 figs., XIV, 232 pages

190 **Neutrino Mass**
Edited by G. Altarelli, K. Winter 2003. 118 figs., XII, 248 pages

191 **Spin-orbit Coupling Effects in Two-dimensional Electron and Hole Systems**
By R. Winkler 2003. 66 figs., XII, 224 pages

192 **Electronic Quantum Transport in Mesoscopic Semiconductor Structures**
By T. Ihn 2003. 90 figs., XII, 280 pages

193 **Spinning Particles – Semiclassics and Spectral Statistics**
By S. Keppeler 2003. 15 figs., X, 190 pages

194 **Light Emitting Silicon for Microphotonics**
By S. Ossicini, L. Pavesi, and F. Priolo 2003. 206 figs., XII, 284 pages